Transport Systems, Policy and Planning
a geographical approach

Transport Systems, Policy and Planning

a geographical approach

R S Tolley

and

B J Turton

To Judith and Margaret

Addison Wesley Longman Limited
Edinburgh Gate
Harlow
Essex CM20 2JE, England
and Associated Companies throughout the world

First published 1995
Reprinted 1997

British Library Cataloguing in Publication Data
A catalogue entry for this title is available from the British Library.

ISBN 0-582-00562-0

Library of Congress Cataloging-in-Publication data
A catalog entry for this title is available from the Library of Congress.

Set in 16 in 12/14pt Times
Produced by Longman Singapore Publishers (Pte) Ltd.
Printed in Singapore

Contents

List of figures

Preface

It is traditional to use a preface to explain why a book was written in the first place. The reasons often include the extent to which the subject in question is undergoing rapid change, or is a vital facet of modern economies, or has climbed to a prominent position on the political agenda and is thus exciting much greater interest than hitherto. All of these apply to the study of transport in general. If one adds the role of transport in the developing environmental crisis and the key part played by communications in the dramatic global economic and social changes of the late twentieth century, the case for a restatement of the relevance of geographical approaches to transport becomes stronger still.

This text has been prepared to meet the needs of undergraduates taking second and third year courses or modules which include elements of transport geography, as well as those of 'professional' students preparing for Chartered Institute of Transport examinations. It provides a review of the major spatial aspects of transport systems, followed by more detailed analyses of transport problems in urban and rural areas and an evaluation of the principal social, environmental and policy issues generated by contemporary transport systems. Throughout the book examples are drawn from industralised nations, the developing world and existing and former planned economies.

The four parts of the book build on each other, so that the most practical way to use the text is to start at the beginning. However, it is entirely possible to appreciate some sections in isolation, providing that the reader has some background knowledge in the subject. The synopses at the start of each chapter provide an overview of contents and the further reading sections guide readers towards a few of the most important and accessible texts in the field. The bibliography at the end of the book is designed to incorporate all references and to allow more specialised enquiry into issues that have attracted the reader's interest.

The authors have led transport geography options in neighbouring North Staffordshire universities for over twenty years and are grateful to successive cohorts of students on whom material has been tested and who have provided feedback, much of which has proved invaluable in the preparation of this text.

Thanks also go, through gritted teeth, to the burglar who, when making off with the word processor, at least left the disks behind. There were times when it seemed as though it might have been better if he had taken them, but the staff at Longman gave us support and encouragement to see the project through. Particular thanks go to the cartographers at Keele and Staffordshire Universities – Andrew Lawrence and Owen Tucker respectively – for their technical expertise, advice and imagination in turning sketches of very doubtful quality into polished finished articles. Most of all our thanks go to our families, for patience and understanding during the long gestation period of this book.

Acknowledgements

We are grateful to the following for permission to reproduce copyright material:

The American Geographical Society for Fig. 3.7 (Taaffe, Morrill and Gould, 1963); Ashgate Publishing Ltd, for Table 8.1 (Newman and Kenworthy, 1989); Blackwell Publishers for Figs 2.10a (Plane, 1981) and 2.15 (Patmore, 1983); Butterworth–Heinemann for Table 13.1 (this was first published in *Transport Geography*, June 1993 Vol 1, No 2 [and March 1993 Vol 1, No 1] and is reproduced here with kind permission of Butterworth–Heinemann, Oxford, UK); Centre for Rural Transport for Fig. 2.6 (TGSG/RGSG, 1985); Dr N. Dennis for Table 5.7 (Dennis, 1993); Devon County Council for Table 8.3 (Reproduced from Devon County Council's Publication *Traffic calming guidelines*, 1991); Elsevier Science Ltd. for Table 9.3 (Reprinted from *Journal of Rural Studies*, 4, S D Nutley, Unconventional modes of transport in rural Britain, 73–86, Copyright (1988), with kind permission from Elsevier Science Ltd., The Boulevard, Langford Lane, Kidlington OX5 1GB); Finance and Development for Table 7.2 (Replogle, 1992); Geographical for Fig. 9.5 (Moyes, 1988); George Braziler Inc. for Fig. 5.14 (McHale, 1969); Guilford Press for Figs 8.1 (1986) and 12.4 (Plane, 1986); HMSO for Fig. 10.11 (Department of Transport, 1977) and Table 11.3 (Department of Transport, 1991) both reprinted with permission of the Controller of Her Majesty's Stationery Office; John Murray (Publishers) Ltd. for Fig. 5.12 (Hoyle *Transport and Development in Tropical Africa*, 1988); Journal of the American Planning Association for Table 12.1 (Reprinted by permission of the Journal of the American Planning Association, 56, 1990); Kluwer Academic Publishers for Fig. 2.13 (*Geojournal* 1(3), 1977); Dr Richard Knowles for Table 7.4 (Knowles and Fairweather, *The Impact of Rapid Transport*, Metrolink Impact Study Working Paper 2, 1991); Lund University Press for Fig. 2.7 (Bunge, 1962); Methuen and Co. for Table 7.1 (Daniels and Warnes, 1980); the author A M Moyes for Fig. 9.5 (Moyes, 1974); OECD for Table 10.3 and Fig. 10.8 (OECD, 1988); Paul Chapman Publishing Ltd. for Fig. 4.2 (Dicken, 1992); Policy Studies Institute for Fig. 11.2 (Hillman *et al.*, 1990); The Public

Health Alliance and the author, THSG, for Table 11.6 (Transport and Health Study Group, 1991); Regional Science Association International for Fig. 3.7a (Lachene, 1965); Routledge for Figs 7.2, 7.3 and 7.5 (Dimitriou, 1990); Singapore Journal of Tropical Geography for 3.7c (Aloba, 1983); Taylor and Francis for Table 4.3 (Mwase, 1987); The Geographical Association for Fig. 3.10 (Tolley, 1973); Transport Geography Study Group for Fig. 3.1 (Hoyle, 1990); Transport Research Laboratory for Table 10.4 (Transport Research Laboratory, 1992); University of Hawaii Social Science Research Institute for Fig. 3.8 (Gould, 1966); Victor Gollancz Ltd. for Fig. 7.3 (Thomson, 1977).

Whilst every effort has been made to trace the owners of copyright material, in a few cases this has proved impossible and we take this opportunity to offer our apologies to any copyright holders whose rights we may have unwittingly infringed.

I Introduction

Transport systems are the response to the ever-growing needs for contacts between individuals and societies and for the movement of commodities as part of national and global economies. A mother taking her children on the daily journey to school or the office worker on the regular journey on the metro to the city centre will see these trips simply as an inevitable start to the day, but each is in fact part of an aggregate and highly complex pattern of personal movements.

Similarly the scheduled deliveries of stock in 5-tonne loads to the hypermarket or of crude oil in quantities of 300 000 tonnes to the refinery both form an integral part of a commercial haulage industry that is an essential element of modern economies.

All these movements of individuals and of goods create demands that are met by the transportation industry, within which various distinctive modes of personal travel and freight haulage can be identified. Some movements will demand far less of the transport system than others: the school trip will often be made by car, driven by the mother and requiring very little from the system other than the use of about 2 or 3 km of road space, but the office-bound commuter relies upon an urban railway network with a heavy investment in track, rolling stock and operating staff. The daily deliveries of milk and dairy produce to individual urban households are made by fleets of small-capacity electric trucks but bulk oil transport involves supertanker operations at the global scale and specialist export and import terminals.

Because these requirements for the movement of people and of goods vary so greatly in terms of distance, frequency of journey and numbers (or tonnage), the transportation industry displays a similar variety in its levels of sophistication, organisation, scale of operations and the specific modes available. The character of this industry's internal structure, the technical, financial and economic aspects of its management and operation, the complex links between transport modes and their markets and the external relationships between transport systems and national governments have all been extensively documented by historians, engineers, economists, geographers and other social scientists. The problems which arise in attempting to meet a given demand with a suitable transport facility have similarly been closely researched, and transport policy and planning are today two of the principal issues in society.

Any introduction to a new transport geography text must provide some explanation of how the geographer has approached studies of the transport industry and of how transport geography as a distinctive discipline may be defined. Ullman's view that 'transportation is a measure of the relations between areas and is therefore an essential part of geography' (1956) still provides a valid standpoint and can be usefully complemented with White's statement made in 1977: 'For the geographer, the importance of transport lies in its being one of the principal factors affecting the distribution of social and economic activity. Thus there is a wide interest among geographers either in transport *per se* as a significant human activity, or indirectly through its influence upon the spatial distribution of other activities.'

Since these two statements were first published there has been a dramatic expansion in the variety and scope of transport-orientated issues which have been addressed by geographers but the essence of Ullman's and White's interpretations survives.

1.2 Transport geography in the 1980s and 1990s

Since the 1970s the interests and activities of transport geographers in several countries have been more strongly focused through the formation of specific research groups, resulting in an increased output of texts, papers and, more recently, journals devoted specifically to this discipline. In the UK the Transport Geography Study Group, as a part of the Institute of British Geographers, was founded in 1972 and its members have made a substantial contribution to the growth of the discipline. The aims and achievements of transport geography, as reflected in the group's activities in its first decade, were reviewed by Williams in 1981, and in 1993 the launch of the *Journal of Transport Geography*, with which the group has been closely involved, provided a further opportunity for an updated appraisal of current themes and developments within the subject. From a purely practical point of view, therefore, the student of transport geography in the 1990s has access to a wider and more comprehensive collection of material published by geographers on transportation than has ever been previously available.

1.3 Approaches to transport geography

The development of transport geography has in many respects followed the path taken by geography in general in terms of its content, its methods and the progress which has been made to place current research within much broader societal perspectives. Where, for example, the inequalities in the distribution and delivery of health and welfare services have concerned the social geographer in recent years, transport geographers have looked in particular at levels of access to surgeries and clinics. Gender-based issues also provide a broad contemporary focus within geography, but the differing degrees of mobility of male and female groups within communities have been the especial concern of the transport geographer.

With the shift of emphasis over time on research directions in transport geography some issues have become of less significance or relevance, but there has been a steady accumulation of material which acts as a valuable source of information and stimulation to the student. The presentation of this material has naturally also varied according to the objectives and intentions of particular writers, and before the authors of this text explain the basis for its form and structure it is appropriate to include a brief review of the principal methods of approach to the subject adopted by earlier contributors.

Many of the earliest studies took what White (1977) described as a 'transport and terrain' approach, with an emphasis upon describing and explaining the relationships between transport routes and systems and the physical form of the areas they traversed. The process of route selection, for example, was seen by geographers as a useful illustration of how adapting the alignment of a road, canal or railway to the landscape provided a compromise between initial costs of construction and subsequent costs of operation.

The historical approach explores the initiation, growth and expansion of specific systems, with each stage of development being considered in the context of the technological, economic and social environments of the time. Many of the attempts to unravel the background to the intricate railway networks built in the nineteenth century in the UK and other European industrial countries have been set in this historical framework. In particular what Williams (1981) describes as the 'coal dust and pounding steam' approach is a feature of many books in which a natural enthusiasm for railways has been combined with the skills of the historian and the geographer.

The 'quantitative revolution' of the 1960s provided transport geographers with valuable new methods and techniques to investigate networks and to describe their form and levels of complexity with much greater precision. Descriptions of the South Wales railway network as 'complicated and interlocking' or of that in West Africa as 'minimal' could now be replaced with much more precise assessments based upon a structured analysis of each system. These statistical approaches were complemented by theoretical modelling techniques, allowing geographers to generate or simulate transport systems within a set of parameters and to compare these simulated patterns with actual systems at various stages of their development.

The more recent application of behavioural principles to transport geography has yielded most benefits in the areas of demand–supply relationships, and particularly in the complex process of the decision-making which precedes personal trip generation. This in turn is very closely associated with mobility, seen by Hoyle and Knowles (1992) as 'a fundamental human activity and need' which they identify as their 'first cardinal principle' in the study of transport.

Many of these approaches rest upon concepts and techniques taken from other subject areas, and it is almost impossible for the transport geographer to work successfully without borrowing from related facts and expertise. This multidisciplinary nature of transport studies is recognized by Hoyle and Knowles as their second cardinal principle.

1.4 The structure of this text

In this text the authors have attempted to provide an overview of what they see as the principal aspects of transport systems which have a geographical dimension and relevance. Not all of the many transport and transport-related issues of current interest to geographers can be dealt with in detail, but it is hoped that this book will help readers to appreciate the more significant developments in infrastructure, planning and policy of recent years.

Transport geography rests upon a continually evolving basis of established concepts and techniques which often emphasises the vital interactions between transport and other essentially spatial processes such as industrial location and urbanisation. These linkages are explored in Part One (Chs 2–4), which focuses upon demand and supply relationships, the form and structure of the principal modes, and the ways in which transport has contributed to the establishment of agricultural, industrial and urban patterns. Chapter 2 deals with how the demand for transport is created and the extent to which it is met, placing a particular stress upon the often intractable problems of mobility and accessibility in both advanced and developing countries. The daily sequence of journeys made by members of an African subsistence farmer's household to tend crops and livestock and to collect water and firewood displays a very simple pattern of movements when compared with the tightly programmed network of freight trips within the retail distribution system of an industrialised nation, but both illustrate how access is such a dominant factor in transport.

Chapter 3 examines the character of the principal transport modes, including a review of the basic elements of network analysis and modelling. The effects of new technologies are briefly discussed, ranging from the electric tramways that proved such a successful innovation in urban areas in the early twentieth century to the air cushion vehicle (or hovercraft) of the 1960s that promised so much with its versatility but failed as a commercial venture. Chapter 4 examines the linkages between transport systems and the world's principal agricultural, industrial and settlement patterns, emphasising how the relative importance of transport costs has changed in many economic activities.

In Part Two the character of transport systems is considered at the international (Ch. 5) and national scales (Ch. 6). The 'shrinking world', reflecting the ever-increasing ease with which distance can be covered, is a well-established and effective device used by geographers to illustrate how air transport or modern electronic communications have strengthened contacts at the global level. This search for the continual improvement of long-distance links has been the impetus behind some of the most impressive transport projects and innovations of the later twentieth century, including the jet airliner and its supporting chain of international airports, the Channel Tunnel between the UK and France and the 'container revolution' for freight haulage.

In Chapter 6 the parts played by the major modes within national transport systems are examined with reference to advanced industrial countries, states which until the late 1980s were subject to centralised economic planning and countries in the developing world. The spread of the limited access expressway, or motorway, throughout Anglo-America and western Europe is indeed dramatic evidence for the dominance of road transport in these continents, but in much of Africa a large proportion of all traffic is still composed of pedestrians and animal-drawn carts moving over seasonal tracks. Such disparities illustrate the diversity of the subject-matter of the transport geographer.

Within Europe the period since the end of the Second World War has been marked by the uneven rates of growth of the economies of states in the West and those which were until 1990 an integral part of the Soviet economic bloc. The collapse of Soviet-based governments in Russia and eastern Europe, and the adoption of more democratic policies, have been followed by attempts to forge closer links with western Europe, a process which will require massive investment in the transport infrastructures of states such as Poland, Hungary, Romania and the former eastern Germany. Many geographers are currently investigating the transport difficulties of these countries of eastern Europe, a region where substantial improvements in road and rail communications will be required to support planned economic developments.

For the greater part of the world's population, travel is a necessary but seldom trouble-free activity. Whether we consider the motorist competing with many other fellow drivers for space within city-centre car-parks or the small-scale commercial farmer in a developing country carting his crops to market over dusty tracks, there is a serious mismatch between demand and supply in the transportation system. Part Three looks at these problems in urban and rural areas and reviews the very wide range of solutions that have been devised. Much of the world's population is concentrated in towns and cities, and in the developing countries the rate of urbanisation is steadily increasing, so many of the difficulties in providing for the demand for travel are in these urban areas.

Chapter 7 considers these problems in cities of advanced and developing nations, showing how the most efficient means of urban travel, namely public transport and cycling, are rarely those which are most heavily used. Solving these problems (Ch. 8) requires the skills of many specialists, and the geographer has a specific role to play in the transport planning process since so many of the difficulties faced by the urban traveller have their roots in the discordance between the locations of the principal types of urban land use and the demands for transport which they generate. For example, the neighbourhood or 'corner' shop still survives to meet locally based small-scale needs of the pedestrian, but for the modern car-based shopper a visit to the out-of-town hypermarket is seen as the more convenient alternative. This chapter describes the standard planning process, accounts for its lack of success in relieving congestion and considers more recent approaches based upon limiting demand rather than catering for supply.

Placing firm limits upon private car traffic, coupled with the revitalisation of urban public transport, is now seen as a priority in the effort to solve traffic congestion. In many European cities the search for better public transport has led to the construction of what are known as light rapid transit systems. These are in effect updated versions of the conventional electric tramway, so that the call to 'bring back the trams' is really advocating the reapplication of a technology first devised a century ago.

Chapter 9 examines transport problems in rural societies and the various approaches to solving them. The difficulties experienced in industrialised countries are often confined to specific groups within the rural population, usually those without the use of a car. In developing nations, where much higher proportions of the populations are rural, transport problems can often affect entire communities. Regardless of degree of development, however, the basic problems are usually those of inadequate levels of mobility and accessibility.

Part Four focuses upon what are seen as the most urgent and controversial issues faced by society in its use and misuse of transport. The first two topics (Ch. 10) of current concern are in many ways interlinked and involve (a) the prodigious amounts of global energy resources which are consumed by the transport industry, often in a grossly inefficient manner, and (b) the destructive effects which transport can exert upon the environment, for example in terms of atmospheric pollution and the impact of new road and rail routes upon rural and historic urban developments. The frequently wasteful consumption of fossil fuels by transport has to be seen in the light of world-wide concern over the depletion of these resources by all users, and supporters of the campaign to revitalise public transport cite the private car as one of the most inefficient modes of transport in respect of its energy demands. Searching for an acceptable balance between transport supply and environment protection is today one of society's most urgent and pressing tasks.

Transport also has an impact on the social environment, and in Chapter 11 an emphasis is laid upon differential accessibility and the fact that the 'transport deprivation' suffered by many women, the disabled and the poor is so often overlooked when recognising the freedom of mobility that the car can offer. However, this freedom must be weighed against the ever-growing price paid in terms of traffic accidents or the unnecessary intrusion of motorised transport into residential communities, which should be given much greater protection against this growing threat.

Finally, Chapter 12 gives an overview of current transport issues and policy. In particular it offers a critical assessment of the contemporary efforts made at different scales and in different states in trying to produce transport policies which are politically, socially and environmentally acceptable. The concept of sustainable transport is introduced and the progress made towards achieving its objectives are reviewed.

Within and outside the transport industry policy making and implementation are often seen as the most important processes which currently affect the relationships between the providers and the consumers of transport services, especially those in the passenger sector such as airlines, railways and bus and coach undertakings. All the issues outlined above can be analysed from a policy viewpoint, although the degree to which policy has actually been effective in bringing about desired changes can vary considerably.

For example, during the 1970s and 1980s the related processes of deregulation and privatisation caused changes in the organisation and operations of large sections of the transport industry in the USA and Europe, reflecting the aims of governments to reduce state control over major undertakings and instead allow them to operate within a free market. Transport geographers have taken an active interest in these changes and have been particularly concerned to identify the effects which deregulation and privatisation of airlines, railways and road passenger transport have had upon the market and upon the overall patterns of movement of persons and freight within the countries involved.

In presenting this text the authors have tried to illustrate the subject-matter and the methods of the transport geographer at various scales, drawing examples from countries at widely differing levels of development. The particular contribution of the transport geographer to furthering the study of transport is in many cases still characterised by the emphasis upon spatial systems and upon processes operating through time. Geographers have also pooled their skills with those working in other disciplines in attempts to discover acceptable solutions to the many problems that beset transportation systems.

Further reading

Haggett P, Chorley R J 1969 *Network analysis in geography* Edward Arnold

Hoyle B S, Knowles R (eds) 1992 *Modern transport geography* Belhaven

Hurst M E 1974 *Transportation geography: comments and readings* McGraw-Hill

Knowles R 1993 *Journal of Transport Geography* **1**: 3–11

Lowe J C, Moryadis S 1975 *The geography of movement* Houghton Mifflin, Boston

Taaffe E J, Gauthier H 1973 *The geography of transportation* Prentice-Hall

Ullman E L 1956 The role of transportation and the basis for interaction. In Thomas W L (ed) *Man's role in changing the face of the earth* University of Chicago Press pp 862–88

Whitelegg J (ed) 1981 *The spirit and purpose of transport geography* Transport Geography Study Group

Part One

The basic framework

2 Transport demand and supply

> **Different locations, different requirements – meeting the needs for links between places**
>
> The demand for transport is generated by individuals, groups and by industry. Patterns of personal movements reflect the influences of mobility and accessibility levels, and freight movements are closely associated with the relative locations of raw material sources, processing plants and the final markets. Different sectors within the transport industry must adjust to changing markets by restructuring and reorganising their services in order to remain competitive.

2.1 Introduction

The 500-seat intercontinental airliner taking off from Charles de Gaulle Airport in Paris and the two-seater cycle rickshaw negotiating the congested streets of Calcutta would seem to have little in common but both, despite the great disparities in their technology, are fulfilling a basic demand for transport.

Many personal travel needs can be met at the simplest level by walking or cycling short distances, but when journeys involve use of the car or public transport we are calling upon the services of a complex system of interrelated activities collectively described as the transportation industry. For the motorist, contacts with this industry will be limited to the fuel filling station or the maintenance garage but the regular user of, for example, inter-city railways, will depend upon an undertaking employing many thousands to operate trains and to staff passenger stations and servicing workshops.

Transport is therefore best understood as a service function, and the geographer is interested in the types of demands for transport and the means by which they can be met, with a particular emphasis on the distribution of the many points where demand arises and is met and on the routes which connect these points.

2.2 *Patterns of demand*

The transport industry is an indispensable part of any economy, and ranges in complexity from the highly capitalised urban and inter-urban passenger- and freight-carrying railways of the industrialised world to the rudimentary but no less important animal-powered cartage underlying the subsistence agricultural economies of many of the less developed countries. Regardless of the levels of sophistication involved, however, a geographical study of any transport system needs to identify where demands are generated and where and how they are satisfied. This in turn can be seen as searching for a set of origins and a set of destinations and discovering the various ways in which the two are linked, thus producing a 'geography of movement' or a 'movement geometry'. The scales of movement will vary considerably, since demand for a particular service can be satisfied in many ways. Groceries available in a neighbourhood shop will usually attract customers from a radius of 1 or 2 km and there may be several visits each week, often carried out on foot. In contrast, the retail hypermarket provides for the needs of car-borne customers travelling 20 or 30 km or more, but such trips will be made at weekly or at less frequent intervals. The deliveries of fuel to city filling stations made each week represent the end of the complex chain of transport which began at the oilfield terminal with the loading of a 300 000 tonne tanker with crude petroleum (Fig. 2.1).

2.3 *Industrial land use and freight traffic generation*

Transport availability and costs have always played a prominent part in studies of industrial location, and the complex nature of freight rates is looked at in Chapter 4. As road haulage became of increasing importance, for both assembly of raw materials to factories and distribution of finished goods, geographers began to investigate the links between specific industries and the commercial traffic which they generate. Many of these studies have concentrated upon large urban–industrial areas and the results show that the amount of inward and outward traffic varies according to the type of manufacturing or service enterprise involved (Box 2.1).

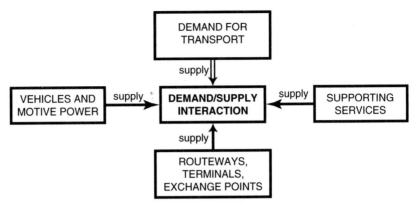

Fig. 2.1
General relationship between the demand and supply aspects of a transport system. *Source:* based partly on Fig. 1.1 in White and Senior (1983).

Box 2.1 Freight traffic generation

Starkie's study of freight traffic was based on industry in the Medway towns of the southern UK. Daily inward and outward journeys were recorded at 77 factories and the volumes were related to the type of industry, employment levels and floorspace at each of the works. The largest number of daily truck trips were generated at building material depots and in particular at 'ready-mixed' concrete plants. High frequencies of trips were also associated with food and drink factories and food product warehouses. Specialist engineering works whose products had a high value-to-bulk ratio required low levels of transport, and the number of inward journeys by vehicles carrying components often exceeded those made for distribution to the market.

    More recent surveys of industries in the north-east UK, the Midlands and the Southampton–Portsmouth area have confirmed Starkie's findings but premises in the service, wholesale and retail categories were also included, showing a high level of transport activity.

Similar investigations have been made in developing countries, and it was found in a sample area of highland Kenya that although agriculture was the dominant land use it was responsible for generating only 10 per cent of all recorded journeys whereas retail and commercial activities produced 30 per cent of the total. Although in these developing countries less information on the links between traffic and different types of economic activity is available, such studies are an essential part of the process for setting priorities for transport investment in these regions.

2.4 Personal travel

Journeys made by individuals and groups are the most important part of the overall pattern of movements in urban areas, and details of personal travel form a significant component in transportation surveys and plans. Deciding on the nature of a particular journey can involve several considerations and the behavioural aspects of trip-making have been investigated in great detail. Before any journey, short or long, is undertaken, choices must be made in terms of destination, mode of transport to be used, route to be followed and timing of departure, and many trips are made to fulfil more than one aim. A regular weekday journey in a large city, for example, would use the family car for carrying children to school and father to the railway station for the rail trip to his city-centre workplace, whilst the wife will continue with the car to her place of employment, the final destination. These complex issues of co-ordination of travel activities are discussed more fully in the urban context in Chapters 7 and 8 (see also Table 2.1).

Many attempts have been made to classify the very wide variety of trips which are commonplace within urban areas, making use of trip purpose or mode of travel as the main basis for differentiating journeys (Box 2.2).

2.4.1 Mobility and accessibility

Any approach to the measurement of demand for personal travel usually rests upon the closely interrelated concepts of mobility and accessibility. The Independent Commission on Transport concluded that 'the real meaning of mobility or the true goal of transport is access' (1974: 106), but a realistic interpretation of these two key terms depends very much upon individual circumstances. Although both terms are used extensively in the description and explanation of movement patterns it is probably impossible to arrive at a precise definition (Box 2.3 and Fig. 2.2).

It is clear from Fig. 2.3 that mobility levels and constraints upon the extent to which particular journeys are made depend upon personal factors such as health and financial resources and upon the range of transport facilities that are available. Any person will experience a life cycle with quite different opportunities and requirements for travel at each stage. The term 'actual present mobility' has been proposed to describe an existing situation in which the ability to travel is determined by personal resources and the perception of transport availability within a framework of locations where demands can be met. If the behavioural element is removed and the potential traveller behaves as 'economic man' (i.e. as in classic studies of industrial location) then trips can be made in a situation of 'optimal present mobility'. If all restrictions upon travel are assumed to be removed then mobility can be completely unconstrained and

Table 2.1
Great Britain: percentage distribution of journeys for different purposes by main modes of transport

	Journey purpose					
Main mode of travel	Work	Education	Shopping	Social/ entertainment	Leisure	All other purposes
Car (driver and passenger)	23	2	18	29	4	24
Local bus services	24	13	29	20	1	13
Rail	51	6	12	15	3	13
Cycle/motorcycle	46	9	11	20	4	10
Walk	13	12	21	19	19	16

Note: Only journeys of over 1.6 km are included.
Source: National Travel Survey, 1985/86.

Box 2.2 Classification of personal trips

Schaeffer and Sclar (1975) have devised a threefold trip classification based upon journey purpose and, to a lesser extent, upon their frequency. Extrinsic trips are those made to fulfil a definite objective such as journeys from home or another origin to the workplace, retail centres or a restaurant or club. Walking, cycling or motoring trips carried out in connection with recreational or leisure activities where no real purpose can be identified are described as intrinsic, the number of such journeys increasing as more time becomes available for leisure pursuits in advanced societies. The third category comprises transport-generated trips, such as car journeys to filling stations and repair garages and train and bus trips to depots during off-peak periods in conurbations.

Hurst (1974) proposed a threefold division of what he defines as 'movement space' based upon the type and length of trip. Most trips are made within a 'core area' such as a major conurbation where travel to work accounts for a large proportion of all journeys; in the UK, a highly urbanised state, 45 per cent of all journeys are less than 4 km in length and 80 per cent of all trips of less than 1.6 km are made on foot. The second category of a 'median area' encompasses less frequently performed journeys including business, social and holiday trips and the 'extensive area' is defined as the total spatial extent within which people travel and interact.

---

journeys can be made at any time, in any direction, by whatever means is seen as most convenient and by all members of the community. This latter situation, offering complete freedom of movement to everyone, does not, of course, exist in reality, but the extent to which individuals should be able to move around without undue constraints is an important concept which is explored more fully in Chapters 8, 9 and 11.

Box 2.3 Accessibility, mobility and the law of constant travel time

For every trip the traveller weighs the expected benefits to be gained at the destination against the costs of getting there, one of which is time, a scarce and finite resource for most people. Although nowadays people are travelling further, to work and the shops for example, it seems that the number of trips and the time spent on them remain constant through time. Hupkes' work in the Netherlands shows that between 1962 and 1972 the number of trips in a year per person and the travel time in hours hardly changed, but the total distance travelled rose from 7156 to 11 478 km. In other words people used the increased speed of travel not to access the same facilities as before and spend the saved time on some other activity, but instead chose to travel for the same amount of time and go further.

Hupkes explains this *law of constant travel time* in terms of the utilities that are derived from travel. In his diagram below, the *intrinsic* utility curve represents the value that travellers derive from the journey itself, such as the opportunity for fresh air or appreciation of the landscape. However, boredom will set in at some point and eventually extra distance will actually give negative utility. *Derived* utility, the value that can be gained from activities that the trip makes possible, will also reach an optimum beyond which either the extra travelling time eats into the time set aside for the activity or is felt to be too high a price to pay for it. Higher speeds do not change the shape of the curve (i.e. people's willingness to spend time on travel) though they may move the optimum further away in distance. This is not only because travellers willingly go to places that they formerly could not reach, but also because higher speeds lead to decentralisation which forces everyone to travel further than they did before in order to carry out the same range of activities. Though they are getting more and more mobility, their levels of access do not improve, while for those without cars, they get much worse.

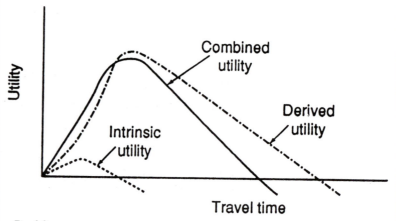

Fig. 2.2
The utility of travel time. *Source:* Hupkes (1982).

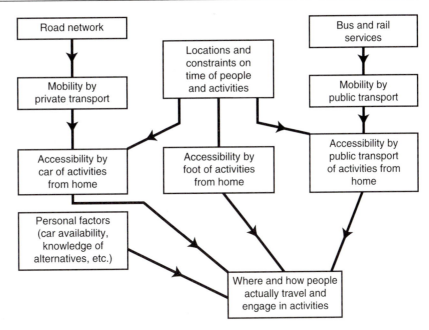

Fig. 2.3
Basic components of the accessibility-mobility-personal travel relationship. *Source:* based on figure in Jones (1981).

Plate 2.1
The village of Serradell in the Tremp basin on the Spanish side of the Pyrenees in the province of Lerida. Although only 20 kilometres from the nearest service centre the village was until recently accessible only by a gravel track, making personal mobility difficult.

JOURNEYS	PURPOSE OF TRAVEL								METHOD OF TRAVEL					TRAVEL WITH			TRAVEL TIME
	Work	School College	Shopping	Visiting People	Enter-tainment	Recre-ation	Meetings	Taking People	Walk	Bus/ Train	Van/ Car	Moped/ M'cycle	Cycle	H'hold Member	Friend	On your own	in minutes
MONDAY																	
1																	
2																	
3																	
4																	
5																	
TUESDAY																	
1																	
2																	
3																	
4																	
5																	
WEDNESDAY																	
1																	
2																	
3																	
4																	
5																	
THURSDAY																	
1																	
2																	
3																	
4																	
5																	
FRIDAY																	
1																	
2																	
3																	
4																	
5																	

Fig. 2.4
Example of a weekly travel diary used for recording trip details by day, purpose, mode of transport and travel time. (Saturday and Sunday not included in this chart.)

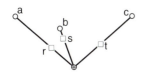

⊕ Location of person requiring particular service (x)

○ a,b,c, Location of three other points where (x) can be obtained

⊕—○ Individual - service location linkages

□ r,s,t, Possible intermediate locations where individual can satisfy demand for services

Fig. 2.5
Basic spatial aspects of accessibility.

> **Box 2.4 The personal travel diary**
>
> Travel diaries are kept for 7 days and in the example opposite groups travel purpose into 8 categories, with 5 modes of travel and space for up to 5 daily journeys. Distinctions between individual and group travel are made and the overall journey times recorded include the walking, waiting (for public transport) and car-parking times incurred for each overall trip.

Present mobility may be measured in terms of one individual's activities in a given period or on the basis of journeys carried out by an urban population as a whole (Table 2.2). National travel surveys often rely for their data upon personal travel diaries, but only a partial picture is presented as they record only those trips actually carried out by respondents (Box 2.4 and Fig. 2.4). Journeys which might be made if fewer constraints were present play no part in these diaries and in consequence the all-important aspect of latent demand for travel cannot be analysed. Interpretations of accessibility usually focus upon the spatial relationship between the point at which a demand originates and the location or destination at which it can be satisfied (Fig. 2.5). The nature of the links which are made between origin and destination is an essential component of accessibility studies, and measurements of accessibility are devised as an expression of the extent to which demands for transport are actually

Table 2.2
Great Britain: distance travelled per person per week for selected purposes by main modes of transport (km)

Main mode of travel	Journey purpose				
	Work	Education	Shopping	Social/ entertainment	Holidays, day trips
Car	22.7	1.3	13.9	37.1	14.6
Rail	4.5	0.3	0.6	1.9	1.4
Local bus services	2.4	1.1	2.4	1.9	0.2
Cycle	0.5	0.2	0.2	0.2	0.2
Private hire bus	0.5	1.0	0.2	0.2	0.2
Pedestrian trips >1.6 km	0.5	0.3	1.0	0.6	0.8
All modes	35.5	4.5	18.9	45.9	21.0

Source: Department of Transport, *National Travel Survey, 1985/86.*

realized. Claims that 'access, not movement, is the true aim of transport' or that 'the problem of modern urban transportation is not congestion or speed but access' underline the significance of accessibility in the study of personal travel patterns.

The association between mobility and accessibility can be illustrated in terms of a specific travel situation. A public library sited within a town centre can be easily reached by the able-bodied pedestrian or by the use of a car or bus, i.e. the demand is met by the book borrower travelling to the facility. Access can, however, also be provided by the use of travelling libraries in both urban and rural areas, allowing the less mobile individual to borrow books without carrying out a lengthy trip. In the UK and many other highly urbanised European states mail is delivered regularly to the household door but elsewhere it often has to be collected from centrally placed post offices, as the cost of delivering post to widely scattered premises in rural areas would be prohibitive. This practice of satisfying demand by bringing a facility or service to the individual or to some mutually convenient location has been extended to banks, health clinics and shops and is examined more fully in a rural context in Chapter 9 (Box 2.5 and Fig. 2.6).

2.4.2 *The measurement of accessibility*

Studies of actual situations where accessibility is to be measured need to take into account both the location of points where services are available and the various trips that are made by individuals to these points (Box 2.6 and Fig. 2.7). In industrialised countries a household's weekly food shopping requirements could be satisfied by one of several different supermarkets or retail malls and the eventual choice will be influenced by factors such as transport availability and opening times. The relative accessibility of these alternative locations which offer retail facilities can be compared and measured in terms of journey time or cost or a composite of these two. For the pensioner on a tight budget the cost of travelling to the shops or to a leisure centre can be more significant than the time involved, whereas in contrast a two-person working household may base their decision on trip time and ease of car-parking.

Measurements of individual accessibility are not easy to make since they involve the behavioural aspects of travel, in particular the personal choice of the services which are required, and the evaluation of the various means of transport available to reach them. As with studies of mobility, measures of accessibility are usually concerned with actual rather than potential levels and research into these areas is very limited.

Box 2.5 Access to hospital services

Medical facilities range from the localised pattern of individual doctors' surgeries to group practices, health centres and clinics and the specialised services available only in the larger hospitals. A survey of travel by out-patients to clinics attached to two major hospitals in South Wales was based upon the definition of areas from which residents could reach a hospital within a specified time making use of public transport. A maximum bus journey time of 45 minutes was defined, combined with a walking time of 15 minutes, and the period during which clinic services were available was also taken into account. A map of accessibility contours was constructed, enabling attention to be focused upon those areas where access to the hospitals presented difficulties.

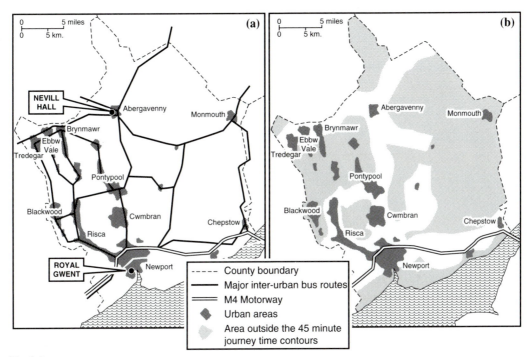

Fig. 2.6
Accessibility to hospitals in South Wales, UK: (a) location of district hospitals, main urban areas and inter-urban bus routes; (b) journey time contours (in minutes) for access by bus to hospitals. *Source:* based on TRRL/Gwent CC Joint Working Group (1981).

2.4.3 'Non-transport' influences upon accessibility

Most of the studies of personal travel patterns focus upon the use made of the various means of transport available, but attention has been shifted recently towards what are described as the 'non-transport' influences upon how people decide where, when and how to travel (Ch. 9).

Table 2.3
Farm and household walking
trips classified by purpose
and family status in two
sample Kenyan villages

| Trip purpose | Family status | | | | | |
---	Father	Mother	Son	Daughter	All	%
Village A						
Sample size	95	127	74	64	360	
Farm:						
Graze animals	5	1	12	2	20	8
Cultivate	18	58	28	33	137	55
Farm transport[a]	1	3	2	3	9	4
Sub-total	24	62	42	38	166	67
Household:						
Wash clothes	—	—	—	2	2	1
Bathe	7	1	16	2	26	10
Collect water[a]	1	28	2	12	43	17
Collect firewood[a]	—	7	—	4	11	4
Sub-total	8	36	18	20	82	32
Total	32	98	60	58	248	99
Trips per day	0.34	0.77	0.81	0.91	0.69	
Village B						
Sample size	132	147	89	83	451	
Farm:						
Graze animals	22	15	21	5	63	15
Cultivate	42	57	40	18	157	38
Farm transport[a]	9	14	2	6	31	8
Sub-total	73	86	63	29	251	61
Household:						
Wash clothes	1	44	1	2	8	2
Bathe	7	—	14	4	25	6
Collect water[a]	—	72	3	28	103	25
Collect firewood[a]	—	15	—	7	22	5
Subtotal	8	91	18	41	158	38
Total	81	177	81	70	409	99
Trips per day	0.61	1.20	0.91	0.84	0.91	

[a] Trips that involve carrying loads.
Source: Kaira (1985).

Although the types of transport available obviously exert a strong influence upon travel, it is important also to consider social and institutional structures when analysing the overall process of trip-making by groups and individuals. A knowledge of the lifestyle of households and their members is an invaluable guide to the reasons underlying travel and to understanding the constraints upon movement imposed by social factors (Table 2.3). For example, the changes in trip-making which take

Box 2.6 Access to essential daily requirements in rural Africa

In rural Africa accessibility is a critical factor affecting trips made to fulfil daily household needs, to carry out subsistence or small-scale commercial farming and to deal at local markets. Essential trips for water and firewood collection in Kenya and Uganda can occupy up to 4 hours each day and travel to and from cultivated areas can account for up to one-third of all journeys made by a household. In many parts of West and East Africa the rural population can spend up to 30 per cent of each day in travel as the result of inaccessible water supplies and the scattered location of agricultural plots. These types of trips are illustrated in Fig. 2.7 and discussed in more detail in Chapter 9.

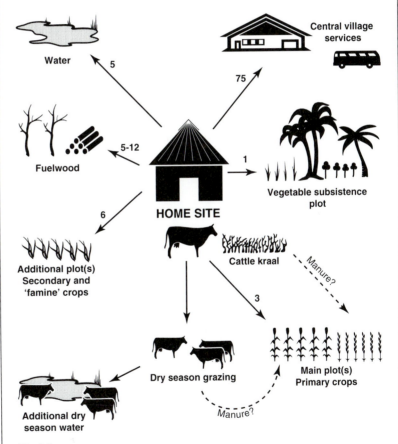

Fig. 2.7
Diagrammatic representation of work trips in an African farming community, showing trip lengths in kilometres. *Source:* based on McCall (1985).

place when a family moves to a new house are often the result of a desire to acquire a larger home, or one in what is seen to be a better residential area, rather than of the need to shorten the journey to work. A survey of people changing jobs in Manchester found that only 13 per cent of respondents cited a need for better access to their workplace as the main reason for change. The whole complex pattern of travel to work is influenced by the need to secure the best possible job and the most suitable house, and employees will have differing views on which of these two requirements is the primary concern in terms of location.

In the UK the progressive decline in the total numbers of doctors' surgeries, pharmacies, post offices, small retail outlets and schools has been accompanied by an increase in car ownership and travel. Access to these various services by car is usually much easier than by public transport and a fall in the number of points at which a particular service is available does not therefore always imply a decrease in the number of persons to whom it is accessible. However, the concentration of facilities such as primary health care at fewer clinics or surgeries is often dictated by a need for increased efficiency within the organisation concerned and questions of accessibility are seldom the prime consideration. For the consumer of these services the most satisfactory situation is achieved when access to the various points of delivery is achieved with a minimum amount of travel. It is the differences between the needs of the consumer and the organisational structure of the undertaking providing the service which are at the root of so many travel problems in both urban and rural areas.

2.5 Types of movement

The various movements which result from satisfying a demand for transport are discussed more fully in subsequent chapters. At this stage it is only necessary to state that a specific trip is a reflection of the combined influences of journey length and time and costs of the various modes of transport available. The relative importance of these factors will vary according to whether or not freight or personal transport is involved, and the attributes of the different types of transport and their suitability for various traffics are discussed in Chapter 3.

2.5.1 Problems of route selection

Since in theory there is an infinite number of specific locations at which journeys may start and finish it is necessary to simplify any description

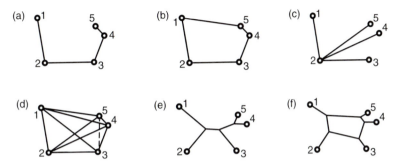

Fig. 2.8
Diagrams to illustrate the various ways in which five locations (1, 2, 3, 4, 5) may be connected by straight-line links. The chosen method of linkage will be influenced by (a) costs of construction and (b) costs of operation. Network (d) for example represents a solution where operating costs are minimal but where construction costs are at a maximum, whereas (e) is described as a 'least length' network where building costs are minimised. Network (c) identifies point 2 as a principal node with radial routes to all other points and example (b) is often termed the 'travelling salesman' network. *Source:* based on Bunge (1962).

of types of movement. If it is assumed that movement is a process which seeks to minimise journey length, time and cost then it is possible for the transport geographer to study personal travel and freight transport in terms of the constraints placed upon movement. The quest for what is described as maximum route efficiency has attracted many mathematicians and has also been pursued by geographers in the context of route networks. Where several locations are to be interconnected the solution to what is known as the shortest-path problem ranges from complete interlinkage, providing direct movement between all points, to more circuitous routeing where the network is of a simple form. Between these two extremes lies the compromise solution where network building and operating cost constraints are combined to create a framework of routes where both direct and multi-stage trips can be made (Fig. 2.8).

Distance can also be considered in terms of time and cost, and the choice of a particular route can be strongly influenced by its ability to offer the cheapest trip between origin and destination. Inter- and intra-urban road networks suffer from various levels of traffic congestion and in consequence car drivers may often divert their journey from a direct but heavily trafficked road to one which offers a shorter trip time although involving a greater distance and higher travel costs. Urban motorists who adopt these measures are seldom popular in the residential areas through which they take 'short cuts', however, and traffic-calming schemes are now introduced to curb this practice.

Diversions from direct lines are often obligatory where traffic management schemes in cities have created pedestrian zones, vehicle-turn prohibitions at intersections and one-way flow streets. Bus operators in particular incur additional running costs when operating in and around pedestrian zones, though this is less of a penalty than having to contend with car traffic in congested conditions.

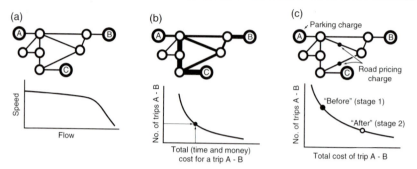

Fig. 2.9
The effects of road tolls and car-parking charges upon route selection and traffic flows in an urban area: (a) the basic road network, with A, B and C as origins and/or destinations for all trips; (b) a flow pattern for the network, assuming that all trips make use of routes that minimise travel time and costs; (c) application of road and car-parking (at A) charges to the network. The changes in travel behaviour caused by these selective charges can then be evaluated in terms of (i) benefits to car drivers, (ii) revenue collected and (iii) changes in operating costs. A new flow pattern can also be simulated in which the number of trips made will reflect the additional cost of each trip. *Source:* based on TRRL (1981).

Inter-urban toll roads, which are common on the motorway systems of the USA, France and Italy, also influence route choice by commercial freight hauliers and bus companies, who have to balance the additional expense of using these expressways against the overall benefits gained in terms of reliability of schedules and total running costs. If the UK and other countries pursue their plans to construct privately financed toll roads then drivers will be able to select what they perceive to be the most economic route in terms of fuel costs, toll expenses and overall journey time.

Within towns and cities route selection by motorists is also influenced by the availability, capacity and location of pay and free car-parks (Fig. 2.9). The introduction of inner urban area vehicle licensing or selective road-charging schemes will create even more complex choices for motorists who must select what they judge to be the most efficient route for their regular trips, especially journeys to work. Experiments in selective admission to city centres based upon odd or even vehicle registrations on alternate days are often seen by motorists as a challenge to their mobility and many ingenious plans such as interchangeable registration plates have been devised. A more detailed discussion of how vehicle restraint can help combat urban congestion will be found in Chapter 8.

2.5.2 *Personal movement patterns*

These patterns are the sum total of individual journeys, and in advanced industrial societies the opportunities for personal mobility have been dramatically improved since the 1950s with increasing car ownership levels. Data on movement patterns are now readily obtained from the surveys at local, regional and national levels which are an essential part of transport planning programmes. In these surveys a specific trip is described in terms of its major purpose, the location of its starting and finishing points, the distance travelled, the route adopted, the means of transport, the time at which it is made and the frequency or regularity with which it is carried out.

Movement patterns are most complex in large urban areas where trips involving several purposes and modes are common. Travel to work is one of the most common personal movement patterns and has been studied in detail, as it involves large numbers of people, requires a substantial investment in transport facilities and presents some of the most intractable problems to the urban transport planner. In many urban areas up to 20 per cent of all trips are to and from work, and these are primarily responsible for the congestion in the two daily peak travel periods (Fig. 2.10). Relationships between places of work and of residence have already been briefly discussed in the context of accessibility, and a strongly developed trend in many urban–industrial regions is the increasing spatial separation of home and workplace as incomes rise and transportation becomes more flexible. The large numbers of cyclists converging upon factory gates that were common in the 1950s have been replaced by the car-borne workers who now live at much greater distances away from the works.

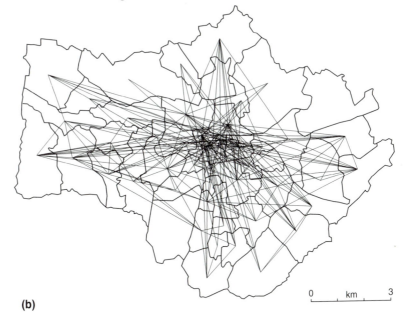

(a)   **(b)**

Fig. 2.10
(a) The principal types of commuting flows in urban areas: (1) within the city core; (2) inward from suburbs; (3) outward from the city core to the suburbs or beyond; (4) within the suburbs; (5) 'cross-commuting' between rural areas and suburbs. (b) Reading, Berkshire, UK: journey-to-work trips showing all zones where 50–149 trips are generated. These accounted for 4 per cent of all journey-to-work trips in the town. *Sources:* (a) based on Plane (1981); (b) based on Lomax and Downes (1971).

Up to the mid twentieth century the concentration of employment in, or close to, city centres produced radial commuting patterns, with morning and evening flows concentrated along rail, tram and bus routes. In the modern city several factors now interact to create a much more complex situation with work trips across cities and from inner to outer areas, all counter to the traditional flows. Decentralisation policies have led to the relocation of many older inner-city enterprises in peripheral industrial estates, a trend aided by the flexibility offered by the car. New employment sources have also been attracted to these outer zones, further reducing the need to travel into urban cores. The implications of these trends are examined in more detail in Chapters 8 and 12.

The daily travel of children and students to schools and colleges produces an intricate picture of movements within suburbs and inner residential areas. In rural districts the progressive closure of small village schools as numbers of pupils fall has often increased the lengths of these essential trips (see Ch. 9). In many countries the sizes of catchment areas for schools are defined by local education authorities, who often have a statutory obligation to provide buses for children travelling longer distances.

Social and leisure trips in advanced industrial nations have become more numerous and extensive as a result of higher disposable incomes, shorter working days and the flexibility provided by car ownership. Journey lengths vary from a few kilometres between home and locally based sporting facilities to day excursions to recreational attractions up to 150 or 200 km away. For longer holidays, however, much of the demand is no longer focused upon the domestic coastal or inland resort but is now increasingly orientated towards overseas centres served by air. Recreation and the associated transport systems are also discussed later at the national and international scales in Chapters 5 and 6.

2.5.3 Freight traffic

Freight movements are largely determined by the distribution of the industries which generate this traffic, and industrial transport may be placed into three categories. The assembly of raw materials and power resources at the production or processing plant is an essential initial transport requirement and is followed by the second type of movement, the transfer of semi-finished goods between plants where more than one location is involved in manufacturing. The distances covered by these trips and the mode of transport used vary with different industries. In the case of petrochemicals almost all the necessary transfers can be made by pipelines within the refinery and between associated plants, and no other transport services are required until the final distribution stage is reached. In contrast, motor vehicle manufacturing usually requires the regular delivery to the assembly plant of components drawn from a widely scattered pattern of ancillary works and both road and rail can be used.

The third category, distribution to the market, is often very complex and the trips made depend upon the nature of the commodity and whether the market is consumer-orientated or one for capital goods. In many industrial nations a large proportion of all goods haulage takes place within urban areas, which contain both the factories and their markets, but raw materials usually require bringing in from more distant sources. Although road transport is now the dominant mode for freight haulage, trucks of widely differing capacities are employed, with transhipment of goods at depots within or on the peripheries of towns and cities.

Rises in the volume of freight traffic in industrial nations, as measured in tonne-kilometres, are often the result of increases in individual trip length rather than in the total amount of goods carried. In the UK larger trucks with capacities of over 20 tonnes have recently taken a greater share of all freight carried (70.7 per cent in 1981 as compared with 42.5 per cent in 1974). The reverse trend has been shown by smaller vehicles, with a reduction in the same period from 57.5 to 29 per cent in their share of all road freight, but an increase from 35.8 to 49.2 km in average trip length. These increases in average journey distance can be explained in terms of changes in the market size and its relationship to the location of factories, storage depots and final retail outlets (Fig. 2.11).

More extensive market areas for home-produced and imported goods, coupled with changes in sales patterns within markets, have both led to longer trips by trucks. A shift in the location of a factory away from the core of its sales area, or the concentration of production and storage activities at a smaller number of sites within an established market, will make longer hauls necessary between factory and warehouse and between the latter and sales outlets.

The displacement of coal as a primary industrial energy source by oil, natural gas and electricity has produced fundamental changes in many fuel movement patterns. Coal haulage by rail used to require movement in small loads of only 10 tonnes, with frequent remarshalling of wagons, together with a time- and space-consuming process of coal stockpiling at or near the consumers' factories. In contrast, energy can now be distributed in a continuous flowline process from production or storage point to the consumer over a fixed and exclusive network. Electricity was the first such energy source to be transmitted through a grid system of cables over or under the surface, but extensive use is now also made of pipelines for the transport of crude and refined oil and natural methane gases (Fig. 2.12).

Freight transport at the international scale provides a final example of the complex nature of goods movements. Transfers between exporters and importers in two countries linked by shipping services will make use of a combination of local, general and specialised cargo ports. Individual locations which form either trip origins or destinations are connected by different modes with freight distribution centres and with the seaports,

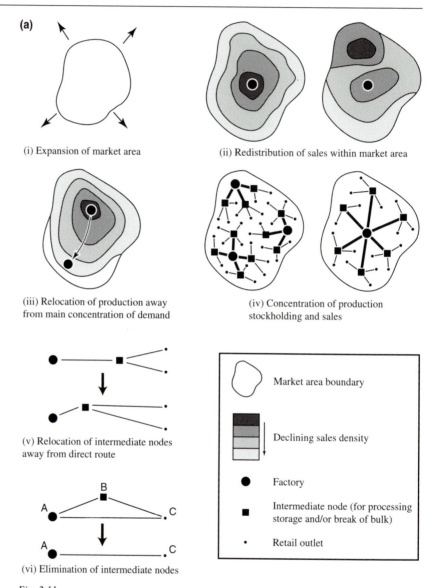

(a)

(i) Expansion of market area

(ii) Redistribution of sales within market area

(iii) Relocation of production away from main concentration of demand

(iv) Concentration of production stockholding and sales

(v) Relocation of intermediate nodes away from direct route

(vi) Elimination of intermediate nodes

Market area boundary

Declining sales density

● Factory

■ Intermediate node (for processing storage and/or break of bulk)

• Retail outlet

Fig. 2.11
(a) Changing relationships between locations of factories and the areal extent of markets and the effects upon journey lengths. (b) The locations of factories, distribution depots, warehouses and retail outlets in relation to journey lengths (facing page). *Source:* based on McKinnon (1983).

and each importer and exporter will decide upon the most appropriate means of transport (Fig. 2.13).

The time dimension can also be an important factor influencing flow patterns, especially where there is agricultural production with a marked

(b)

Channels of distribution

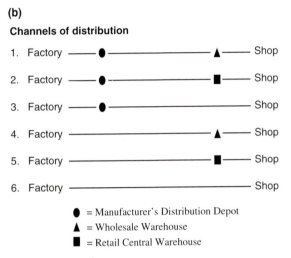

● = Manufacturer's Distribution Depot
▲ = Wholesale Warehouse
■ = Retail Central Warehouse

Fig. 2.11 continued

seasonal regime. In Nigeria the harvest and subsequent dispatch of cotton and groundnuts from the northern savanah and of palm products from the rainforests to the coast takes place at specific times of the year, and demands for haulage can often place a severe strain upon transport services which at other times are underoccupied. A similar concentration of freight flows into a particular period occurs in southern African states such as Zimbabwe, where the tobacco crop is sent by rail for export through South African ports.

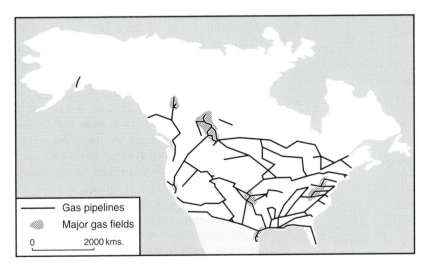

Fig. 2.12
Principal natural gas fields and gas pipelines in Canada and the USA.

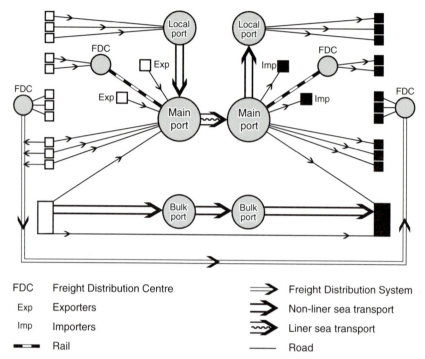

FDC   Freight Distribution Centre   ⟹   Freight Distribution System

Exp   Exporters   ⟹   Non-liner sea transport

Imp   Importers   ⟿   Liner sea transport

▬▬   Rail   ——   Road

Fig. 2.13
Illustration of a cargo distribution system involving exports from a hinterland to a foreland
via main, local and bulk commodity ports. *Source:* based on *Geojournal* **1**(3).

2.5.4 *Commodity flow analysis*

Attempts have been made to classify freight flows in terms of time, distance
and combinations of these two variables, and Hay's groupings (1976)
recognised international, hinterland and hierarchical patterns (Fig. 2.14).
International flows occur within large trading blocs where the individual
trading states may be either within the same contiguous region or
dispersed throughout the world. Examples of closely knit blocs include
the European Union (EU) and the former Comecon organisation, and
existing flows in these two areas are likely to be substantially altered
as the political changes of the 1989–91 period are followed by the
extension of Western trading links into eastern Europe. All such international blocs are distinguished by the dominance of internal trade links
and external commerce is usually at a low level (Box 2.7).

A second type of flow pattern is that confined within a hinterland,
where one major centre is the attracting and/or dispersal point for freight.
Seaport hinterlands with flows of exports and imports typify these patterns
but major industrial cities can also generate similar movements within
their spheres of influence. Finally, hierarchical flows are associated with
the assembly and distribution movements in manufacturing industry

Box 2.7 Commodity flow analysis

The concept of commodity flow analysis helps to explain the linkages between the various components of the world's production and exchange systems. Ullman (1956) used the term 'complementarity' to define the basic relationship between supply and demand locations and the movements which this generated. Where a demand could be satisfied by more than one source of supply then 'intervening opportunity' was created, and the term 'transferability' describes the costs involved in transport between locations of demand and supply. These basic concepts have been elaborated using gravity modelling techniques to compare predicted and actual flows of goods within a region, relating specific movements to the various modes of transport that are available.

The complementarity index compares actual flows with predicted flows calculated using a gravity modelling approach. If data on all inward and outward flows for all centres and regions within a state are available it is possible to prepare a transaction flow analysis which yields a salience score. The value of this score expresses the relationship between predicted and actual flows and this type of analysis has been applied to regional flows within Nigeria. Forms of regression analysis have also been used in a West African context to identify towns receiving more or less than the predicted amount of traffic.

$$Fe_{ij} = \frac{F_{i*} \cdot F_{*j}}{F_{**}}$$

Fe_{ij} is expected flow from location i to location j

F_{*j} is total flow in to j from all other regions

F_{i*} is total flow from i to all other regions

F_{**} is total flow in system

already discussed, and examples of all three types are looked at again in Chapters 5 and 6.

2.6 Changes in demand

Deserted railway lines, derelict docks and abandoned canals are evidence that the demand for transport fluctuates both from place to place and in time. Changes in the nature of industrial activities or within urban communities will be followed by changes in transport requirements, and advances in transport technology will in their turn alter the capability of different transport modes to respond to the demands of industry and society.

In the Western world the most obvious example of a shift in demand is provided by the decline of public transport in the face of steadily rising car ownership. In a similar way the flexibility of commercial road

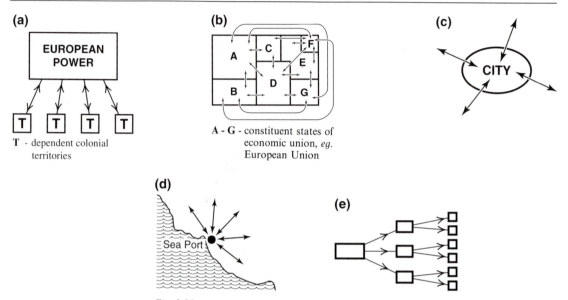

Fig. 2.14
Basic types of flow pattern: (a) trading bloc, e.g. European power–colonial territory pattern; (b) trading block, e.g. economic links between states within a continental bloc; (c) hinterland pattern, e.g. daily inward and outward commuting flows; (d) hinterland pattern, e.g. import and export flows; (e) hierarchical pattern, e.g. distribution of goods from warehouse to retail units.

transport, offering a direct service between consumers, has been responsible for the fall in the relative importance of rail and inland waterways as carriers of freight. Where an established demand for the services provided by a particular transport mode is reduced or even eliminated then adjustments to this changed market have to be made by the affected mode if it is to survive. The extent to which these responses are necessary depends upon the flexibility of the service which is losing traffic and upon the degree to which alternative markets may be identified and exploited.

At one extreme are the railways which were built to cater for a specific demand, such as the haulage of mineral ore from mine to processing plant or export outlet, but with exhaustion of the resource the link becomes totally redundant and is abandoned. In contrast, air passenger services may be easily diverted from a declining to a rapidly expanding market, and bus and coach hire enterprises have a similar degree of flexibility which enables them to adjust to changing demands far more easily than a static facility such as a railway is able to do.

In the UK the canals, like the railways which succeeded them, were unable to match the advantages of road haulage for industrial traffic and after 1920 they rapidly declined as commercial routes (Box 2.8). However, with the growth of leisure activities in the second half of the twentieth century the attractions of these redundant waterways were soon recognised and they now satisfy a growing demand for towpath walking and canal

---

Box 2.8 The decline of the British canal system

At its peak in the 1830s the British canal system had a total length of 48 000 km and was the principal means of transport for both bulk raw materials such as coal, iron ore and limestone, manufactured goods and much of the agricultural produce destined for the urban market. With the expansion of railways after 1840 and the later growth of road transport, demand for inland water transport was progressively reduced and by the mid twentieth century much of the canal network was disused. Commercial traffic is now restricted to a few waterways where the needs of heavy industry or the convenience of moving freight to and from ports still justifies the use of barges. By 1991 total inland waterway traffic had declined to 5.4 million tonnes and only the Thames and Aire–Calder systems carried over 1 million tonnes annually.

---

cruising, having already been exploited on a large scale for fishing. The British Waterways Board have improved many channels and locks, and several disused canals have been restored to a navigable condition by private organisations to meet the need for a more extensive network of 'cruiseways'.

Similar adaptations may be seen on many of the rural branch railways abandoned by British Rail during the programme of widespread contraction in the 1960s. Several of the lines described as uneconomic and subsequently closed now enjoy a flourishing and profitable life as privately run concerns, fulfilling a popular demand from tourists for steam train services along scenic routes in upland England and in North Wales (Fig. 2.15). The Severn Valley Railway and several similar enterprises have now been reconnected to the British Rail network, enabling the demand for recreational travel on these preserved lines to generate additional journeys on the main line system. Other abandoned lines have been adopted as cycleways and walkways in popular touring areas such as the British National Parks.

Railways, which are especially vulnerable to challenges for their market from more flexible means of transport such as road and air undertakings, must attempt to maintain revenue levels by concentrating upon the freight and passenger traffic for which they can still offer a competitive service. The US railways in particular provide an example of how a rapidly declining demand for inter-urban services has been countered by new initiatives. As in most industrialised states, long-distance freight haulage in America was a monopoly of railways and inland waterways in the nineteenth and early twentieth centuries, but after about 1940 there was growing and vigorous competition from oil pipelines and the trucking industry. By the late 1980s the railways had become the minority partner in freight carriage, road haulage and pipelines having expanded to capture roughly equal proportions of the national market (Table 2.4).

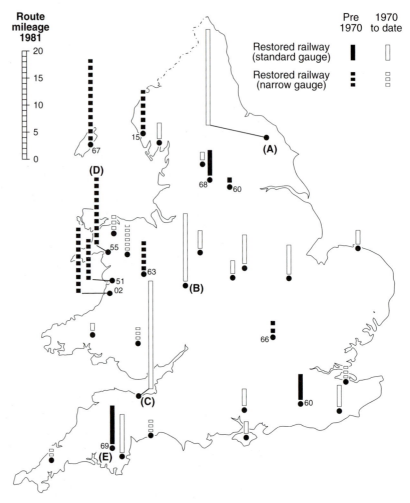

Fig. 2.15
England and Wales: railways closed to traffic by British Rail and later reopened as private undertakings for passenger services. A, North York Moors Railway; B, Severn Valley Railway; C, West Somerset Railway; D, Ffestiniog Railway; E, Dart Valley Railway. Dates (e.g. 60 is 1960) refer to year of opening as a private undertaking. *Source:* based on Fig. 2.9 in Patmore (1983).

The American railways' share of the national passenger market also fell rapidly after 1940, with the spread of domestic air services and the increasing use of the private car for long-distance trips on the express highway network. The proportion of all inter-city passengers carried by rail fell from 77 per cent in 1930 to 7 per cent in 1970, the year in which the National Railroad Corporation (Amtrak) was created by the federal government to try to ensure the survival of nationwide rail services. Amtrak operates its own trains over the privately owned rail systems,

(a)

(b)

Plate 2.2
(a) Reuse of former British Rail lines closed as uneconomic. The Severn Valley line, closed in 1962, has been reopened as a private railway between Bridgnorth and Kidderminster and attracts a flourishing tourist traffic in summer. Limited services are also operated in the winter. View of the station at Arley. (b) Goathland station with diesel multiple unit on the railway between Pickering and Grosmont in North Yorkshire. This line was closed by British Rail in 1965 but was reopened in 1973 as the private North Yorkshire Moors Railway.

Table 2.4
Changes in US inter-city
freight and passenger traffic
1970–90

	1970	1980	1990
(a) Freight traffic by mode (percentage of total tonne-km)			
Rail	39.8	37.5	36.3
Road	21.3	22.3	27.7
Inland waterway	16.5	16.4	13.2
Oil pipelines	22.3	23.6	22.7
Domestic airways	<1	<1	<1
(b) Passenger traffic by mode (percentage of total passenger-km)			
Private cars	86.9	83.5	80.6
Domestic airways	10.1	14.1	17.6
Bus/coach	2.1	1.7	1.1
Rail	0.9	0.7	0.7

Source: Statistical Abstracts of the United States.

which themselves now concentrate upon freight haulage, and the only lines owned by the corporation are the Boston–New York–Washington and Chicago–Detroit railways, both of which still retain substantial passenger traffic (Fig. 2.16).

Inter-urban Amtrak services are now run over 44 000 km of track, serving 500 stations, and during the 1980s the total number of passenger journeys was stabilised at between 20 and 21 million. Federal financial support was reduced by 46 per cent between 1981 and 1989 and Amtrak revenue increased over this period, being derived both from the long-distance trains and the high-density traffic in the north-east corridor, including Boston commuter services. The Amtrak experiment has survived, despite political hostility from some quarters and vigorous competition from long-distance buses, and although some express trains have been withdrawn Amtrak has displayed an overall stability in terms of the number of services operated, and the timekeeping of its long-distance trains has been maintained (Table 2.5).

In order to maintain and expand its share of the passenger market it has been suggested that Amtrak should concentrate on services between cities no further than 500 km apart, where there is the prospect of generating traffic with a minimum expenditure on infrastructure. In California, noted as a state dedicated to air and car travel, federal and state support for improved Amtrak services on the Los Angeles to San Diego line resulted in a rise in annual traffic from less than 400 000 to almost 1.5 million in the 1980s, although this amount is of course only a very small proportion of California's total passenger traffic. Furthermore much of Amtrak's market, and that of similar state-supported ventures in Canada, lies in the leisure and tourist sectors, with trains such as the California Zephyr or the Southwest Chief being promoted to exploit the

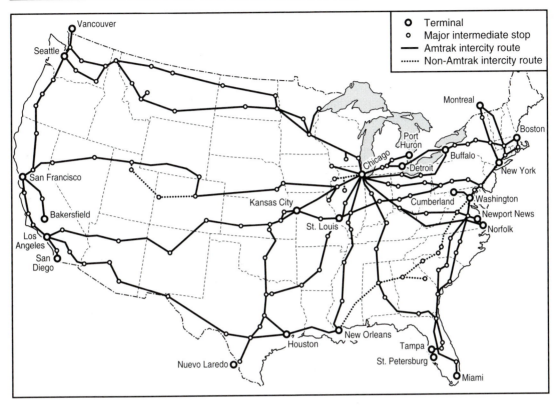

Fig. 2.16
Principal passenger services operated by Amtrak (National Railroad Passenger Corporation)
over the United States railway network.

Table 2.5
Amtrak traffic and
performance 1975–87

Year	Inter-city passengers (millions).	Passenger-miles (millions)	Stations served	On-time performance (% all trains)
1975	17.4	3939	484	77
1979	21.4	4915	571	57
1983	19.0	4246	497	82
1987	20.4	5221	487	74

Source: Nice (1989).

growing market in 'nostalgic travel'. To encourage this trend Amtrak is
advertised abroad with period rail passes giving access to either the entire
system or regional networks.

2.7 Concluding summary

In the sense that it meets a demand for movement the transport industry is properly described as a service activity, but its range of activities and the complex relationships which it has established with its markets are equalled in few other economic systems. As the organisation of society and the economies upon which it depends have become more sophisticated so the challenges for the transport industry have increased.

The opportunities for personal travel at all scales in many countries of the industrialised world have been expanded with increasing car ownership and air transport, but these advances have widened the 'mobility gap' between these states and those of the developing world where so many trips are still carried out on foot.

Road transport has steadily expanded to eclipse the railway as the principal means of movement for both people and freight, and the railway industry has been obliged to restructure its operations and make use of new technology in a bid to remain in the market. In cases where survival has proved impossible railways have relinquished their traditional role as freight carriers and scheduled passenger train operators and have exploited new markets such as leisure and recreational activities.

Further reading

Lowe J C, Moryadis S 1975 *The geography of movement* Houghton Mifflin, Boston
Hay A 1976 *Transport for the space economy* Macmillan
Jones P L et al 1983 *Understanding travel behaviour* Gower
Burns L D 1979 *Transportation, temporal and spatial components of accessibility* Lexington Books, Lexington, Mass.

3 *Transport form and structure*

> **Destinations, directions and decisions – the multimodal choice for travellers and industrialists**
>
> This chapter examines the morphology of transport systems and discusses network analysis and modelling as approaches towards the understanding of the features of different modes. The principal modes are then considered in terms of their capacity, market attraction, and flexibility in changing economic circumstances.

3.1 Introduction

Linear transport features such as railways or motorways form one of the most prominent and easily recognisable elements in the landscape, and many of the early geographical studies of transport focused upon the relationship between routeway and the physical environment. In the UK many landowners were unwilling to allow what they regarded as intrusive lines of railway to be laid across their estates, and the pioneer railway engineers were often obliged to make use of cuttings or tunnels so that views from country mansions were not obscured. This opposition to what is seen as a discordant feature of the landscape continues today in the campaigns waged against motorway alignments in rural UK.

This chapter examines the principal aspects of transport form and structure, drawing upon illustrations from the main modes such as roads, canals, railways, maritime and air services. Several terms are now in common usage for the analysis of transport form and it is appropriate at this stage to identify these as they are used extensively throughout this text. These definitions provide a common basis for the description of transport systems, allowing comparisons to be made between systems in different regions.

A **transport system** may be defined as the assemblage of components associated with a specific means of transport. Thus a railway system will be composed of the tracks and signalling equipment, rolling stock and motive power, freight depots, passenger stations and maintenance workshops. Road transport systems are more loosely co-ordinated and consist of the road network, private and commercial vehicles, fuelling stations, haulage depots and other ancillary features such as car-parks and motorway service stations. A more fundamental distinction between railway and road systems is that the former are usually owned and controlled by a single organisation whereas the latter is made up of many individually owned and operated components. Collectively all the individual transport systems within a country are referred to as the national transport system, with the relative importance of each mode varying from state to state.

The term **network** is applied to the framework of routes within a system and can be used to describe both tangible and visible communications such as railways or motorways or intercontinental air and sea corridors. A **route** is simply a single link between two points which is a part of a larger network.

Nodes and **terminals** are points on a network where several routes converge, and often act as the focus for transport services or for the exchange of traffic between two modes of transport. Such focal points often have their location predetermined by the industrial or settlement geography of the region which the transport system was built to serve. In the UK, for example, most important towns and cities in existence in the early nineteenth century became railway centres. In other cases, however, a transport system has created its own nodes in response to the particular requirements of the traffic which it carries. Major motor intersections often occupy rural locations and important railway junctions are not always associated with major population centres. In the UK the towns of Crewe and Swindon both owe their origins to their focal positions on the growing railway network of the 1840s.

Traffic interchange points can vary in size and importance from an inland freight container depot or urban railway station to an individual bus stop or small car-park, but all share the common characteristic of allowing the transfer of freight or persons from one mode of transport to another.

3.2 Transport morphology

In Chapter 2 the responses of transport systems to different types of demand were examined. Each transport mode caters for a variety of demands, and as it has evolved to serve its market so it has assumed a characteristic form. Both canals and railways were built primarily to meet the demands of industrial Europe for more efficient means of transport, but the geographical extent of inland waterway networks was much less than that of railways because of constraints such as gradient imposed upon canal building. The motorway systems of the late twentieth century were built specifically to meet the demands of large volumes of motorised transport and so differ substantially in their form and layout from the conventional roads they were designed to supplement.

Many sections of the railway network in industrialised countries have become relict features of the landscape as road competition has made them redundant, and the once flourishing passenger terminals in major seaports have also been converted to other uses with the capture of their traffic by air transport (Box 3.1 and Fig. 3.1).

Changes in technology exert a strong influence upon transport form. The steam locomotive railways released the industrialising countries of nineteenth-century Europe from the limitations of horsepower, and the twentieth-century development of air transport has in its turn aided the

---

Box 3.1 Use of redundant transport land

Land no longer required for operational transport purposes can often represent a valuable asset in both urban and rural areas. There is a basic distinction between the linear form of abandoned lengths of railway and canal and the more compact form of land formerly in use as major rail terminals or docks, which in urban areas has frequently been the site of new commercial, industrial and residential developments. Former railway lines have been incorporated into road bypasses and converted into cycle- and walkways. Greater Manchester had 180 km of disused railway track within its boundaries in the mid-1980s and proposals for a network of cycleways using both railway lines and canal towpaths were made, stressing the advantages of these traffic-free routes for recreational and journey-to-work purposes. The same city also converted the redundant Central Railway Station into a major exhibition complex and the Salford Docks on the Manchester Ship Canal are also being redeveloped for industrial and commercial uses.

In the USA and Canada many large tracts of redundant docks and wharves in major seaports have been the subject of ambitious schemes for alternative urban functions under the general description of waterfront regeneration. In Vancouver the Burrard Inlet shore is now the site of marinas, hotels and convention centres and the False Creek waterfront became the venue for the 1986 Exposition, with an associated development of marinas, retailing and housing (Fig. 3.1). The Lake Ontario port of Toronto has had changes in waterfront use carried out in a co-ordinated manner by the Toronto Harbour Commission, with the construction of hotels, housing, amusement centres and commercial buildings in Harbour Square and adjacent areas. In this case developers were able to make use of both former harbourside land and land reclaimed from Lake Ontario. ▶

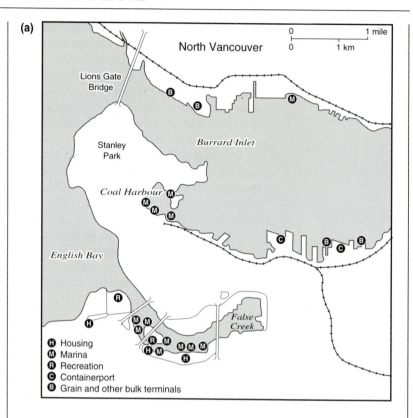

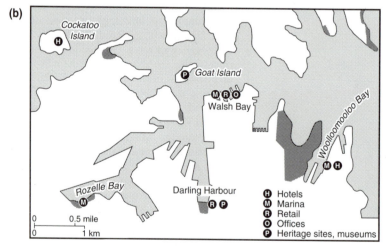

Fig. 3.1
The redevelopment of redundant transport land: (a) central Vancouver port area, showing redevelopment of False Creek waterfront and part of Burrard Inlet southern waterfront; (b) actual and proposed redevelopments in Sydney Harbour. *Source:* based on Hoyle (1990).

growth of mining and industrial communities in remote parts of Canada, Australia and Russia, where adverse climate and topography have severely inhibited land communications.

Although electricity is now an indispensable part of modern life it has had only a selective application as a form of motive power in transport, because of its high capital costs of installation when compared to the use of steam or diesel power. Electric traction is therefore largely confined to railway systems carrying high-density freight or passenger traffic, where levels of revenue can justify the expense of electrification.

Transport systems can also provide interesting examples of the delayed application of technological advances. The civil engineering techniques required for the construction of long railway tunnels under estuaries or through mountain barriers were available in the late nineteenth century, and were used in the building of tunnels through the Alps of Europe and through the Rockies in Canada. Road traffic in these areas, however, suffered the inconvenience of detours and difficult, often snow-blocked, high-level mountain passes until the 1960s, when the volume of road traffic had increased to levels high enough finally to justify the expenditure on tunnels.

The initial form of a transport system will be strongly influenced by the amount of capital available for construction, and this in turn will depend upon the perceived benefits to be derived from the proposed road, railway or canal in terms of revenue and the potential for economic growth in the region to be served by the system. During the nineteenth century privately raised finance was the principal source of capital for railway construction, but contemporary projects such as motorway networks or major airports are often built with public sector funds and the immediate financial benefits are thus not always a critical issue. The Channel Tunnel between the UK and France offers an example of a complex funding operation, with the tunnel itself being a product of private enterprise but the eventual building of a high-speed link between London and Folkestone being a matter for negotiation between the British government, British Rail and private financiers (Box. 3.2).

The original capacity of a transport system was often constrained by limited financial resources, but as the undertaking prospered additional capital became available for improvements and extensions. For example, international steamship routes were considerably shortened with the opening in 1869 of the Suez Canal and in 1914 of the Panama Canal. Both schemes were expensive to build but the tolls paid by the shipping companies who profited from them made these canals a worthwhile investment. However, with the increasing size of ships in the second half of the twentieth century the canals' advantages were reduced as they were unable to accept large vessels, and a significant proportion of international maritime trade no longer uses either the Panama or the Suez routes (Table 3.1).

Many mining and industrial regions within the developed world have

Box 3.2 Investment and the Channel Tunnel

The opening in 1994 of the Channel Tunnel linking the UK and France was the culmination of over 200 years of political discussions, civil engineering advances and negotiations over the raising of the necessary capital. Many of the earlier schemes were to have been financed by British and European railway companies but the Anglo-French governments' initiative of the 1970s involved much public investment. This project was abandoned partly because of controversy within the UK over the cost of a new high-speed railway between London and the tunnel terminal at Folkestone, which would have been a call on public funds. The existing tunnel was financed by private capital and will eventually be linked to all major European cities by networks of motorways and new railways built to the requirements of high-speed trains such as the *train à grande vitesse* (TGV). However, proposals for the construction of the urgently needed high-speed railway connecting London with the tunnel have involved both private consortia and the public sector, in the form of British Rail. Agreement has yet to be reached on the final route, and on the financing of construction, and the situation is further complicated by the forthcoming privatisation of British Rail. As one of the major civil engineering projects of the twentieth century the Channel Tunnel has been justified in terms of its market potential and it encountered very few physical building difficulties; it is the investment problems behind the London–Channel rail link that have proved controversial. The fact that the UK still lacks this vital new rail link illustrates the difficulties in co-ordinating the funding of different components of a major transport project.

highly complex railway networks, often with duplicated routes between major cities, and these patterns frequently reflect the competitive environment in which these lines were initially operated. Regions generating large quantities of lucrative freight traffic, such as the expanding coalfields of the UK in the mid nineteenth century, would attract several pioneer railway companies and by 1900, when most of the British network was completed, many examples existed of route duplication between leading industrial centres and also within manufacturing regions such as south Lancashire or the west Midlands. Passengers were offered several different services between the main provincial cities and London and could take advantage of the 'fare wars' waged between rival railway companies.

Transport form has also been strongly influenced by political and strategic motives, often stemming from the aims of a state to gain control over areas seen to be of economic or military significance. The Roman road network of western Europe is an early example of a military transport system, echoed in the same region in the 1930s by the vigorous programme of motorway building in Germany initiated by the National Socialist regime (Fig. 3.2). Designed ostensibly for commercial motives

			Maximum vessel size (m)		Traffic in 1990/91	
Table 3.1		Maximum depth			No. of vessels in	Total net register
Capacity of Suez and Panama canals	Canal	(m)	Length	Beam	transit	tonnage
	Suez	19.5	No restriction	70	18 221	419 429 000
	Panama	12.8	289	30	12 572	162 695 000

Source: Lloyd's Maritime Directory (1993).

Fig. 3.2
The Roman road network in western Europe.

and to relieve unemployment, one of the major aims was in fact to provide mobility for armed forces as part of the plan for territorial expansion of the Third Reich.

Political fragmentation has also affected the structure of transport systems in Europe. After the partition of the Austro-Hungarian Empire in 1919 the railway system focusing upon Vienna and Budapest required substantial alterations to meet the needs of the newly created states of Yugoslavia and Czechoslovakia and a much-reduced Austria. The political changes of the late 1980s in eastern Europe, resulting in the partition of Yugoslavia and Czechoslovakia, have also had repercussions upon road, rail and air transport networks.

In North America the construction of transcontinental railways from east to west coasts as undertaken to consolidate the territorial units of the USA and Canada. The Canadian Pacific Railway was built from Montreal westwards across the Prairie Provinces to strengthen the hold of the government upon a productive region whose natural lines of communication, principally the Red River valley, led southwards, enabling contacts to be made more easily with the USA rather than eastern Canada until the railway was opened. On a more ambitious scale the Trans-Siberian Railway was constructed to strengthen Russian influence east of the Urals through Siberia to the Pacific seaboard at Vladivostok.

Similar motives lay behind railway building in the African colonies of European powers in the late nineteenth century. In West and East Africa the layout of railways in the former colonies of Senegal, Ivory Coast, Cameroon, Kenya, Tanganyika (now part of Tanzania) and Mozambique was designed with the dual aims of carrying primary resources to the coast for export and establishing territorial security.

The relationships between the physical landscape and transport systems and the methods adopted to overcome constraints imposed by relief have attracted the attentions of many geographers. When these constraints are translated into building and operating costs the completed transport project can often be seen as a compromise, where economies in construction can involve higher costs of operation, or where a heavy investment in building will be rewarded by lower operating costs. If, however, the initial demand for the proposed new route is at a high enough level then the necessary capital investment will be made available to overcome the more serious physical obstacles. Where the anticipated traffic is light and financial returns modest then the amount of building capital will usually be less and the transport system provided will not be so easily adapted to the terrain.

3.3 Route selection

The relative effects of these various factors on the form of a transport system can be assessed in the context of route selection, the process by

Box 3.3 Route selection processes

Black's (1993) recent discussion of route location is based upon a theoretical economic approach designed to aid our understanding of the alignment of existing routeways and applied to railways in the US Midwest. Identifying a least-cost routeway involves making a series of assumptions including the following:

1. Construction costs will vary directly with distance.
2. Construction costs will be kept to a minimum.
3. Demand for flows between points on the routeway will be constant.
4. The region in which the routeway is located has only two potential termini for this route.

Using these assumptions the actual alignments of three railways in Indiana, all having the state capital of Indianapolis as one of their termini, are compared with the alignments generated on a theoretical basis (Fig. 3.3). In all three cases strong similarities between the actual and generated alignments can be seen and the differences that do exist result from local deviations that are not allowed for in the theoretical framework.

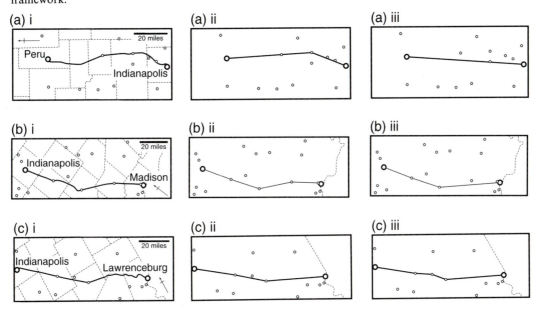

Fig. 3.3
Three examples of railway route selection in Indiana, United States with all three routes radiating from Indianapolis (a) north to Peru (b) south-east to Madison and (c) east south-east to Lawrenceberg: (i) the railway as constructed; (ii) graphical representation of route; (iii) route generated using population data and a shortest-path algorithm. *Source:* based on Figs 7, 8 and 9 in Black (1993).

which the most acceptable alignment for a road, canal or railway is chosen (Box 3.3 and Fig. 3.3). Wellington's pioneer work (1887) explored the relationship between the aims of a transport undertaking to maximise traffic on the one hand, and to minimise construction and operating costs on the other, expressing these objectives in terms of the extent to which a projected route is deflected from a straight line. The early canal engineers tried to avoid costs of lock construction by following a level line as far as possible, although this practice often resulted in very circuitous 'contour' routes.

The Leeds and Liverpool Canal was the pioneer trans-Pennine navigation in the UK, and diversions to make use of one of the lowest passes through the uplands resulted in an overall length much greater than the direct distance between the two terminal cities. In the north-west USA builders of the railway from Spokane across the Cascade Range to the Pacific coast considered several planned alignments, and the final route represented a compromise solution in this difficult upland terrain. In contrast, the Perth to Port Augusta railway in southern Australia encountered few physical obstacles and includes a straight length of 530 km across the Nullabor Plain.

This approach can be applied to many analyses of contemporary route selection but many other factors in addition to the terrain can now act as 'deflectors' from a direct alignment. New roads can be diverted from the original planned route to ensure the protection of environmentally sensitive areas such as National Parks or nature reserves. Existing routes may be realigned to meet the needs of new or improved transport modes. In France the original trunk railway from Paris to Lyons has now been replaced by a new route designed primarily to accommodate the TGV express services, and similar express railways have been built in Italy and Germany. In the UK the controversial Union Railway planned to connect London with the Channel Tunnel terminal at Folkestone has encountered strong opposition from environmental groups but it is likely that the final route will not diverge significantly from the direct alignment, even after objections have been met.

3.4 Quantitative approaches to transport form

The application of topology and network analysis to transport systems enables the spatial arrangement of the lines and points which make up a network to be described in clear and precise terms. Two attributes which are of particular interest to geographers are the complexity of a network and the relative accessibility of points within it. Network analysis, as adopted by geographers, is based upon the simplification of a transport system into a graph where the routes connecting nodes are usually indicated as straight lines. Complexity can be measured by a series of connectivity indices, which are based upon the numbers of lines and nodes in a network and indicate the progression from a rudimentary network to one where all points are directly connected with each other. The centrality values are used to describe the degree of accessibility of nodes on a network and can be arranged in matrix form for a direct comparison of the values of individual points (Box 3.4 and Fig. 3.4).

Network analysis becomes more complex, but at the same time more valuable to the transport geographer, when specific values are assigned to lines and points within a framework. A valued graph is produced when each line is described in terms of its length, or the time taken to travel along it or the cost of travel. A directed graph is one where certain lines permit movement in one direction only, so that access between pairs of points will not be the same in both directions. Both valued and directed graphs are of use in the description and interpretation of urban street networks, where the movement of both private and public transport can be affected by restrictions such as one-way streets.

The effects of changes in transport systems, such as the closure of parts of a railway system or the addition of new highways to a road network, can be measured using these techniques. The connectivity of a primary highway system will be enhanced with the building of additional links, and the relative accessibility of individual points on the system will also be altered. The deviation of an actual route from the direct alignment between two points can be measured by the detour index, which is the ratio between this actual distance and the straight-line distance. Lösch applied the physical law of light refraction to route alignment where a routeway is planned across two or more areas, each with different transport costs depending upon the terrain. He suggested that the optimal route would be identified by an equation involving transport costs and the angles of entry and refraction at the boundary between each type of terrain (Fig. 3.5).

Box 3.4 Some basic measures in network analysis

At its simplest level a network consists of edges (or routes) which intersect or terminate at vertices (or nodes). A tree or branching network is one where no loops or circuits are present, whereas a circuit network contains one or more loops. Where not all the edges and vertices are interconnected a series of subgraphs exists and this distinction between partially and completely interconnected networks influences several of the indices used to measure the characteristics of networks.

1. Measures of connectivity – an application of the gamma and beta indices is illustrated in Fig. 3.4.
2. Measures of centrality – the associated number and the accessibility index are used to describe the varying levels of accessibility to individual vertices on the network.
3. Measures of shape and compactness – the diameter of a network is defined as the number of edges included in the shortest possible path between the two most remote vertices.
4. Measures involving values of each individual edge. ▶

(a) (i)

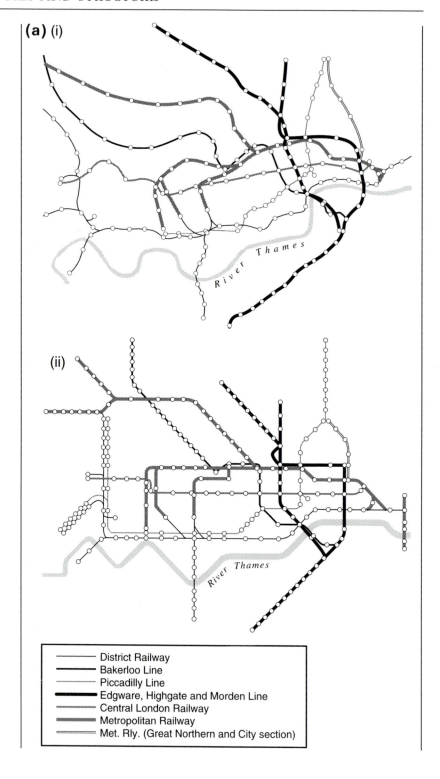

(ii)

	District Railway
	Bakerloo Line
	Piccadilly Line
	Edgware, Highgate and Morden Line
	Central London Railway
	Metropolitan Railway
	Met. Rly. (Great Northern and City section)

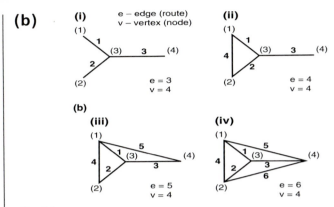

Fig. 3.4
(a) The London underground railway system as shown in (i) a map of the 1920s and (ii) H Beck's graphical interpretation of 1931. (b) An illustration of how a simple network with four vertices increases in complexity as the number of edges is increased. The level of connectivity increases as the number of direct links (edges) between each of the four vertices increases progressively from example (i) to example (iv), where a maximum number of six edges is present.

Connectivity may be measured by two indices:
    Beta index = e, which is 0.75 for example (i) and 1.5 for example (iv), where maximum connectivity has been achieved. Values of between 0 and 1 indicate branching or disconnected networks and values greater than 1 indicate networks of increasing complexity.
    Gamma index = e, which is 0.5 for example (i) and 1 for example (iv). The gamma index expresses the ratio of the observed number of edges on the network to the maximum possible number of edges that would be present if all vertices were directly interconnected. The gamma value varies between 0 and 1 in a fully interconnected network.

3.5 Transport nodes and interchanges

On a graphical network a node (or vertex) can be defined as the intersection of routes or edges or as a terminal point. In a wider sense nodes are recognised not only as focal points within a particular network but also as locations where two or more different modes make contact. In both cases there are usually facilities for the reorganisation or exchange of traffic and many key nodes have developed into important commercial locations such as seaport cities. At the simplest level, as already noted, bus stops, taxi ranks and bus stations provide examples of nodes where pedestrians make the transfer to motorised transport. Major rail stations are points of interchange for rail passengers, travellers using urban taxi, bus and suburban rail services, cyclists and pedestrians. Airports provide examples of nodes with the most rapid rates of growth in terms of traffic

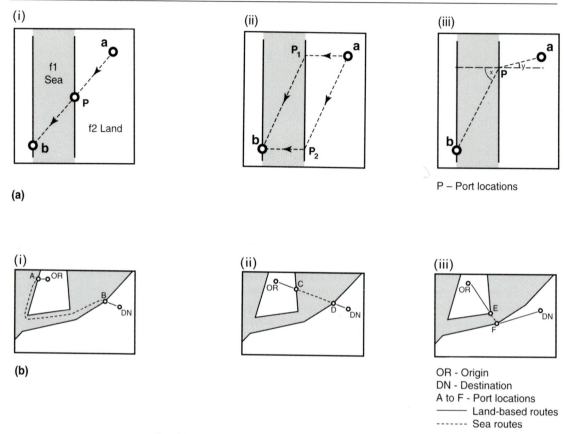

Fig. 3.5
Examples of route refraction in land and sea environments. (a) Lösch's application of the law of light refraction to routes across sea and land surfaces with transport costs f1 and f2 respectively. In (i) the direct route ab is adopted irrespective of transport costs. In (ii) the route aP1b minimises land and maximises sea transport costs, whereas route aP2b maximises land and minimises sea transport costs. Diagram (iii) illustrates the choice of an optimal route defined by the formula:

$$\cos x \; f2 = \cos y \; f1$$

(b) Three routeways between an origin OR and a destination DN making use of (i) the nearest ports A and B to the origin and destination, (ii) a direct alignment via ports C and D and (iii) the short sea crossing via ports E and F.

in the period since 1950, and this expansion has necessitated the building of many new roads and railways to connect with city centres (Fig. 3.6).

Within towns and cities the location and distribution of traffic nodes are the key elements in transport infrastructure, and the improvement of the efficiency of such nodes is a major concern of urban planners (see Ch. 8). Up to the mid twentieth century the bus terminal, usually sited within or close to the central business district, was the major focus of daily travel in many west European cities but today the multi-storey car-park can act as a much more important node. Policies aimed at

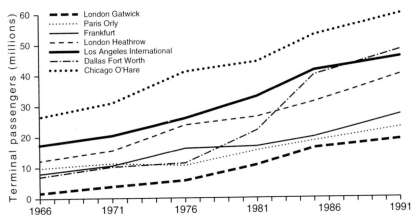

Fig. 3.6
The growth of terminal passenger traffic at selected major world airports between 1966 and 1991. *Source:* United Nations Statistical Bulletins (1966–92).

integrating the different urban transport modes have also produced multi-purpose exchange centres bringing together rail and bus stations and car-parks. Rail goods depots in urban areas have lost much of their significance with the capture of freight traffic by road transport, and many of the traditional railway goods marshalling yards have now been supplanted by inland road–rail container terminals.

Plate 3.1
The central station in Amsterdam is an important interchange between ferry, metro, tram, bus and rail traffic, offering commuter services to surrounding towns and long-distance services to other cities on the European rail network.

The development of seaport nodes has been traced from their origins as primitive land–sea traffic exchange sites to modern complexes forming the nucleus of major cities such as Hamburg, Cape Town or Vancouver. Bird's pioneer analysis (1971) identified successive growth stages in terms of ship-berthing, cargo-handling and freight storage facilities, with the later phases of expansion being marked by dock construction and the provision of specialised terminals for petroleum, iron ore and other bulk freight and, most recently, for containers (see Ch. 5).

As with seaports, the airport can vary in size and complexity, from the simple dirt or grass landing strip, still a vital element in the air transport of many less developed countries and in interior Australia and Canada, to the modern terminal capable of handling up to 300 daily aircraft movements and over 30 million passengers each year. Since passengers provide the bulk of air transport revenue airports must be located close to major cities, although such sites are not always entirely satisfactory from the aircraft operators' viewpoint. In terms of operational requirements the most acceptable location is often remote from population centres, but from the viewpoint of traffic generation the opposite is true and most major airport nodes normally occupy a compromise site. The evaluation of alternative locations for new airports is a technically complex and politically sensitive task particularly in view of the environmental impact of such facilities and the operation.

3.6 Models and transport form

Several conceptual models have been devised as aids to the understanding of the development of transport systems and of their component parts, and there is often an emphasis upon identifying salient stages of growth. In their hypothetical approach Ekstrom and Williamson (1971) recognise an **initial phase**, with the introduction of a new transport mode, followed by a **spread phase** with spatial diffusion of the network and a **co-ordinating phase** where the new and existing modes become integrated. These three may be followed by a **concentration phase**, involving an emphasis upon certain flows along selected routes, and the processes that can lead to the decline or demise of a system are included in what is termed the **liquidation phase**.

The Lachene (1965) model is deductive and examines the development of a transport system upon a hypothetical isotropic plain, which in practice could be a mid-continental region such as the American Midwest. It may be compared with Lösch's approach to the evolution of an economic landscape, progressing from an initial network of paths and trails arranged in a grid pattern to the selective growth of towns and villages at favoured intersections, and culminating in a smaller number of high-order settlements connected by high-grade routes such as railways and motorways. Throughout the period of growth the aggregate length of the network is gradually reduced from the complete grid of pathways in the early stages to the selective framework of major routes in use in the final stage (Box 3.5 and Fig. 3.7).

The Taaffe, Morrill and Gould (1963) approach is inductive, descriptive and based upon observations of colonial transport development in West Africa. The model traces the progressive establishment of links between the sea coast and the interior of an undeveloped region, with the emergence of dominant and subordinate coastal and inland nodes connected by lateral routes and feeder branches. In the final stage high-grade links are, as in the Lachene model, provided between selected ports and interior towns.

Both the Taaffe and Lachene models emphasise the fact that progressive changes in a network of nodes and links can produce a pattern in which movement usually becomes increasingly concentrated along a few favoured routes. Neither model, however, caters for the behavioural or probabilistic element in transport growth but other studies have attempted to incorporate this approach. Kansky's (1963) simulation of the growth of the Sicilian railway network involved the prediction of internodal links on the basis of selected geographical characteristics, and Black (1967) adopted a similar approach in his simulation of the expansion of the US state of

---

Box 3.5 Applications of the basic Taaffe, Morrill and Gould (TMG) model

The basic TMG model has, since it was first published in 1963, attracted the attentions of other geographers who have attempted to apply its concepts to other regions of the less developed world. Hoyle (1983) offers an explanation of the growth of the East African transport network from the fifteenth and sixteenth centuries, identifying the Arab trading phase, the colonial pioneer railway phase up to 1914, the railway elaboration period of the 1920s and 1930s and the post-1949 phase during which several isolated inland lines were connected by coastal links. This application does illustrate the significance of links between the evolving railway system and the distribution of mining and commercial agriculture, but Hoyle also emphasises the more recent importance of roads as routes between the interior and the ports.

In his application of the TMG model to rural road expansion in the Ile-Ife cocoa-growing region of Nigeria, Aloba (1983) recognises five periods of road growth between 1910 and the 1980s during which progressive increases in network complexity led to improvements in the rural economy with the gradual expansion of cocoa cultivation into areas of subsistence farming. However, this example involves an interior area dominated by one large urban centre so that not all components of the TMG model can be realistically applied.   ▶

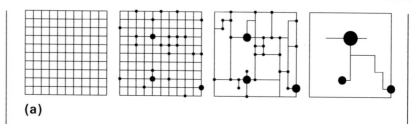

(a)

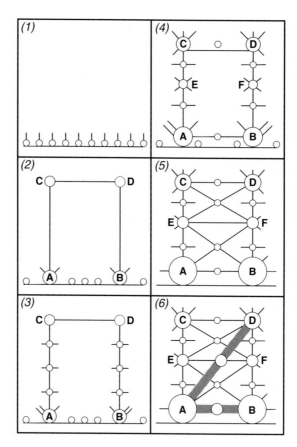

A, B – Ports
C, D – Inland centres
E, F – Intermediate centres

(b)

Fig. 3.7
(a) The basic sequence of network development and settlement growth according to Lachene. (b) The basic Taaffe, Morrill and Gould transport model of 1963. (c) An application of the Taaffe, Morrill and Gould model to a rural area of the interior of West Africa (facing page). *Sources:* (a) based on Lachene (1965); based on Taaffe, Morrill and Gould (1963); (c) based on Aloba (1983).

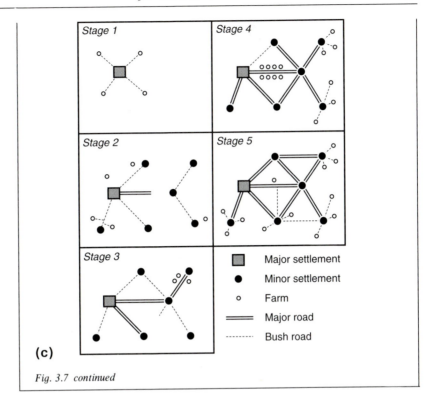

(c)

Fig. 3.7 continued

Maine's railway system between 1840 and 1851. His theoretical 1840 pattern was very similar to the actual network in terms of nodal connectivity (Box 3.6 and Fig. 3.8).

Box 3.6 Gould's spatial exploration model

This behavioural model was proposed in 1966 as an alternative to the Taaffe, Morrill and Gould concepts of transport development. It incorporates a random approach and is based upon a simulation of search theory, with the development of a transport network within an area which contains resources and hazards, or constraints, indicated by isoriths of environmental quality. The developer aims to tap the resources of a previously unexploited area, depicted as a square, by building roads from a port on the coast which forms one side of this square. As road building proceeds so the developer will encounter the resources and the constraints, such as mountains or rivers, within the environment. In Stage One capital is invested in roads which diverge from the port in straight lines. In Stage Two information on the nature of the resources or of the hazards encountered by the advancing roads is fed back to the developer, who may react in one of two ways. The resource already tapped may be exploited by investing in all-weather roads, or the search may be continued for other resources by extending the road network. Stage Three comprises the construction of further links following the principles outlined in the first two stages.

Realism is added to this exercise by incorporating rules governing interest rates, capital budgets, repayment time limits and rates of road depreciation. Hazards are also added in the form of 'shocks' such as floods or droughts and variations in the market prices of the resources that are being exploited. In constructing this model Gould demonstrates the importance of random occurrences and of the behavioural element, showing that bad route planning and poor alignments are often traceable to insufficient survey information or incorrect interpretation of the data.

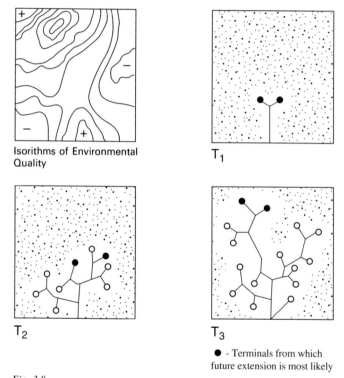

Isorithms of Environmental Quality

T_1

T_2

T_3

● - Terminals from which future extension is most likely

Fig. 3.8
Gould's simulation of spatial exploration in time periods T_1, T_2 and T_3. *Source:* based on Gould (1966).

3.7 Transport modes

Each principal means of transport varies widely in its form and in the arrangement of its component parts. These differences reflect the types of traffic that each mode was designed to carry, and in this section transport systems are examined with reference to (a) their capacity for particular types of passengers and freight, (b) their ability to attract various segments of the market in terms of service quality and reliability and (c) mode flexibility, that is, the extent to which a transport system can adapt to changing markets.

The broad view of transport modes presented by White and Senior (1983) considered five basic factors which influence the shape of transport systems and the ways in which changes take place. These factors act in various combinations to determine the contemporary features of transport modes and are summarised in Box. 3.7.

3.7.1 Capacity

With the notable exception of the pipeline, all the principal transport modes involve the movement of unit loads of passengers or freight in purpose-built vehicles along routeways which, for inland communications, are clearly defined. Vehicle design and tractive power have been progressively improved to move larger unit loads at higher speeds or at lower costs, and there has been a particular increase in the development of vehicles designed for specialised traffic such as chemicals and foodstuffs. In the road transport sector vehicle size is determined by the types of traffic to be carried. In urban areas there is the need for frequent movement of low-capacity trucks and vans carrying out regular collection and delivery services, whereas on inter-urban highways long-distance

---

Box 3.7 The basic factors shaping transport systems (after White and Senior, 1983)

1. The historical factor – this involves the location and pattern of systems, technological development, institutional development and settlement and land-use patterns.
2. The technological factor – the technological characteristics of each major transport mode are considered together with a discussion of the effects of technological advances.
3. The physical factor – this includes physiographic controls upon route selection, and geological and climatic influences.
4. The economic factor – the structure and nature of transport costs are examined, together with service quality and methods of pricing and charging.
5. Political and social factors – these include political motives for transport facilities; government involvement in capital, monopolies and competition, safety, working conditions and co-ordination between modes; transport as an employer and the social consequences of transport developments.

Table 3.2
Cargo vessel size by
dead-weight tonnage
capacity, 1992

Type of vessel	Capacity (dead-weight tonnage) (no. of vessels)				Total no. of vessels
	Below 25 000	25 000– 49 999	50 000– 99 999	100 000 or more	
Oil tankers	4 208	675	644	815	6 342
Chemical carriers	1 581	242	36[a]	—	1 859
Bulk dry carriers	1 209	2161	811	371	4 552
General cargo carriers	16 860	156	—	—	17 016
Roll-on/roll-off cargo carriers (including passenger/ cargo carriers)	3 509	36	—	—	3 545
Container vessels	834	496	84[b]	—	1 414

[a] 50 000–89 999 tonnes.
[b] 50 000–69 999 tonnes.

Source: Lloyd's Register of World Fleet Statistics, December (1992).

trucks with payloads of up to 40 tonnes and average speeds of 80 km per hour are commonplace.

Many large railway undertakings now concentrate upon the bulk freight market, hauling minerals, oil, coal, chemicals and other commodities in scheduled train loads of 20 to 40-tonne wagons, and the 10-tonne all-purpose truck requiring constant marshalling into different trains is now used for only a very small proportion of rail traffic. The development of large specialised freight vehicles has reached its peak in the maritime sector, where massive rises in oil consumption since 1950 have stimulated the introduction of 'very large crude carriers' (VLCCs) for the carriage of crude petroleum. Similar advances have occurred in the transport of iron ore from North and South America to western Europe and from Australia, India and South America to Japan, using ships of up to 200 000 tonnes capacity (Table 3.2).

The growth in numbers of these high-capacity ships has led to the eclipse of the trampship, an all-purpose vessel of less than 10 000 tonnes dead-weight which could carry most commodities and enter almost all commercial seaports, and which until the mid twentieth century was responsible for the carriage of most maritime trade. Increases in the size and carrying capacity of vehicles on inland routes are often limited by loading gauges in the case of railways and by weight restrictions on bridges on road networks. Air transport is less constrained in these respects but the length and load-bearing capabilities of runways do exercise some control over aircraft payloads. Intercontinental planes such as the Boeing 747 series can accommodate over 400 passengers on flights of up to 9000 km, but on short-haul routes, such as north American domestic or northern Europe–Mediterranean services, aircraft with between 100 and 200 seats are more common.

Table 3.3
Service attributes of urban
transport modes

Mode of transport	Maximum capacity in persons per hour	Average speed (km per hour)	Interval between access points (km)
Bus on conventional road network	9 000–10 000	16–24	0.2–0.5
Bus using reserved lane on express highways	20 000	56	0.8–1.6
Urban railway	50 000–60 000	32–48	1.6
Light rapid transport system	40 000	26–38	0.5–1.3
Private car on conventional road network[a]	1 000	19–40	—
Private car on urban motorway network[a]	3 000	72–80	—

[a] Assuming 1.5 occupants per vehicle.
Source: World Bank reports.

The pipeline is unique in its method of freight transport, acting as vehicle and routeway combined, but this advantage of offering a continuous flow system is achieved only at a high capital cost of installation: pipelines with diameters of over 1 m can handle an annual throughput of petroleum of 50 million tonnes. Traffic is still largely confined to crude oil and its refined derivatives and to gas products, but coal in slurry form is pumped from colliery to power station in the USA and experiments with other commodities in capsules within the pipeline have been made.

Vehicle speed, rather than capacity, is often a critical factor in the inland passenger transport market, with variations from less than 10 km per hour for the cycle rickshaw in south-east Asian cities to over 350 km per hour on the French TGV railways. Inter-city journey times have been considerably reduced where electric rail traction is available and over distances of up to 350 km rail can offer shorter city-centre to city-centre timings than domestic air services (see Ch. 6). The relative significance of speed and capacity in the passenger market is best demonstrated within urban areas, where there is a great variation between modes in terms of the numbers of passengers passing through a given distance per unit of time (Table 3.3). Surface and subsurface electric railways can carry up to 65 000 persons per hour whereas a conventional road has a capacity of only 1000 persons per hour when the private car is used. These remarkable disparities and their implications for urban transport are discussed in more detail in Chapters 7 and 8.

3.7.2 Service characteristics

A discussion of the basic structure of operating costs and freight and passenger rates is provided in Chapter 4, and these economic factors have

(a)

(b)

Plate 3.2
Striking contrasts in vehicle speed in vessels carrying tourist traffic on Lake Garda in the Italian Alpine region. The 'Freccia' hydrofoil (a) offers rapid transits between lake-side resorts whereas the paddle-steamer (b) provides much slower and more leisurely travel, catering for the nostalgia for steam-powered ferries.

an important influence upon the types and amounts of traffic carried on different transport modes. However, there are many other features in addition to rates charged to the consumer which can be considered under the general heading of service quality and which influence the final choice of mode that is made. Common to both freight and passenger systems are issues of speed, security and convenience, whilst the passenger market also involves levels of service reliability and comfort. White and Senior (1983) also emphasise the importance of government intervention in terms of controls over monopoly, levels of competition, safety of passengers and working conditions for transport employees. Freight transport systems must also provide specialist services for perishable goods and other traffic for which particular care in transit is required.

These considerations are today often as significant as the purely economic issue of costs of transport, and consumers do not necessarily select those services offering the lowest rates. Levels of reliability, frequency and quality are now often of more relevance in the selection process, and an example of the importance of these attributes is provided by the vigorous marketing campaign conducted by English Channel ferry companies during the period prior to opening of the Channel Tunnel in late 1994. This campaign was based upon the attraction of the more relaxed and comfortable, if longer, transit or 'mini-cruise' across the Channel provided by a conventional ferry, as compared with the service offered by the tunnel 'shuttle' trains.

3.7.3 Mode flexibility

Many changes in the structure of passenger and freight markets have taken place in the second half of the twentieth century and these have highlighted the varying abilities of transport modes to adapt and modify their systems to meet either increased or reduced demands. The ease with which a particular mode can face challenges from its competitors and succeed in maintaining its share of the market depends to some extent upon its physical form, and some systems have suffered considerable losses in revenue because of their inflexibility. For example, an oil pipeline or a railway with high-capacity oil tanker wagons can usually offer much lower rates than the equivalent road vehicles. However, if there are any changes in the location of oil sources and markets then new directions of flow are created, and the inflexible nature of pipelines and railways makes it difficult to respond to the new situation. On the other hand road transport can be adapted much more easily to any new patterns of oil flow providing that adequate roads are available, although the true costs of road-borne oil carriage can be much higher than those by rail or pipeline.

Public transport modes in urban areas also vary greatly in their physical flexibility and vulnerability to challenges to their traffic. In the 1890–1910 period railways in major European cities attempted to combat competition from the newly introduced electric trams by opening additional stations and more frequent services. However, the tram networks were able to penetrate inner urban areas far more effectively than the

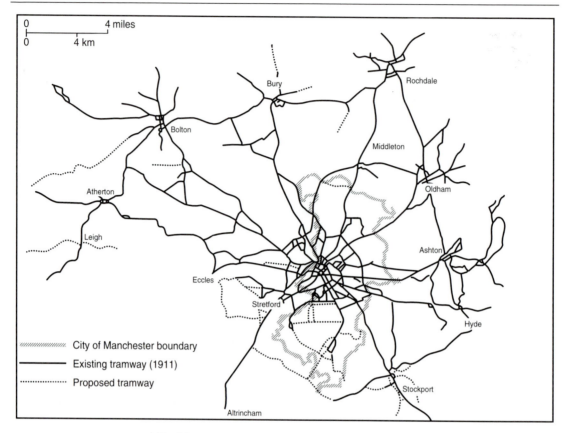

Fig. 3.9
The electric tramway system of the Greater Manchester area in 1911. *Source:* Knowles, personal communication (1994).

railways, which in consequence lost much of their city traffic (Fig. 3.9). In the 1930s the electric tramways in their turn were superseded by the motor bus, whose routes could be easily altered where necessary to serve the growing needs of the expanding suburban districts. Maximum flexibility in towns and cities is of course provided by the private car, although such freedom of movement has generated many problems of congestion and environmental deterioration, as discussed in Chapters 7 and 10.

In the less developed countries road transport is often the only means of freight haulage and personal travel, although its advantages are often reduced by the limited length of all-weather roads and, in the case of the car, by the low purchasing power of much of the population. The use of all-purpose trucks and minibuses carrying people and their goods is common in both rural and urban areas (see Chs 7 and 9 and Table 3.4).

Experiments in transport technology are frequently aimed at developing new levels of flexibility, both in terms of transhipment between modes

Table 3.4
Ranking of farmers'
preferences, based upon
vehicle attributes, in Bomo
state, northern Nigeria. The
scores in rank columns 1, 2,
3 and 4 are based upon
survey data.

Factors		Rank				Total score	Overall rank
		1	2	3	4		
Speed	Frequency	196	54	56	19		
	Aggregate score	196	108	168	76	548	1
Load-carrying capacity	Frequency	56	228	29	12		
	Aggregate score	56	456	87	48	647	2
Reliability	Frequency	21	8	226	70		
	Aggregate score	21	16	678	280	995	3
Cost	Frequency	51	37	13	224		
	Aggregate score	51	74	39	896	1060	4

Load-carrying capacity: explained as the capability of the vehicle to carry the user and all his loads.
Reliability: explained as the ready availablity of the vehicle at whatever time required.
Cost: explained as the vehicle's capability to render required services with reasonably low total (i.e. both initial and operating) costs.
Aggregate score: this is obtained by multiplying the frequency of the rank order.
Source: Barwell et al (1985).

Box 3.8 The air cushion vehicle (ACV)

The air cushion vehicle (ACV) or hovercraft was only able to exploit its amphibious capabilities in those commercial sectors where (a) traffic could be attracted from other modes by the speed of service and (b) passengers were prepared to pay a premium fare for this advantage of a shorter transit time. In southern England a hovercraft service was introduced in 1965 between the coastal city of Portsmouth and the small town of Ryde on the Isle of Wight, separated from the mainland by a 7 km channel. Here the AVC offered a transit time of 20 minutes compared with the 35 minutes on the ferry and was used by commuters and tourists. Similar services were introduced on the Clyde, Humber and Thames estuaries but did not survive. Larger craft competed with conventional ferries on the Dover–Calais route across the English Channel and by 1970 the use of the ACV as a passenger ferry had extended to the USA, Canada, Scandinavia and the Mediterranean, i.e. all locations where either commuters or tourists were prepared to pay premium fares for the service on offer. However, the high fuel consumption rates and maintenance costs as compared with conventional ships gradually exposed the commercial vulnerability of the ACV and since the 1970s the number in use as passenger ferries has fallen sharply. Moreover the hydrofoil, which has lower operating costs than the ACV but similar speeds, has captured much of the traffic, especially on routes in the coastal waters of Italy, Japan and the USA and on rivers such as the Volga in Russia, a country which in the 1980s had over 1000 hydrofoil ferries in operation.

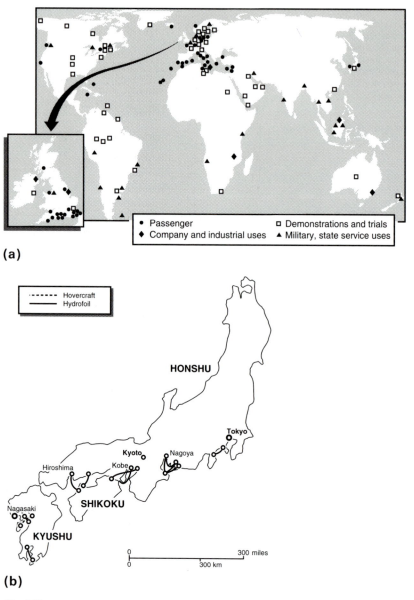

(a)

(b)

Fig. 3.10
(a) World air cushion vehicle operations in the 1960s. (b) Air cushion vehicle and hydrofoil services in Japan in the 1970s. *Source:* (a) based on Tolley (1973).

and entirely new means of transportation. The air cushion vehicle (ACV) or hovercraft introduced in the late 1950s was a successful technical advance which was initially seen as having useful applications in both the industrial world and the developing countries (Fig. 3.10). Fully amphibious versions of the ACV are driven by aircraft-type propellers and travel over

Plate 3.3
This Sikorsky 56 helicopter is operated by British Airways in Scotland and its manoeuver-
ability and small capacity is well suited to services to islands where runways for conventional
aircraft are difficult to construct. Similar helicopters are used extensively to service the North
Sea oil industry from bases on the north-east coast of Scotland.

any land or sea surface on a low-pressure cushion of air, whilst the marine
version is powered by conventional immersed screws. The ACV thus offers
a novel alternative to conventional wheeled transport since no prepared
routeways are required and a direct transition between land and sea
surfaces is possible. This advantage of operational flexibility over different
types of surface is only achieved at the expense of high fuel costs, however,
and the first commercial ventures were limited to short sea transits with
heavy passenger traffic (Box 3.8). Many other services were ephemeral,
with the combination of premium fares and poor travel conditions in
heavy seas proving fatal to commercial success. However, ACVs are still
employed for military purposes, for exploratory mining ventures and for
similar activities where high operating costs are justified by the advantages
of mobility in difficult terrain unsuited to conventional transport.

   The helicopter as an innovative form of air transport has been much
more successful than the ACV, offering maximum flexibility in terms of
its vertical take-off and landing capability in a limited area, although again
incurring heavy operating costs. It is used widely in both military and
passenger-carrying sectors and is also employed extensively for crop
spraying, aerial surveillance and rescue operations by police and fire services.

The transfer of freight between sea and land-based modes has been greatly improved with the introduction of the roll-on/roll-off system for loading and off-loading trucks and other vehicles on to purpose-built ferries. The lighter-aboard-ship (LASH) system for maritime freight transport also has advantages over conventional land–sea interchanges where inland waterways are available, and a similar principle is used in the 'piggyback' method of carrying road trucks on railway flat cars. The greatest advances, however, have been made with the world-wide use of standardised containers for freight, involving the construction of specialised ships, railways wagons and road trucks and the installation of complex handling equipment at coastal and inland locations for the rapid interchange of containers between modes (see Ch. 5).

3.8 Concluding summary

Many of the world's contemporary transport planning problems have their roots in transport infrastructures dating back to the eighteenth and early nineteenth centuries, and an understanding of the basic features of road, rail, maritime and air transport systems is necessary in order to appreciate the current issues of concern to policy-makers and planners. Certain modes can only offer a limited range of functions: the railway, for example, is able to provide long-distance bulk freight transport at competitive rates but cannot match the flexibility over shorter journeys that road can offer. Seaports have been displaced by airports as the major nodes for long-distance passenger travel and certain maritime terminals have ensured their continuing prosperity by specialising in particular commodities such as oil or bulk minerals.

 Subsequent sections in this volume consider many of the general points made in this introductory chapter in more detail, emphasising their relevance at the international, national and local level and in urban and rural environments.

Further reading

Haggett P, Chorley R 1969 *Network analysis in geography* Edward Arnold
Hoyle B S, Knowles R 1992 *Modern transport geography* Belhaven
Lowe J C, Moryadas 1975 *The geography of movement* Houghton Mifflin, Boston
Taaffe E J, Gauthier H 1973 *Geography of transportation* Prentice-Hall
White H P, Senior M 1983 *Transport geography* Longman

4 *Transport and spatial structures*

The role of transport in location: crops, jobs and homes

This chapter examines the relationships between transport and development in the industrialised and less advanced nations, with an analysis of how transport costs and rates can influence industrial location and agricultural patterns. The significance of transport in settlement geography is discussed with a particular emphasis upon urban areas.

4.1 Introduction

Explanations of industrial distributions, agricultural patterns and settlement structures rest upon a wide range of contributory causes and constraints but transport in most cases plays a very significant part. Nineteenth- and early-twentieth-century theorists concerned with manufacturing location, for example, placed much emphasis upon transport facilities as a location factor, although contemporary data and other evidence in support of their claims were seldom available.

Urban growth in the nineteenth and twentieth centuries owed much to the introduction of suburban railways, tramways and later motor buses, and although in contemporary cities road traffic is often seen more as a problem than as an asset, it undoubtedly contributed to spatial expansion. At the level of the world economy transport has made a vital contribution to the expansion of Western capitalist influences into colonial possessions during the nineteenth and early twentieth centuries. Railways in particular were of major importance in consolidating the political and economic control of colonial powers within their overseas territories whilst the steamship provided the essential long-distance links.

During the last quarter of the twentieth century transport and the associated communications industry have undergone substantial changes and have become more closely integrated into the global economy, typified by the transnational corporation and satellite technology. Advances in maritime freight handling and carriage have dramatically cut the cost of international haulage, and modern transport has been aptly described as one of the 'enabling technologies' contributing to the growth of industrial organisations at the world scale and to the establishment of what has been identified as the 'new international division of labour'.

4.2 *Transport and development*

Improvement and innovation in the transport sector, a fundamental reorganisation of industrial production and operational techniques and the perfection of electronic communications technology have all played a vital part in transforming much of the world's industrial capacity from a nationally to an internationally orientated scale. New materials, new processing methods and above all new products in the industrial sector have been introduced in new locations, often remote from their markets, and demanding economic and efficient transportation of goods and quick and reliable transfers of personnel and information. These developments and the corresponding changes in the transport industry have proceeded together and any discussion as to which was the more significant will involve similar arguments to those surrounding the role of transport in the Industrial Revolution of the eighteenth century.

4.2.1 *The industrialised world*

The part played by the transport industry in supporting existing and promoting new industrial growth may be examined under several headings. In the industrialised Western world the railway has continued as an important carrier of raw materials and manufactured goods, aided by the introduction of specialised high-capacity bulk wagons and container trucks. A large proportion of twentieth-century industrial growth, however, owes more to the flexibility of road transport than to the railway, and with the expansion of motorway networks within and between towns and cities the high-technology industries and the larger regional distribution warehouses have been attracted to sites alongside these routes, as in California or southern England, and at highway intersections (Fig. 4.1). Within cities in Europe and North America large office complexes and

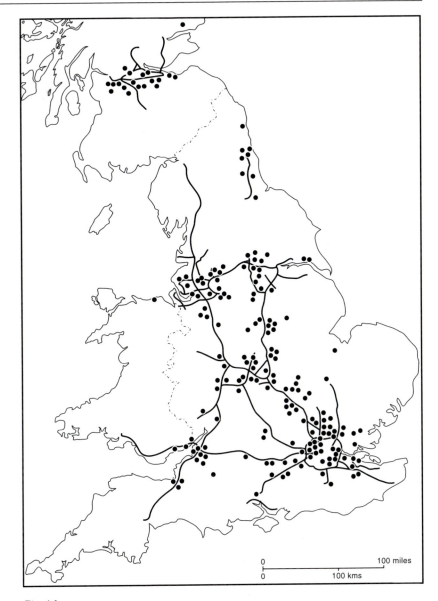

Fig. 4.1
The British motorway network in 1990 and the location of central warehouses of 50 multiple retailers. *Source:* McKinnon (1991).

business parks have been built in suburbs at focal points on local rail or road networks.

Many large-scale manufacturing operations have abandoned the 'just-in-case' approach, involving the use of extensive storage facilities for components, in favour of the 'just-in-time' concept, a move which has resulted in changes in transport requirements. Contemporary assembly and distribution movements are much more closely integrated with production processes, and complex scheduling of transport is necessary to meet the needs of modern industrial firms which seek to minimise investment in warehousing space for raw materials and finished goods.

Modern business, finance and consultancy firms in the service sector employ staff who require a high degree of mobility, and proximity to major international airports is often an important consideration when selecting premises. For some highly specialised technologically based industries whose products, such as electronic goods, have a high value per unit weight, access to air freight services can be a significant locational consideration. In particular the expansion of the electronics industry in the countries of the Pacific Rim has been aided by the provision of regular air services between the Far East and the markets in North America and Europe.

Most long-distance international freight, however, is still carried by sea, and the continued increase in capacity of oil tankers, accompanied by the introduction of bulk freighters for dry cargoes and container ships, has been an essential factor in the growth of the transnational manufacturing corporations. These specialised vessels serve the needs of the new global enterprises in the same way as their less sophisticated predecessors, the all-purpose steamships, met the demands for transport within the various European colonial empires. Similarly, the nineteenth-century telegraph which was the nerve centre of this earlier transport system has been superseded by today's complex satellite- and electronically-based global communications.

4.2.2 The developing world

The comparatively low levels of economic activity in many of the less developed countries is often reflected in the modest scale of their transport systems. Early-twentieth-century studies of colonial growth frequently assessed opportunities for agricultural or industrial expansion in the context of road or rail transport availability, implying that until adequate means of communication are provided little or no growth can take place. At this time transport was indeed accepted by many geographers and economists as the principal factor promoting economic growth, but in the second half of the century a much more cautious view of transport has been taken and it is seen now as a permissive rather than as a deterministic factor. Acceptance of the fact that transport alone cannot generate or encourage economic growth is of particular significance in the context of the developing independent nations of Africa and South-east Asia, since investment in transport improvements can now normally only be achieved through international loans requiring rigorous feasibility studies of the proposed schemes before funds are made available.

The building of a new road, railway or seaport in a less developed country can occasionally fail to meet the objectives set by its promoters and financiers. In many cases the region in which transport improvements have been made may prove incapable of further economic growth because of adverse climate, soils or geological features, and thus the old concept of transport as a 'magic wand' capable of overcoming all adverse factors can no longer be accepted.

A lack of sufficiently reliable or accurate forecasts on the traffic potential of a region to be served by a new road or railway may mean that the latter is unable to cope with the amount of freight generated after completion of the scheme and that additional capacity is required. The costs of using new road or rail services may be beyond the reach of small-scale farmers or local industrialists, and unless a subsidy is provided by regional or national government the service will be underused and thus not justify the initial investment. In some cases the new facility may not be operated efficiently because of inadequate local management or technical skills and a lack of spare parts for vehicles.

Of all these factors one of the most critical is the economic potential of the area for which a new scheme is planned. Any exploratory forecast of the effects of a transport improvement must assess the nature of existing agricultural or industrial enterprises in the area and the extent to which they could benefit from it. Translated into financial terms, the fundamental question that has to be asked is whether the transport sector is necessarily the most suitable recipient for any financial aid which is available for promoting economic growth in a specific region. Where claims on finite investment funds exceed the amount available then the allocation of resources to the transport sector as a first priority could well prove to be an inappropriate decision, and a more satisfactory return could have been achieved by initially directing the funds to agricultural improvement programmes or schemes for rural light industry based upon local resources. Alternatively, the supply of new railway equipment or road vehicles to be used on existing routes might in some cases prove to be just as effective an investment as providing entirely new railways or roads.

Analyses of the impact of new transport facilities suggest three possible results in terms of how a local or regional economy may be affected. In the most favourable circumstances there will be a measurable increase in agricultural or industrial output which can be positively attributed to the benefits brought by the new road or railway. Alternatively, a neutral effect may be identified, whereby the investment in transport does not bring about any discernible change in the local economy. Finally, a negative effect may be recognised whereby the introduction of a new transport facility may actually be detrimental to the economy. This latter situation may be produced when investment in a new national airline, port extensions, roads or railways may prove uneconomic and where the finance could have produced more beneficial results if allocated directly to the agricultural sector.

4.3 New communications technologies

The perfection of the electric telegraph in the nineteenth century and the introduction of radio in the early twentieth provided the first examples of communications on the global scale. They enabled the transport industry to operate more efficiently by co-ordinating movements of shipping and railways and providing consumers of transport facilities with a more reliable service. In the late twentieth century the transport of information has been transformed with the use of satellites and fibre optics for the high-speed electronic transmission of oral or printed messages, in the form of the 'fax', throughout the world. This development in turn has enabled the demands of transnational industries and other global organisations for transport to be met more effectively.

Satellites orbiting the Pacific and Atlantic Oceans have a much greater capacity for transmitting information than cable networks and are used for both telephone traffic and television links. Fibre optic networks have proved their efficiency at the national scale and can be connected to satellite systems to provide world-wide cover, being used for both commercial and private communications (Fig. 4.2). The transmission of information has also been radically improved with the introduction of information technology, which draws upon the advantages of the computer and the new communication techniques.

Existing telecommunications networks display the 'hub and spoke' pattern, with focal points in the UK, the USA and Hong Kong and interconnect the world's older industrial nations and those of the Pacific Rim. Many major transnational companies control their own networks: Texas Instruments communicates with its 50 factories in 19 countries by means of over 8000 terminals and 140 computers. The complexity of these new communication systems at the regional and global scales reflects their need to meet the needs of modern industrial societies. In early 1994 a major US telephone company announced plans for what was described as an 'electronic superhighway' across the continent in conjunction with other communications interests and the computer industry. The project, to be completed by the end of the century, will deliver both commercial information and entertainment programmes to homes and offices using fibre optic technology.

4.4 Transport costs and rates

Transport systems are owned and operated by both state undertakings and private enterprise, and range in size from the haulage firm with two or three trucks to national railway companies and international airlines. Where the state has assumed control the profit motive is not always uppermost and rates charged to consumers can often be equivalent to or less than the actual cost of transportation. The shortfall is usually covered by a subsidy, which is often applied as part of a deliberate policy to encourage the use of a particular service which may be uneconomic but is seen by the state as worthy of support on social grounds. Such practices are common in many major cities and in some rural areas and are discussed at length in subsequent chapters.

Where the transport industry is in the hands of private enterprise operating in a competitive environment then costs must be covered by rates charged to the consumer. An undertaking will try to attract as much of the available traffic as it can cater for and therefore endeavours to provide as efficient a service as possible. Each of the major modes can offer specific advantages in terms of ease of handling, speed, cheapness of haulage, security of goods in transit and availability of vehicles of suitable capacity, but surveys of transport users indicate that where a choice of transport facilities is available to the consumer the final selection is not always necessarily made on the basis of the lowest rate.

The type of traffic on offer and the location of the initial loading and final unloading points have an important effect upon costs. Freight transport companies will try wherever possible to secure outward and return loads, even lowering rates below costs to attract custom, but in many cases an empty back-haul journey is unavoidable and the rate charged to the consumer must reflect this situation. Among the more common examples of one-way-only carriage are the transfer of coal from colliery to power station or of iron ore from coastal import terminal to smelter and the collection of agricultural produce from farms and delivery to a deep-freeze or canning factory. The further expansion of the container trade between industrialized nations and the developing world is often constrained by the latter's inability to provide full, or even partial, return loads for containers which arrive at their ports with manufactured goods and which have to be returned empty to their points of origin.

Within the larger urban areas operators of rail and bus commuter services must provide sufficient vehicle capacity to cope with the peak-hour traffic demand but are not always able to make economic use of their fleets at other times. They must therefore ensure that the revenue received at peak times is sufficient to cover costs of operation during other periods. This obligation is a critical factor in perpetuating urban transport problems, as Chapter 7 shows.

In some areas there are physical constraints upon transport operations, and limits are placed upon the times during the day or months during the year within which services can be provided. Inland waterway transport can be hindered in this way, especially in tropical regions such as the Niger basin, where the dry season produces river regimes in which navigation is impossible at low water. Shipping on the St Lawrence Seaway–Great Lakes system and on the rivers of northern Russia is disrupted by icing in winter, and in all these cases where climate is a limiting factor the rates charged during the navigable season must yield sufficient revenue to cover the period when vessels are unable to operate.

The costs incurred in providing a transport service fall into two main categories. Fixed costs are those necessary to maintain the routeway on which vehicles run and to provide the traffic-handling and, where necessary, storage facilities at terminal and exchange points. Variable or line costs are the costs of actually operating vehicles on freight or passenger-carrying services and include the expenses of fuel and crews' wages. The proportions which these fixed and variable costs represent of

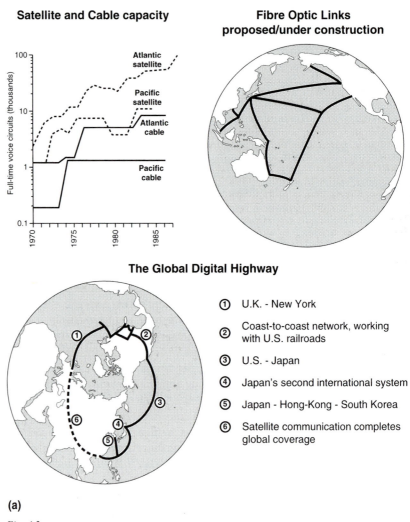

(a)

Fig. 4.2
(a) Satellite and fibre optics communications at the global scale. (b) Telecommunications networks for a major US transnational corporation (facing page). *Source:* based on Figs 4.5 and 4.6 in Dicken (1992).

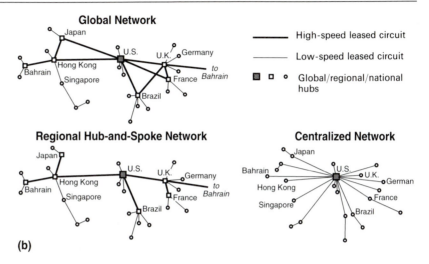

Fig. 4.2 continued   **(b)**

total costs differ from mode to mode and within each mode and are one of the most distinctive features of a transport system.

Railways, which are one of the few transport undertakings to own their routeways and have exclusive control over them, incur very high fixed costs since railway tracks and modern complex signalling equipment require regular maintenance. To these costs must be added those of providing passenger stations and freight depots. In contrast road hauliers, who make use of public road networks, have lower fixed costs which comprise capital outlay on vehicle purchase, vehicle licensing and insurance expenses and the costs of maintaining, where necessary, loading and unloading facilities. Maritime transport usually has the highest fixed costs and the lowest variable costs, and the substantial expense incurred when a vessel is in port has been one of the principal incentives for the development of new cargo-handling techniques designed to minimise loading and unloading times and to ensure that vessels are kept at sea earning revenue for as large a proportion of their working life as possible.

The generalized relationships between fixed and variable costs for road, rail and sea transport are illustrated in Fig. 4.3 which indicates the broad zones within which a particular mode offers the lowest transport charges. Detailed comparisons between these three modes are difficult, however, as operating costs will depend upon the volume of goods transhipped, the nature of the cargo, the lengths of haul and the frequency of the journeys (Box 4.1 and Fig. 4.4).

The rate payable for a particular transport service will normally be higher than the operator's cost, but the precise level will be determined by the factors outlined in the preceding paragraph. In practice several different types of freight rate are levied according to the demands of the consumer. Much depends upon the distances involved, since with increasing length of haul the fixed costs, which are generally independent of journey length, can be spread over a greater distance and the rate

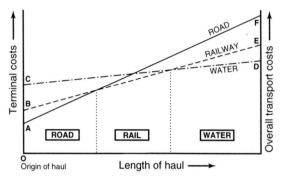

Fig. 4.3

Fixed (terminal) and variable (line-haul) costs of three major transport modes. Terminal costs are indicated on the vertical axis (OA, OB and OC for road, rail and water transport respectively) and increases in variable costs are shown by the gradients of the lines AF (road), BE (rail) and CD (water). *Source:* based on diagram in Smith (1971).

Box 4.1 Oil cargo freight rates

Freight rates for a commodity carried by several modes may be compared by using a cost coefficient. For example, if a 200 000 tonnes dead-weight oil tanker is assigned a cost coefficient of 1 then a road oil tanker and trailer will have a coefficient of at least 30. Between these two extremes comes a large-bore pipeline with flows of over 40 million tonnes per annum, with a coefficient of between 2 and 4, and a 4000 tonnes dead-weight inland waterway barge where the coefficient will be between 6 and 8. For oil transport in rail tanker wagons the value will be between 8 and 16, depending upon wagon capacity.

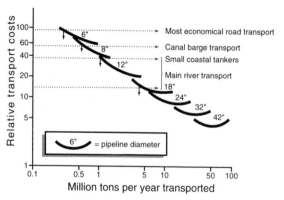

Fig. 4.4

Relative oil transport costs by pipeline, sea and land. *Source:* Shell Information Services.

Plate 4.1
The Botlek in Rotterdam where grain imported in bulk is transferred to smaller 1500 tonne barges for transport up the Rhine to river ports in France, Germany and Switzerland.

chargeable can therefore be less than when much shorter trips are made. The simple cost–distance rate is based upon a standard charge per unit of distance covered, but it may often be modified to a tapering rate in which charges are reduced with increasing length of haul, reflecting the fact that the operator's costs will diminish over longer distances. The cost–distance structure may also be modified by grouping several lengths of haul from a given origin into one zone and charging the same amount for all journeys within this class. This is known as the zonal or blanket-rate system and is characterised by a stepwise increase in charges from zone to zone. An extreme example of this approach, and one which is generally limited to postal services, is the 'postage stamp' rate where a standard national charge is levied regardless of distance and where only the weight of the consignment determines the final cost (Table 4.1).

Charges will also vary according to the nature of the cargo as well as its volume. Certain commodities, such as high-value manufactured goods, perishable agricultural products and livestock, require more attention during transit than low-value cargoes such as coal, iron ore or basic chemicals and in some cases purpose-built vehicles are essential. Where a consumer is able to offer the transport contractor regular consignments of goods over a long period a contract may be negotiated which involves what is described as a commodity rate. If the goods are in bulk and of low value and move between fixed origins and destinations then the charges are likely to be reduced further. The shipping of iron ore and

Table 4.1
Summary descriptions of
principal types of freight rates

1. *Uniform, flat* or '*postage-stamp*' rate. The rate levied is irrespective of the distance involved, but the costs to the consumer do increase with increasing weight of the consignment.
2. *Cost–distance* rate. The cost of haulage is related to the distance over which the consignment is carried, with the rate levied reflecting the terminal and line-haul costs involved. Many public transport passenger undertakings also adopt this type of charging.
3. *Zonal* rate. A variation of the uniform rate charging system in which haulage distances are grouped into zones, within which the rate is uniform.

Tariffs, or rates charged, will also reflect:

(a) the class of commodity transported, with bulk freight being cheaper than goods requiring more care in haulage;
(b) competition between hauliers, with lower rates being levied on routes carrying heavy traffic and higher rates being charged on lightly trafficked routes.

crude petroleum at relatively low cost over distances of several thousand kilometres is only possible because of these types of contract. Thus the basic cost–distance rate is only applied in a limited number of situations and it is the character and amount of cargoes which often determine the rates levied.

Two examples are now examined in more detail to show how freight rates can be affected by specific circumstances. The first is based upon changes in charging policy following the introduction of containerisation and the second, drawn from Southern Africa, illustrates how an improvement in the transport infrastructure may not necessarily bring about cheaper movement costs.

4.4.1 Container transport costs

The structure of maritime tariffs is extremely complex and its basis is beyond the scope of this text. However, certain aspects of the pricing system are of relevance to the transport geographer as they involve choice of port location and of shipping route. Before the introduction of containers, goods carried by sea could be subject to the 'equalisation principle', whereby the rate paid by the consumer using the service from a particular port was the same irrespective of which overseas port the cargo was ultimately shipped to. Since the consumer is normally responsible for the costs of moving exports or imports to or from the coast the nearest convenient port would be chosen in order to minimise these land-based expenses and thus the overall costs of haulage. With the adoption of containers this equalisation of costs principle could not be retained, as not all the ports used by conventional shipping were equipped to deal with the new unitised system and thus the choice of ports became much more limited. Longer journeys between points of export or import

Table 4.2
Freight costs of container
transport

(a) Comparison of distribution costs of a container from four different
ports in the UK

Container port	Distribution wholly by freightliner	Distribution wholly by road haulage	Distribution by least-cost mode
Liverpool	£105	£110	£101
Tilbury	£151	£160	£146
Southampton	£152	£161	£148
Felixstowe	£162	£172	£155
Liverpool saving	£46–£57	£50–£62	£45–£54

(b) Costs of diverting different sizes of
containership from the North America–
northern Europe route via the English
Channel to Liverpool[a][b]

Vessel size (21 knots)	Daily costs at sea	Diversion cost per box[c]
1000 TEU	£24 000	£60
1500 TEU	£28 000	£70
2000 TEU	£34 000	£85
2500 TEU	£39 000	£98
3000 TEU	£43 000	£108

[a] The figures apply to a 500 nautical mile diversion from the
English Channel (Bishop Rock) to Liverpool and return.
Steaming at 21 knots over this distance equates to a 24-hour
diversion.
[b] Daily cost figures updated to recent cost levels.
[c] Calculated on the assumption of a container exchange of 400
boxes.
Source: Pearson and Fossey (1983).

and the final inland origins or destinations became necessary and
land-haul costs rose in consequence. The equalisation principle was
therefore replaced by what was termed 'absorption pricing'. With this
system the container land-haul costs payable by the consumer were based
upon the distance to the nearest port rather than upon the distance to the
actual container port that was being used, enabling more competitive
rates to be offered.

Within the UK this principle was applied on a 10 km by 10 km grid
square basis, the cost to the shipper being based upon the distance
between the centre of the grid square within which the consignment was
loaded and the nearest port, irrespective of whether this possessed
container-handling facilities or not.

Actual transport costs for containers vary with the mode adopted, being cheaper per unit distance by road than by rail. Maritime rates depend upon length of voyage and are usually expressed in costs per 100 twenty-foot-equivalent-unit (TEU) miles, with the lowest costs for trans-oceanic voyages being achieved with a 3000 TEU capacity vessel at an average speed of 19 knots and the highest with a 1000 TEU vessel operating at 25 knots (Table 4.2).

4.4.2 *Land-locked states in southern Africa*

Inland states such as Zambia, Malawi, Zimbabwe and Botswana often encounter problems in gaining satisfactory access to the sea and much of the railway building programme in the colonial period was concerned with export routes to ports on the Atlantic and Indian Ocean coasts. Following the independence of Zambia in 1964, the Chinese-financed TAZARA (Tanzania and Zambia Railway) link was opened in 1976 to provide an additional export route for Zambian copper, other minerals and agricultural produce. Prior to completion of the TAZARA, Zambian copper was railed through Angola to Lobito, on the Atlantic coast, or through Mozambique to Beira on the eastern seaboard, both journeys being of about the same length. As western Europe was the leading copper market Lobito, which is 5600 km closer by sea to Europe than Beira, would appear to be the favoured port but under international shipping agreements the costs of sea transport for the copper were the same from both ports despite the differences in distance (Fig. 4.5).

With completion of the TAZARA link to Dar es Salaam an alternative route became available and choice of the most suitable clearly depended upon the freight rates on offer. In common with other southern African independent states, Zambia has a shortage of foreign currency, which favoured the maximum use of its own railway system when selecting an export route. Thus, although the Lobito route provided the shortest overall distance to Europe, and initially offered lower freight rates than the TAZARA, the latter involved a shorter land haul between copper belt and coast and almost one-half of the journey was on Zambian rails, compared with only 1 per cent when Lobito was used (Table 4.3).

Copper can also be exported via South African ports, making use of the Zimbabwe and South African railways, and therefore Zambia has a wide choice of routes to the sea. Freight rates on the TAZARA have to produce sufficient revenue to cover all fixed and variable costs and also ensure repayment of all outstanding loan charges to China by the early twenty-first century. Much of its traffic is carried at preferential rates which do not always meet actual costs and its copper rates were, during the first decade of operation, usually lower than those of any of the competing routes. The main difficulty which the new line faced was to attract as much traffic as possible by offering competitive rates and also ensure that sufficient revenue was gained to cover costs and loan

Fig. 4.5
Rail routes between the Zambian copper belt mining centres and principal coastal outlets.

Table 4.3
Freight charges for copper between Mufulira (copper belt) and four ports

Port and distance (km) from Mufulira over each system	Railway system and freight charge (US$ per tonne)					
	ZR	TAZARA	BR	RR	RR/SAR	Total charge
Dar-es-Salaam 　1860 TAZARA 　1265 ZR	11.00	51.96	—	—	—	62.96
Beira/Maputo[a] 　925 ZR 　1547 RR	31.97	—	—	36.54	—	68.51
Port Elizabeth 　925 ZR 　2373 RR/SAR	31.97	—	—	—	106.84	138.81
Lobito 　157 ZR 　2365 BR	12.26	—	33.17	—	—	45.42

ZR = Zambian Railways, TAZARA = Tanzania and Zambia Railway, BR = Benguela Railway, RR = Rhodesian Railway, SAR = South African Railways.

[a] Distances between Mufulira and Beira and Maputo are different but the 'chargeable' distances are similar.

Source: Mwase (1987).

repayments. If the Southern African Development Community's objectives of improving the other copper export railways to Lobito and Beira are eventually achieved then Zambia will be able to reconsider the pattern of transit routes in terms of the freight rates on offer.

4.5 Transport costs in agricultural geography

When von Thunen (1826, in translation 1875) published his ideas on the relationship between specific types of agricultural activity and distance from market the only transport modes available were horse-drawn wagons and canal or river barges. Contemporary views on the location of agricultural activities incorporate a modified version of von Thunen's approaches, using the concept of 'economic' or 'location rent'. This term is used to identify the use of land which will provide the highest return per unit area and is calculated with the aid of data on market prices, production costs, crop yield, distance from market and transport rates:

$$\text{Location rent} = \text{yield (market price} - \text{production cost)}$$
$$- (\text{yield} \times \text{transport rate} \times \text{distance})$$

Since the market price of a specific crop is dependent upon the supply–demand balance, and also the costs of transport between farm gate and selling point, then producing areas which are closer to the market will in theory enjoy an advantage over more distant areas. This is expressed in the location rent equation above, where the initial profit is reduced by the costs of transport. At a certain critical distance from the market these costs will exceed the profits derived from a particular crop, and an alternative crop will then be produced whose market price will be greater than the transport costs (Box 4.2 and Fig. 4.6).

This is illustrated in Fig. 4.6, in which three crops are considered, each with their production area determined by the location rent general

Box 4.2 The von Thunen theory and agricultural marketing costs

The von Thunen pattern of concentric agricultural production zones around a central market assumed a homogeneous environment with no spatial variations in climate or soil fertility. The relative costs of transporting different commodities to the central market determined the distance from this market at which they could be produced. Where a waterway could offer cheaper carriage than road transport the producing zones become elongated alongside the river. Von Thunen also accepted that where a smaller market existed within the hinterland of the major centre this would in turn generate its own minor zonal pattern of farming types. ▶

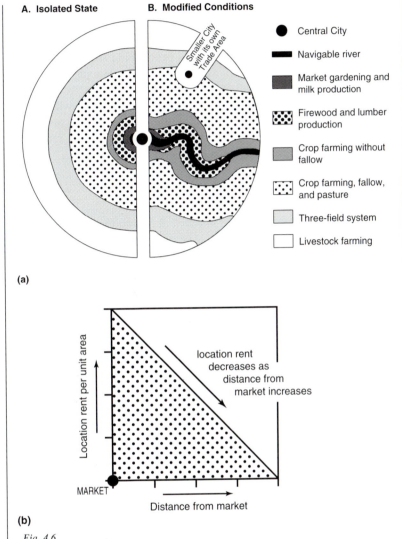

Fig. 4.6
(a) Diagram to illustrate the basic von Thunen concept of agricultural land use, showing (left) zonation of different agricultural activities and (right) the modifications made when a navigable river reduces transport costs for produce marketed at the central city. (b) The relationship between location (or economic) rent, distance from market and transport costs. *Source:* based on Fig. 2.13 in Lloyd and Dicken (1977).

equation. Farmers will be willing to pay a higher rent for land closer to the market in order to reduce transport costs and there will be competition for this more accessible land, a process which has been compared to the urban land 'bid rent' principle proposed by Alonso (1963). Each product will have its own 'bid rent' curve, and if the various curves in the two-dimensional diagram are rotated around the location rent axis a series of concentric zones will be produced which bear similarities with von Thunen's original system.

These ideas presuppose that all agricultural activities are taking place within areas of homogeneous physical conditions focusing upon just one centrally located market centre. Actual conditions of climate, soil fertility and other physical characteristics will vary greatly and subsidiary marketing or competing markets will be present, so that many adjustments to the basic location rent pattern are required.

4.6 *Transport costs in industrial geography*

Considerations of transport have always occupied a prominent place in theories of industrial location. The emphasis on the broader scale has changed from the least-cost location factor basis to more recent approaches involving a behavioural element, and issues connected with the global organisation of capital and manufacturing capacity. However, any detailed analysis of industrial activity still requires a knowledge of transport costs and of their significance within total production costs.

Weber's pioneer theory (1909) involved the search for the least-cost location at which the industrialist would be able to maximise profits (Box 4.3 and Fig. 4.7). Many of Weber's ideas have been incorporated in modified form in later industrial location concepts and, with increasing access to reliable data on transport costs, his theory has been recast to take account of variable freight rates and the transfer of semi-processed goods between producing plants (Box 4.4). Although transport was seen by Weber as the dominant factor to be considered in industrial location analysis, he did accept that in some cases labour costs could be of more importance as a determinant of location. D. M. Smith (1971) has developed this latter issue with the aid of isodapanes, or lines of equal transport costs (Fig. 4.8). Locations established on the traditional basis of least transport costs are compared with alternative possible sites where labour costs are cheaper than at the original point, and critical isodapanes are then identified within which labour rather than transport costs will be the principal influential factor.

When transport costs were considered by the pioneer workers on industrial location it was assumed that transport was 'bought in' and that freight rates were therefore payable by manufacturers to these outside contractors. Contemporary industrialists can take advantage of a much more varied range of transport services, including fleets of road vehicles owned by the manufacturing company, long-term contract leasing of vehicles from transport firms, short-term hire arrangements or a straight-

Box 4.3 Weber: transport costs and industrial location

Weber grouped costs of industrial production into three categories: transport costs, labour costs and costs associated with either concentrated or dispersed patterns of location. A least-cost location was identified as one where the total costs of transport of raw materials from sources to plant and of output from plant to market were at a minimum, and Weber used the product of distance and weight of materials as his basic index for calculating costs. A range of different location patterns, increasing in complexity from one market and one material source to multiple sources, were investigated and Weber illustrated his least-cost concept with a 'locational polygon' where each corner represented either a market or a material source, weighted according to costs and where the plant site with minimum transport costs lay within the perimeter.

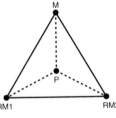

Weight losing industry

(a)

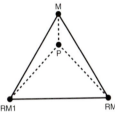

Weight gaining industry

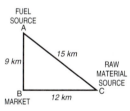

(b)

Fig. 4.7
(a) The basic Weber diagram illustrating relationships between plant location, material sources and transport cost: M, market; RM1, RM2, raw material sources; P, plant location. (b) Diagram to illustrate how plant location can be influenced by costs of raw material assembly and costs of final product distribution. Each tonne of product requires 2 tonnes of fuel and 4 tonnes of raw material. Location of plant at raw material source C: fuel transport costs 30 tonne-km; product transport costs 12 tonne-km; total transport cost 42 tonne-km. Location of plant at fuel source A: raw material transport costs 60 tonne-km; product transport costs 9 tonne-km; total transport costs 69 tonne-km. Location of plant at market B: raw material transport costs 48 tonne-km; fuel transport costs 18 tonne-km; total transport costs 66 tonne-km. Based on these costs the optimum location as defined by minimum transport costs is at point C, the raw material source.

Box 4.4 Some post-Weber theories on transport costs and industrial location

1. Hoover (1948) saw transport and production costs as the main determinants of industrial location and devised isotims (lines joining points of equal delivered price) to define an optimum plant location where the combined costs of raw material assembly and output distribution would be at a minimum.
2. Greenhut (1956) suggested that most plants would display a market-orientated location, except where raw materials are perishable or where the costs of hauling materials from source to plant are far in excess of distribution to market costs. For example, fruit and vegetable processing factories would be sited in or very close to crop-producing areas and iron smelting and some chemical works would also be located close to raw material sources or import points.
3. Isard (1956) grouped industrial location factors into three classes. Transport costs were seen as generally varying regularly with distance between plant and material source or between plant and market. Other production costs, such as those of labour, power and loan interest, varied independently of distance. The third group included those costs associated with either concentrated or dispersed industrial patterns.

---

forward use of a transport contractor for specific trips. The relative importance of transport costs in the overall production process is therefore now much more difficult to identify, given this range of transport services.

A manufacturer's costs of delivering products to the final market may be recouped from the consumer in a number of ways. The simplest pricing system disregards the distance between factory and customer location. with the latter paying what is known as a uniform delivered price for the product. The alternative method involves the customer bearing the purchase price of the chosen commodity plus the cost of its carriage from the factory gate, an element which of course varies with distance and creates a variable pricing pattern. This 'free on board' pricing system will affect the spatial extent of the market, as an increasing length of haul will produce a higher total cost for the purchaser and demand will therefore be affected. In practice many pricing systems incorporate elements of both the uniform delivered price and free-on-board structures and tariff rates for commodities are often very complex.

The cost structures of industrial concerns are of particular significance in the context of regional economic development, where plans to revive decaying manufacturing activities must consider the most effective means of reducing costs and thus aid the survival of existing firms and the attraction of new enterprises to the region. Economic planners recognise the importance of industrial transport costs and the extent to which investment in new or improved road and rail links can reduce them. The

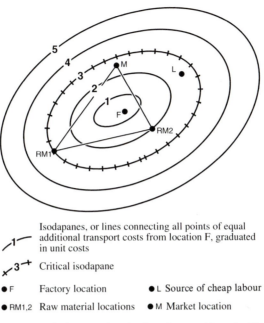

Isodapanes, or lines connecting all points of equal additional transport costs from location F, graduated in unit costs

Critical isodapane

● F   Factory location   ● L Source of cheap labour

● RM1,2 Raw material locations   ● M Market location

F represents the least-cost location for a factory with market M and raw materials drawn from RM1 and RM2

Fig. 4.8
The effects of cheap labour sources upon the Weberian minimum-transport cost approach to plant location. F represents the least cost location for factory serving a market M and with raw materials transported from RM1 and RM2. If cheap labour is used from location L, labour costs at F are reduced by 3 unit costs for each item produced. Labour source L is closer to the factory than the isodapane for 3 unit costs of transport so that if the factory is moved from F to L the overall costs of production could be lower than at F. The isodapane for 3 unit transport costs was identified by Weber as the critical isodapane as it encloses the area within which the cheap labour location (L) provides lower overall costs than does the minimum-transport cost site (F). *Source:* based on Fig. 8.2 in Smith (1971).

effectiveness of this policy will clearly depend upon the proportion of an industry's total production expenses accounted for by assembly and distribution costs, and many ambitious transport improvement programmes have been subject to strict appraisal in terms of their potential effectiveness in reducing these costs. Decisions made by manufacturers to rationalise their pattern of production units and storage depots will also take into account their transport requirements and an example drawn from the UK is now examined in more detail.

4.6.1 Industrial transport costs in a depressed region

The county of Lancashire in north-west England relied heavily upon the textile and engineering industries for its economic prosperity during the

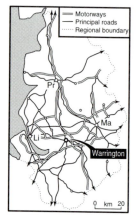

Ma – Manchester
Li – Liverpool
Pr – Preston

Fig. 4.9
The location of Warrington
with respect to motorways
and principal roads in
north-west England.

nineteenth and early twentieth centuries. Subsequent industrial decline has been countered by a series of policies and plans designed to stimulate revival and attract new enterprises to the region. In particular Lancashire has seen a very substantial investment in transport infrastructure since 1960, with the electrification of the main west coast railway between London, Liverpool, Manchester and Scotland, the construction of the M6 and M62 trunk motorways and completion of a network of inter-urban motorways linking Merseyside with the Manchester conurbation and other industrial centres. A programme of New Town building was also undertaken and the industrial promotion campaigns mounted by these centres place particular emphasis upon the advantages to be gained by industrialists from locating within a region so well-endowed with motor-ways. Warrington has been described as 'Crossover City' in terms of its location close to the intersection of three motorways and two trunk roads, and although it is undeniable that the town is an important node within the north-west region it is ranked at only twelfth place on the national motorway network in terms of accessibility (Fig. 4.9).

The influence of the M62 motorway upon industrial costs of firms along its route was assessed in order to see whether or not the transport cost advantages of being sited close to the new road would generate any benefits such as opportunities for additional employment. The results of the study indicated that the impact of the M62 was to cut total manufacturing costs in the corridor by about 0.33 per cent, a negligible reduction which was very unlikely to create new jobs in the area. Given that the transport costs incurred by most UK industries are less than 2 per cent of total costs, any benefits accruing from a motorway location could be more than outweighed by more general changes produced by fluctuating economic conditions. It is also significant to note that many of the enterprises attracted to new industrial estates adjacent to the north-west region's motorways are primarily distribution rather than manufacturing firms, and the levels of employment generated in these warehouses and depots are much less than those in production units. It is indeed possible that motorways in north-west England have actually reduced employment potential by allowing markets in the region to be served for the first time from depots in other parts of England. Although the electrified rail network in the north-west has been accompanied by the introduction of scheduled container services these again have had little effect upon the regional economy, since the benefits of rail freight transport are often only secured over distances in excess of 240–320 km with train-loads of over 30 containers per day – requirements which are unlikely to be met by industrial centres in the region.

There is undeniable evidence that improved roads can lower transport costs for manufacturing plants and distribution depots, but such reductions are rarely significant in terms of their contributions to economic growth. The advantages of 'motorway corridors' are widely publicised by those responsible for attracting new industry to depressed areas, and regions with poor transport infrastructures clamour for motorways in order to improve industrial accessibility and overall performance. The image of such roads as generators of economic revival is, however, not always supported by reliable evidence.

4.6.2 Transport costs in modern industrial economies

Two major conclusions to emerge from recent research are that (a) transport costs are today not so significant a component of overall manufacturing cost structures as suggested by industrial location theorists, and (b) road or rail improvements alone are unlikely to create substantial reductions in these costs. The significance of transport investment schemes within regional economic planning programmes will therefore depend upon the extent to which those industries where transport costs still form a large part of total costs, and would therefore benefit from an improved transport infrastructure, are present in the region. In the UK, for example, such industries include quarrying, coal-mining and iron and steel manufacture, but these are of declining importance in many regions where the transport infrastructure has been improved.

Economies in the transport of freight can also be achieved through a more efficient use of existing roads and vehicles, since about 70 per cent of total haulage costs are fixed and unaffected by distances travelled or the condition of the road network. Improved distribution systems involving computer-assisted delivery programmes and the use of more fuel-efficient trucks can both help to reduce costs, although within the European Union legal restrictions on drivers' hours and constraints on loading and unloading in congested urban areas also have to be taken into account when formulating movement patterns and timings. Modern manufacturers who have adopted the 'just-in-time' methods of production depend upon precisely timed goods delivery and distribution services, and the ability of transport contractors to meet these demands can be of more importance than the issue of who can offer the lowest freight rates.

It has been suggested that the levels of accessibility provided by existing roads, and to a lesser extent railways, in some regions is already at a maximum and that these regions have attained their optimum transport capacity in terms of infrastructure. Further reductions in industrial freight haulage costs are therefore only likely to be made by adopting some of the improvements in distribution organisation outlined above. The contemporary trend is towards larger-capacity manufacturing plants, such as the Japanese motor works at Washington, in north-east England, or at Burnaston, near Derby in the Midlands (Fig. 4.10), and improved distribution methods have enabled these plants to extend their markets at regional, national and global scales.

The concept of equity has also been introduced into this debate on transport costs and regional economies. Although a concentration of state

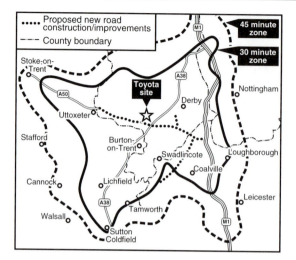

Fig. 4.10
The location of the new Toyota motor vehicle plant at Burnaston, Derbyshire, UK in relation to existing and proposed motorways and principal roads. *Source: Evening Sentinel,* Stoke-on-Trent, 30 June 1989.

investment in roads in industrially depressed regions can be interpreted as a subsidy towards the encouragement of new development, it has been argued that freight transport costs could be more effectively lowered in such areas by the granting of direct subsidies to the haulage costs of individual firms, in the same way that fiscal concessions are often available to manufacturers to stimulate expansion.

4.7 Transport and the geography of settlement

The relationships between transport and settlement are most complex at the urban scale, but planners and geographers are also concerned with the role of transport in rural areas, particularly with respect to the absence or decline of village facilities and the problems of accessibility to essential services, issues discussed in detail in Chapter 9. Christaller's central place theory (1933) attempted to explain settlement evolution in terms of urban services centres and rural hinterlands, with an emphasis upon the construction of a hierarchical system of centres based on the range over which certain functions are discharged. These patterns were built up on a landscape where transport opportunities were assumed to be uniform throughout but Christaller, like many of the classical economic theorists, did recognise that certain settlements would benefit if they lay on a major transport route and modified his theory accordingly.

The revolution in inter-urban transport brought about by canal and railway building confirmed the industrial strengths and the existing nodality of many towns, but also conferred a new importance on others by virtue of their strategic location on these routeways. Trade at ports such as Liverpool and New York expanded steadily during the railway age, and many African and South American ports such as Cape Town, Mombasa and Buenos Aires owed much of their initial growth to the rail links with their hinterlands.

Most rural settlements in the industrialised world received few immediate benefits from the railways, and the impact of transportation facilities upon economic and social life in the countryside has been at its greatest in the twentieth century, with the provision first of motor-bus services and subsequently the expansion of access to the private car. In the less developed countries rural communities have received only limited benefits from modern transport technology and continue to rely to a large extent upon animal power and walking.

4.8 Transport and urban geography

The significance of transport within urban areas may be explored in several ways. The introduction of public services by rail and later road played a significant part in urban expansion in North America and western Europe during the nineteenth and early twentieth centuries and several models of city growth emphasise the importance of transport (Box 4.5 and Fig. 4.11).

Construction of railways and new roads in cities also influenced the morphology of the urban area, these routes often acting as barriers to growth and as physical boundaries to communities. Contemporary research is strongly focused upon the interaction between urban land use and intra-urban transportation and on the problems associated with personal travel within the city, and these issues are examined in Chapters 7 and 8.

When most people had no choice but to walk to carry out their essential daily tasks, cities were of necessity compact. People lived at their workplaces or close by, producing high-density living environments and cities that were small in size and functionally integrated. Rarely did such 'foot cities' achieve populations of more than 50 000 or have diameters of more than 8 km. Only in the aftermath of the Industrial Revolution did vehicles of relatively high capacity and speed, as compared with walking pace, appear, allowing greater distances to be travelled and larger quantities of goods to be exchanged. This relaxed the restrictions on city size and established the interdependence between transport technology and urban

Box 4.5 Urban growth models and transport technology

Several of the growth models devised by urban geographers emphasise the significance of transport facilities. Burgess (1925) cites the importance of the advent of cheap public transport in his model of concentric urban zones, with its inner low-status housing belt and the outer upper-class residential areas and the peripheral commuter zone. Although the sectors in Hoyt's (1939) model were defined mainly on the basis of housing types, in locating his high-status residential area within the urban framework, he acknowledged the significance of the access provided by major routeways Similar links between the alignment of road and rail transport routes and housing areas can also be seen in Harris and Ullman's (1945) multiple nuclei model.

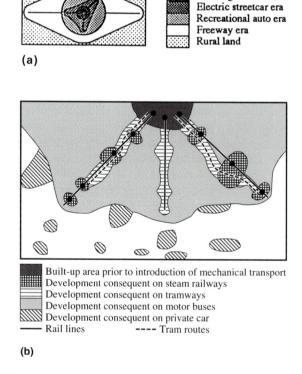

Walking-Horsecar era
Electric streetcar era
Recreational auto era
Freeway era
Rural land

(a)

Built-up area prior to introduction of mechanical transport
Development consequent on steam railways
Development consequent on tramways
Development consequent on motor buses
Development consequent on private car
—— Rail lines   ---- Tram routes

(b)

Fig. 4.11
(a) The sequence of urban growth in a United States mid-West city in relation to the development of transport technology. (b) Urban development and changing transport modes in the European city. *Source:* (a) Muller (1989); (b) White, personal communication.

form. Thus, as this technology developed, so did the built-up area of the city, reflecting the changing accessibility of various locations. The 'foot city's' compact circular form evolved into more extensive and star-shaped urban areas, reflecting the greater accessibility along the lines of railways and trams (or streetcars). These two new forms of transport were critical in separating home and workplace and encouraged functional specialisation within the city, as well as promoting the penetration of the surrounding countryside by fingers or 'beads' of development.

Herbert and Thomas' (1982) examination of the development of urban systems underlines the particular importance of improvements in the technology of public transport. However, as Dennis (1984) emphasises, it is essential to understand exactly why investments were made in tramway and suburban railway undertakings in the nineteenth century. The complex issues of urban transport demand are examined by Daniels and Warnes (1980), who consider that although public transport greatly assisted urban growth, and especially the process of suburbanisation, the provision of these services did not in itself initiate this growth in most cases.

In Europe and North America it was the introduction of horse-drawn and later electric tramways, coupled with improvements in working hours and pay, that made it possible for a higher proportion of urban workers to move into housing areas sited away from their places of employment. The widespread investment by entrepreneurs and municipal authorities in electric tramways after 1880, and the extension of these networks into the pioneer council housing estates in the twentieth century, is seen by Dennis as one of the first instances of transport and urban expansion proceeding simultaneously.

In the UK and the USA, railway and tramway construction was often closely linked with the efforts of property developers to attract wealthier city workers to the benefits of life in the rural areas beyond major cities. The Metropolitan Railway, which served the areas north-west of London, formed a company which constructed many housing estates close to its stations, establishing a 'rural suburbia' known affectionately as Metroland.

Towns and cities in the post-industrial stage display the effects of the decline of public transport and the dominance of the car as a means of obtaining a level of intra-urban mobility which the bus and train had never been able to match. Non-tracked forms of transport, at first the bus and then the car, have been responsible for the infilling of wedges of open land between earlier fingers of urbanisation, and have massively extended the range of daily movements of people, services and goods. The sprawling suburbs of the North American city of the late twentieth century thus reflect the dominance of the internal combustion engine in much the same way as the high-density pre-industrial European city was a product of the reliance upon walking as the major mode.

Traffic congestion in inner cities and the construction of urban motorways offer the most obvious evidence of the domination by the car, but the decentralisation of retailing, office and industrial activities and the continuing growth of suburbs are processes which have also been encouraged and aided by mass car ownership. These are issues which have profound environmental and social consequences and are discussed in Chapters 10 and 11.

4.9 Concluding summary

Transport has played an indispensable part in the growth of modern industrial systems and urban societies. During the first half of the twentieth century transport, and in particular the railway, was seen as asserting a powerful influence upon industrial location and upon economic growth in the colonial world, and transport costs in particular were recognised as the determining factor in siting manufacturing industry. Urban geographers also saw rail and road public transport as a significant influence upon town and city growth after 1850.

A much more moderate view is now taken of transport, particularly in terms of its contribution to the process of economic growth in the less developed countries, where national policies and investment feasibility studies often assign a much lower priority to transport than in the past. In the Western world the car, which for so many years has been popularised as the symbol of urban mobility, has played a major part in shaping mid- and late-twentieth-century urban expansion but its disadvantages are now becoming increasingly apparent.

Further reading

Brotchie J et al 1985 *The future of urban form: the impact of new technology* Croom Helm

Dennis R J 1984 *English industrial cities in the nineteenth century* Cambridge University Press

Dicken P 1992 *Global shift* 2nd edn PCP

Owen W 1987 *Transportation and world development* Johns Hopkins

Pearson R, Fossey J 1983 *World deep sea container shipping* Gower

Watts H D 1987 *Industrial geography* Longman

Part Two

Spatial systems

5 *International transport*

International connections and global transport networks

Many changes have encouraged movements on the international scale since the mid twentieth century. Technological advances have provided us with high-capacity jet airliners, ships which carry over half a million tonnes of crude oil over thousands of kilometres, and the capability of handling freight in more economic ways. Movements at the global scale are now within the reach of ever-increasing numbers of people and commercial enterprises, and the flows of tourist and business passengers and of raw materials and manufactured goods by sea and air have created a demand for more efficient terminal handling facilities.

5.1 Introduction

This chapter considers the international movements of freight and passengers at various scales and by different modes. Transport systems which are operated at the global scale are the expression of the need for links between both individual nations and trading blocs, and have complex spatial networks. The levels of trade which produce the demand for international transport are obviously influenced by economic conditions but it is important to remember also the roles of governments and global organisations of the major shipping and airline companies. For example, the failure to negotiate an acceptable conclusion to the General Agreement on Tariffs and Trades (GATT) talks in the early 1990s was a constraint upon the expansion of trade. Decisions made by the various world shipping conferences and by the International Air Transport Association must also be taken into account when examining flows of freight and passengers at the world scale. Policy is thus now a significant factor in international transport.

New processing techniques have aided the manufacture of lighter and less bulky consumer goods which are easier and cheaper to transport, so encouraging the establishment of larger production plants serving distant and more extensive market areas. The rapid expansion of transnational manufacturing companies in particular has been responsible for much of the increase in international traffic in the late twentieth century.

A literal interpretation of the term 'international transport' would involve an unmanageable pattern of cross-frontier movements, so in this chapter attention is focused upon long-distance freight and passenger flows, and upon smaller-scale movements which are of significance within trading blocs such as the European Union. The growth of multinational industrial corporations, the increasing dependence of many industrial nations, such as Japan, upon external sources for basic raw materials, and in particular the demands for petroleum, have been met by the creation of a fleet of highly specialised cargo vessels and, in the case of oil, by a network of pipelines crossing national frontiers.

Almost all long-distance passenger travel is now by air and the expansion of tourism has produced a demand for many additional scheduled and charter services, with distances of thousands of kilometres now being a commonplace element in package holidays. Technological advances in aircraft and shipping design and construction have been essential to meet the growth in international transport from 1950 to 1990. Freight rates and passenger fares have been reduced as much higher-capacity ships and aircraft have been brought into service, and this in turn has had an important effect upon the size, functions and relative importance of seaports and airports.

5.2 *Freight transport*

Maritime transport carries over 75 per cent of all world trade by tonnage and 80 per cent of this is in the form of bulk traffic. The unscheduled small-scale freighting performed by the ubiquitous tramp steamer of the early twentieth century has been largely replaced by long-distance liner services using high-capacity vessels which often specialise in one type of commodity and are confined to specific routes and a limited number of ports.

Containers and other forms of unitised traffic are also increasing their share of world trade at the expense of general cargoes and, as a result of these two trends, the proportion of all maritime freight carried by conventional shipping has declined considerably. The four major categories

Table 5.1
Structure of maritime
cargoes 1990/91

Cargo	Million tonnes dead-weight loaded	Percentage of all cargoes
Crude petroleum	1 276 082	31.9
Oil products	468 555	11.7
All other cargoes	2 250 091	56.4

Source: UN Monthly Bulletin of Statistics (1992).

of freight are as follows:

(a) crude petroleum, which in 1991 accounted for 32 per cent of all
    international tonnage;
(b) dry bulk cargoes, principally iron ore, chemicals and grain;
(c) container and other unitised traffic;
(d) conventional freight, usually described as general cargo (Table 5.1).

Shipping companies in many maritime nations have increasingly adopted
the practice of registering their fleets under 'flags of convenience' in order
to reduce operating expenses, especially the costs of crews. Panama,
Liberia and Cyprus are the principal nations to offer this type of
registration and together accounted for 112.4 million gross tonnes, or 26
per cent, of world shipping in 1990. The former USSR followed Western
practice in 1990 by placing many vessels under the Cypriot flag of
convenience, thus gaining access to finance from Western countries to
build new ships as well as securing lower operating costs.

5.2.1 Petroleum

World trade in crude petroleum is characterised by the long-distance
separation of the zones of supply and demand. The major production
areas are the Middle East, the Caribbean and northern states of South
America, North Africa, the CIS and South-east Asia, whereas the principal
regions of consumption are western Europe, North America and Japan
(Fig. 5.1). Development of inland and offshore indigenous oil resources
by the UK and Norway in the 1960s has reduced dependence upon more
distant supplies to some extent but the bulk of petroleum entering world
trade is still shipped over distances involving thousands of kilometres.

During the first half of the twentieth century much petroleum was
refined at or near its source, but with the continued growth in demand
many major oil-consuming states adopted the policy of providing their
own refining facilities so that the bulk of petroleum is now shipped in
crude form. Modern oil-carrying vessels are now the largest vessels afloat
in terms of their dead-weight tonnage, and the progressive increase in
capacity has been made in order to minimise freight carriage costs.

Building larger vessels has produced the desired economies of carriage

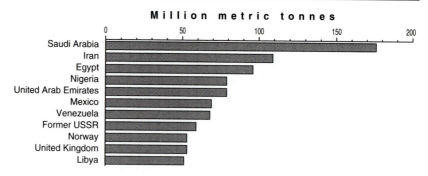

Million metric tonnes

Fig. 5.1
Countries exporting over 50 million tonnes of crude oil in 1990. *Source: UN Monthly Bulletin of Statistics* (1993).

Box 5.1 Oil tanker capacity

In 1938 tankers had an aggregate capacity of 16.6 million tonnes dead-weight and no vessel exceeded 25 000 tonnes. Twenty years later, when the world fleet capacity had risen to 55.7 million tonnes, one-third of oil traffic was carried in larger ships and by the late 1960s only one-fifth of the 135 million tonnes in world trade was transported in 25 000 tonne vessels. The first 100 000 tonne carrier was brought into service in 1963 and by the early 1980s vessels of 550 000 tonne capacity were in operation.

A large fleet of smaller-capacity tankers still exists but much of it is now deployed on coastwise voyages redistributing crude or partially refined oil from terminals served by the very large crude carriers (VLCCs). A 550 000 tonne dead-weight tanker is capable of carrying 3.5 million tonnes of petroleum annually from the Persian Gulf to terminals in western Europe at an average speed of 15–17 knots. During the recession of the 1980s many of these supertankers were put into reserve as it is cheaper to lay up capacity in shipping than to operate vessels at uneconomic levels.

As the size of oil tankers increases so does the potential for disaster become greater. Fully laden tankers wrecked on or close to the shore can cause immense ecological damage to coastal environments, as the break-up and oil spills from the *Torrey Canyon, Amoco Cadiz* and *Exxon Valdez* disasters have clearly shown (see Ch. 10).

---

but has created serious problems associated both with routeing patterns and choice of unloading terminals. Use of the Suez and Panama Canals is restricted to smaller vessels and passage through the former was also severely disrupted during the Arab–Israeli conflicts in 1956 and 1967. Adoption of the Cape of Good Hope route for Middle East–western Europe tanker traffic during the period that the Suez Canal was closed stimulated the design and construction of larger-capacity vessels, and ships of up to 250 000 tonnes had the advantage, exploited after reopening

Table 5.2
Registration of oil tankers by flag in 1992: 12 leading states ranked in order of gross tonnage registered

State	No. of oil tankers registered	Total gross tonnage (millions)
Liberia	445	27.44
Panama	528	16.45
Greece	306	10.87
Bahamas	178	9.81
Norway	124	8.47
Japan	1062	7.166
USA	159	5.493
Cyprus	105	4.67
Singapore	294	4.182
Malta	145	3.085
Russia	276	2.435
Italy	175	2.11
World total (all states)	6342	138.148

Source: Lloyd's Register of World Fleet Statistics, December (1992).

of the Canal in 1967, of being able to follow the Cape route loaded and the Suez passage in ballast on the return voyage to the Gulf (Box 5.1).

In 1990 oil tankers accounted for 33 per cent of global gross shipping, with the largest fleets being registered under the Liberian, Norwegian and Panamanian flags (Table 5.2).

Land-based pipelines now carry a substantial proportion of all international petroleum tonnage and can either be integrated with maritime oil tanker routeings or be confined solely to terrestrial trade. Pipelines built to connect the Iranian and Iraqi fields with the Levant coast of the Mediterranean Sea eliminated the costs of using the Suez Canal for vessels bound for western European ports, but are vulnerable to sabotage during periods of political unrest. Within Europe a network of pipelines links coastal refineries in France, the Netherlands and Italy with inland industrial centres.

One of the most extensive pipeline systems conveys petroleum from Russian wells in the Urals to markets in Poland, Czechoslovakia, Hungary and the former East Germany. The 5327 km link from Druzbha to Kuibyshev was opened in 1965 and by the early 1970s the USSR had 47 200 km of pipeline in operation (Fig. 5.2).

5.2.2 *Dry bulk cargoes*

Dry bulk cargoes now account for about 56 per cent by tonnage of world trade and many of the changes which have occurred in their transport are broadly similar to those associated with the petroleum trade. Few industrialised nations with important iron and steel industries now rely

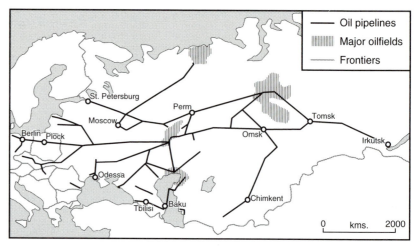

Fig. 5.2
Major oilfields and oil pipelines in territories of the former USSR and in eastern Europe.

upon domestic resources to any large extent and contemporary deposits of iron ore are often located far from the main points of consumption. Although modern iron ore carriers do not approach the size of the supertanker the same principle underlies their construction, namely that a high-capacity vessel can offer lower freight rates and make voyages of over 5000 km between ore deposit and iron smelter a feasible proposition. In 1970 the ore fleet had only 45 ships capable of loads of over 100 000 tonnes but by 1990 320 vessels were of this size or larger. For the UK the principal iron ore sources are Labrador, North and West Africa and Brazil and for the USA South America is a major supplier. The reconstruction of the Japanese iron and steel industry in the post-war period has been founded to a large extent upon ore imports from Australia.

Other ores such as bauxite, raw chemicals, coal and grain have also benefited from advances in specialised bulk ship construction, and wheat and other cereals are also carried across the North Atlantic in this way. Dry bulk carriers of all types now account for about 30 per cent of all world shipping, a proportion only slightly less than that of oil tankers.

The economic depression of the late 1980s and early 1990s led to a decline in world maritime trade and many oil tankers and other dry bulk vessels, formerly on regular charters, were laid up awaiting an upturn in long-distance traffic (Table 5.3).

5.2.3 Containerisation

The introduction and expansion of maritime container traffic have produced some of the most spectacular changes in vessel design and port

Table 5.3
Cargo carriers recorded as
laid up in December 1992

Type of ship	Number	Dead-weight (millions of tonnes)	Average age
Oil tankers	47	4.9	22
Bulk dry/oil carriers	14	1.9	18
Bulk dry carriers	33	1.6	18
General cargo carriers	163	0.7	26
All cargo carriers	441	—	—

Source: Lloyd's Register of World Fleet Statistics, December (1992).

Table 5.4
Container ship capacities
1992

TEU capacity	Number of vessels	Total TEU capacity
Below 1000	630	304 616
1000–1999	435	511 953
2000–2999	218	550 790
3000–3999	96	324 263
Over 4000	35	150 063
World total	1414	1 957 600

TEU = Twenty feet equivalent unit.
Source: Lloyd's Register of World Fleet Statistics, December (1992).

organisation, but it is important not to exaggerate the effects of this innovation. Precise data on container tonnage are difficult to obtain, as the proportion of each container which is loaded varies considerably, but it is unlikely that containerised cargoes account for more than 7 per cent of total world tonnage. Although the gross tonnage of container ships doubled between 1980 and 1990 the total is still less than half that of general non-bulk cargo vessels (Table 5.4).

However, in certain areas and on particular routes the importance of container transport in terms of freight by value is often much higher, and several established ports, such as New York and Hong Kong, have acquired a new significance as container centres, whilst others, such as Tilbury, near London, have been created to deal almost exclusively with this type of traffic.

Efforts to reduce the time spent by shipping in port stem from the fact that maritime transport is at its most efficient when periods spent at sea are maximised, since unloading and loading operations are essentially non-revenue-earning activities. Early experiments in unitised handling of cargoes involved the use of pallets and similar equipment but in the 1960s agreement was reached on the introduction of an international system. This makes use of standard-sized metal containers, suited to road, rail and sea transport, which can be loaded or unloaded as near as possible to their origins and destinations. Since this system involves a heavy investment in new purpose-built vessels and dockside handling equipment, the routes and ports chosen for traffic were determined by the type and volumes of cargo judged to be best suited to containerisation. The totally mechanised handling of containers at terminals can reduce the proportion of a ship's working life spent in port from 66 to 25 per cent, with a consequent increase in revenue-earning capacity, but a very careful appraisal of the viability of this system of cargo handling is essential before the necessary investment is made.

Much of the trade between the UK and Europe is handled in this way, using road container ferries on routes across the North Sea and the English Channel, but with the opening of the Channel Tunnel in 1994 a share of this traffic was diverted to direct rail services. The principal international container routes connect the world's more highly industrialised regions, which can guarantee a regular volume of freight suited to this method of handling. A large vessel carrying 3000 containers operating on a weekly service would require an annual total of 100 000 containers to warrant a regular schedule, and only a few transoceanic routes serve ports that can offer this level of freight (Box 5.2).

Box 5.2 Container shipping

Purpose-built ships include the fully cellular vessel, which carries only containers, the semi-container ship, which can accept a combination of containers and conventional cargo, and the ferry, which is equipped with vehicle ramps and can convey both wheeled containers and deck cargoes. The fully cellular ships, many of which can carry over 1000 standard containers, are operated on the main long-distance trade routes and are capable of eight round voyages in a year between their regular terminal ports. Semi-container ships are used more on routes, such as those between Europe, West Africa and South America, where there is insufficient container traffic to provide a full load and where other space is allocated to conventional cargo. Roll-on/roll-off vessels are usually found on the shorter sea crossings where a road container vehicle is used for the entire journey between shipper and consignee. Vessels may be loaded or off-loaded either by cranes or by the roll-on/roll-off method which makes use of trucks and trailers.

A substantial proportion of the world's fully cellular container fleet is used on services between the east coast ports of North America and those of north-west continental Europe, which can generate this level of traffic (Fig. 5.3). A route of similar importance is that between North America and the Far East, with the growing industrial significance of the Pacific Rim manufacturing states such as South Korea and Taiwan. These two shipping corridors, together with the Europe–Far East route, account for about four-fifths of the world's fully cellular vessel capacity.

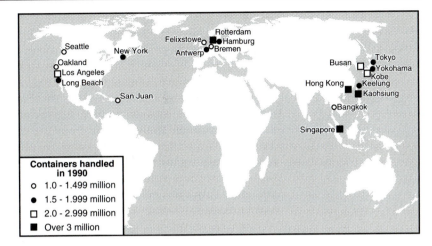

Fig. 5.3
Principal container ports 1990. *Source: Containerisation International* (Dec 1991).

During the decade 1970–80 container traffic grew at an annual rate of 18 per cent but since then the pace of expansion has been reduced. A rapid take-up of container facilities on many of the more prosperous trading routes in the period up to 1980 has meant that the heavy initial investment in containerisation has been able to meet the demand, and the potential for extending the system to other shipping corridors is less than originally envisaged. In addition the global economic slump which began in the late 1980s decreased the demand for some shipping services on which containerisation may otherwise have experienced a continued expansion.

Any future introduction of the system to new trade routes, and any increase in traffic on routes already served by container vessels, will depend upon several economic and technical factors. Economic trends in the USA and Japan in particular will continue to influence rates of container traffic growth since these two industrial nations are involved with such a large proportion of this trade. The spread on a large scale of container traffic to the developing countries is more problematic since so few of these nations have a manufacturing sector which is sufficiently well advanced to offer an acceptable level of suitable traffic.

At present the main routes carrying containers between a less developed state and Western industrial nations are those linking Europe with West Africa and with the Caribbean. Many of the primary agricultural and mining products which form the major part of the exports from developing countries are not suited to unitised cargo handling, and this has hampered the export of manufactured goods in containers from the Western world to developing states since the latter cannot offer any return cargoes.

On the established container corridors linking Western industrial states there may be a possible increase in traffic if commodities currently handled in bulk freighters could be more economically conveyed by container. Some basic chemicals and foodstuffs may in future lend themselves to containerisation on some of the longer routes between the Americas and western Europe but little progress has been made.

Developments in land-based international container traffic have been on a much smaller scale than their maritime counterparts, although rail-borne services between western Europe and the Far East by way of the Trans-Siberian Railway date from the early 1970s (Fig. 5.4). The distance by sea between Rotterdam and Yokohama via the Panama Canal is 23 200 km; by using the USA land-bridge from an east-coast to west-coast port the distance can be reduced to 20 240 km. If the Eurasian rail link is employed, however, the journey length is only 13 770 km, with a trip time of 35–50 days compared with an average sea timing of at least 60 days. Moreover rail freight rates for container traffic can be 20 per cent lower than maritime rates. The land–sea transhipment port of Nahodka on the Sea of Japan is ice-free all the year round and is equipped with container-handling facilities, and the use of this Trans-Siberian route would appear to offer several advantages over the alternative maritime service for European–Japanese traffic.

The disadvantages are that the volume of eastbound traffic is less than that passing from Japan to European markets and that the railway is of limited capacity and must cater for CIS domestic freight as well as international traffic. In the late 1970s the annual container capacity of the Trans-Siberian Railway was estimated at 70 000 units, including 60 000 on the Europe–Far East link, but with the opening in 1980 of the alternative Baikal–Amur line this capacity was probably doubled. Train-loads of 108–116 TEUs are now operated and although most of the route is electrified there are still some lengths of single track and weak bridges. The change of track gauge at the Russian frontier with Poland also causes delays and some freight companies prefer to transfer containers from rail to road at Brest, within Poland. The annual traffic is currently estimated at 80 000 units, but this is only 3 per cent of the total Far East–Europe container market.

With the dismantling of the Soviet Union, and the emergence of plans in some member states of the CIS to expand their trade with Western economies, it is likely that this Eurasian land-bridge will carry an increased container traffic in the future. The opening of the Channel Tunnel to freight in 1994 has also strengthened the attractions of this continental connection between Europe and the Far East.

5.2.4 Air freight

The high costs of air transport, as compared with sea or land modes, limits regular air freighting to fragile or valuable commodities such as electronic equipment or fashion clothing or to perishable goods such as

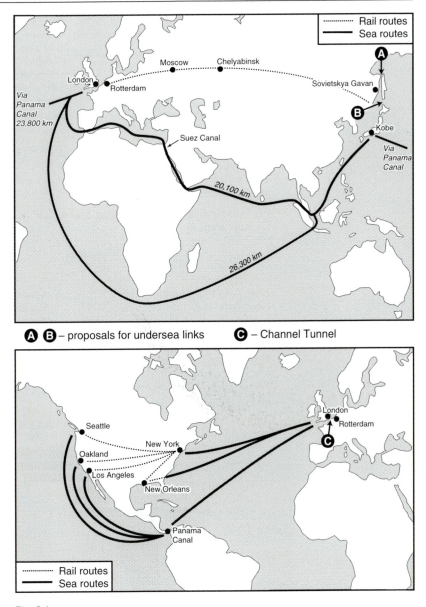

Fig. 5.4
Land-bridge container routes in North America and Eurasia. *Source:* based on Hayuth (1982).

food and flowers. The expansion of high-technology industries in the Pacific Rim, in North America and in western Europe has generated a growing demand for air cargo services, and the highest growth rates have been recorded on the US–Far East routes. World tonnage increased from 639.1 million in 1987 to 740.2 million in 1991, but this total is still insignificant when set against maritime freight (Table 5.5 and Fig. 5.5).

Table 5.5

(a) Air freight imports and exports by value and weight, 1987 and 1991

	Total value (£ millions)		Total weight (tonnes)	
	1987	1991	1987	1991
World imports	14 836	21 465	335 046	425 128
World exports	16 232	20 593	304 085	315 084
European Community imports	2 954	3 997	46 503	45 417
European Community exports	4 293	5 754	56 982	54 117
USA imports	5 194	7 412	111 722	157 165
USA exports	4 164	5 225	90 506	71 890

(b) International airports handling over 0.5 million tonnes of freight (loaded and unloaded) in 1991

	Thousand tonnes		Thousand tonnes
New Tokyo International	1361	Chicago O'Hare	749
New York JFK	1207	London Heathrow	698
Frankfurt	1084	Seoul	631
Los Angeles International	1025	Singapore Changi	621
Miami International	908	Paris CDG	618
Hong Kong International	802	Amsterdam Schiphol	585

Source: Civil Aviation Statistics of the World (1991).

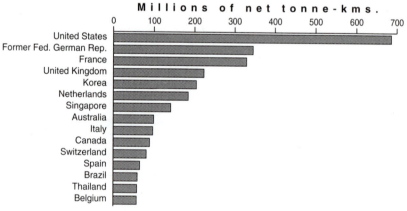

Fig. 5.5
Principal origins of international air freight traffic, 1990. *Source: UN Monthly Bulletin of Statistics* (1992).

Freight is carried in over 850 purpose-built aircraft and the leading international freight centres are now New Tokyo International, J. F. Kennedy at New York, Frankfurt and Los Angeles, with twelve world airports now handling over 0.5 million tonnes each year.

5.3 *Passenger traffic*

International passenger traffic displays great variations in terms of motives for travel, distances involved and the mode of transport selected. Long-distance business travel continues to rely upon scheduled air services but the growth of leisure travel, and in particular the expansion of the packaged holiday programme, has promoted a dramatic increase in chartered air services. In terms of income generated, tourism now ranks second to petroleum in world trade and people are now travelling over increasingly longer distances in search of less developed areas in which to enjoy their leisure periods. Both commercial and leisure travel exploit the fact that air transport can offer a shorter transit time than any other mode and thus enables the greater part of the period spent abroad to be devoted to business or recreation, but there are still examples where sea transport is favoured as a means of access to popular tourist regions.

5.3.1 *Air transport*

In 1991 a total of 1.8 billion passenger-kilometres was recorded on international air services, the leading nations involved in this traffic being the USA, the UK, Japan and France. Transatlantic and transpacific routes carry the highest numbers of passengers, and the traffic carried by scheduled airlines is also most significant on these corridors (Fig. 5.6). Routeing patterns were substantially modified in the late 1980s with the deregulation of airlines in many countries and with the relaxation of 'overfly' restrictions imposed by the former Soviet Union.

Deregulation has often been accompanied by privatisation, as in the case of British Airways, and the strong influence once exerted by many governments over their national airlines has been reduced. Air transport has since the 1940s operated under a set of international agreements, or 'freedoms', which regulate the rights of airlines to operate in the various national markets. In particular the right of the airlines of two nations to carry most of the traffic of those two nations was safeguarded and outside competition was severely restricted by agreement. The 1980s period of deregulation has permitted more competition and, as Sealy (1992: 235) writes, the 'political geography of the airline world is in a state of flux'. For example, the majority of the traffic between the UK and Australia has been carried by the two national airlines of British Airways and Qantas, with intermediate stops in the Far East. This traffic is now being attracted to new services operated by Thai, Cathay Pacific and SIA airlines, who are offering a greater flight frequency and very competitive fare structures.

The granting by the USSR in 1990 of 'overfly' rights to several European and Far Eastern airlines has dramatically changed the pattern of the London–Tokyo and similar routes. In the late 1980s British Airways

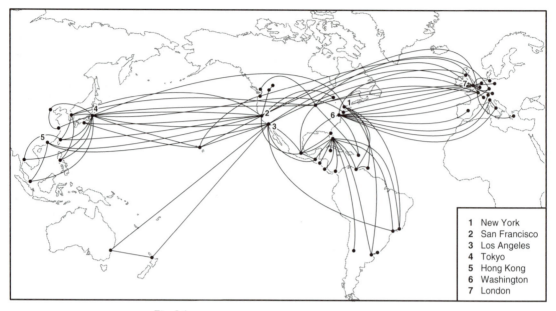

Fig. 5.6
International air passenger services; routes operated by United Airlines, 1992. *Source:* United Airlines current timetables.

1	New York
2	San Francisco
3	Los Angeles
4	Tokyo
5	Hong Kong
6	Washington
7	London

and Japanese Airlines offered 14 flights each week over the North Pole with an 18-hour journey time and a stop at Anchorage in Alaska. In 1991 the flight was rerouted over Russia, reducing the time to 12 hours and, with the attraction to this route of the All Nippon and Virgin Atlantic Airlines, 42 flights per week are now available. Tourist travel has been the dominant growth sector on many long-distance routes and a more detailed study of leisure traffic patterns can usefully illustrate how international air transport has been adapted to meet the growing demands for these movements.

The availability of satisfactory transport facilities is seen to be one of the five major aspects of tourism as a spatial process, the others being the location of areas of attraction, the existence of accommodation, the level of supporting facilities such as shops and personal services and the infrastructure within holiday areas. In the early phase of international tourism in Europe the majority of movements were between the industrial nations of Scandinavia, France, Germany and the UK on the one hand, and the Mediterranean coastlands on the other. The short-distance air links established to meet this demand are still intensively used; although many European tourists are now patronising more remote resorts in East and Southern Africa, South-east Asia, Australia and the Americas, the bulk of the leisure traffic is still on the Europe–Mediterranean corridor.

Plate 5.1
The Tri Star jet-propelled aircraft, here seen in British Airways livery, typifies the medium-range planes used for both scheduled and charter passenger traffic between major international airports and in particular for tourist travel between the United Kingdom and Mediterranean coastal destinations.

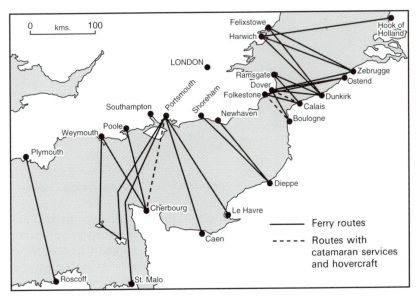

Fig. 5.7
English Channel ferry services between the United Kingdom, France, Belgium and the Netherlands, 1990.

Table 5.6
Passenger traffic between the
UK and the rest of Europe

(a) Accompanied passenger vehicles (thousand vehicles)

To and from:	1981	1986	1991
France	1 402	1 944	3 329
Belgium	591	478	514
Holland	259	325	399
Dover–French and Belgian Ports	1 260	1 539	2 080

(b) Passenger traffic (thousands) by sea

	1981	1986	1991
Dover	12 461	14 377	15 990
Folkestone	1 787	1 192	1 313
Plymouth	318	382	699
Portsmouth	1 032	1 885	2 827
All Thames and south coast ports	19 200	21 030	25 137

(c) Passengers by air on scheduled and charter flights in 1991 (thousands)

From UK to:	1991	
France	5 916	
Germany	5 115	
Greece	3 459	
Italy	3 077	
Holland	3 159	
Portugal	2 417	
Spain	11 653	
All EC countries	41 146	(57% total)
All international air passengers	71 888	

Source: Transport Statistics Great Britain (1992).

5.3.2 United Kingdom–Europe tourist traffic

The pattern of tourism between the UK and continental Europe offers one example where sea transport still accounts for a significant share of total traffic (Table 5.6 and Fig. 5.7). When time is at a premium the British tourist will almost always accept the air charter packaged holiday programme, but increasing numbers whose main objective is to take a touring holiday in one or more European countries make use of the car ferry services across the English Channel. In the six summer months of 1990 15.92 million passengers crossed to France, Belgium and the

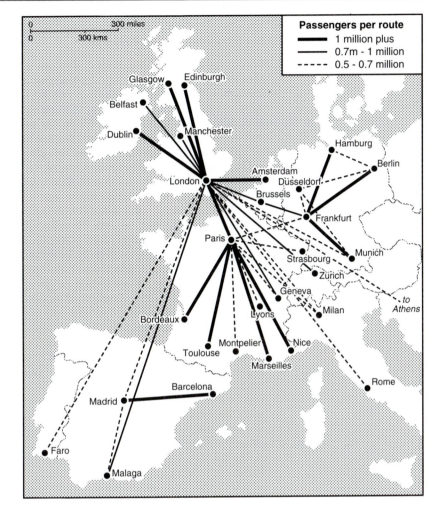

Fig. 5.8
Principal flows of national and international air passenger traffic in western Europe, 1989.
Source: based on *Financial Times* 'Transport links supplement 5', April 1990.

Netherlands by sea compared with 29.22 million using air transport (Box 5.3).

The opening of the Channel Tunnel in 1994 has begun to have an impact upon this cross-Channel ferry traffic and also upon many of the air services linking the UK with Europe (Fig. 5.8). Given that the tunnel shuttle service for road vehicles provides a transit time, inclusive of customs and passport formalities and clearance, of about 60 minutes, compared with about 140 minutes by ferry on the Straits of Dover route, it will undoubtedly prove an attractive alternative to many passengers. Additionally, the introduction of high-speed rail services from London to Paris via the tunnel will eventually cut the present 7-hour journey time

Box 5.3 English Channel ferry traffic

For travel between the UK and destinations in Europe and in the Mediterranean, including North African coastlands, the use of sea ferries fell from 46 per cent of all passengers in 1965 to 38 per cent in 1991, but the actual numbers travelling on cross-Channel passenger ships has increased from 6.59 million to 25.13 million over the same period. This short-sea traffic is largely responsible for the fact that maritime transport still accounts for one-third of all international travel between the UK and the rest of the world. Whereas in the 1950s cross-Channel services were largely confined to ships between Dover, Calais and Boulogne, the contemporary routes involve the use of ports from Ramsgate west to Plymouth in the UK and from Zeebrugge west to Roscoff in continental Europe.

This extension of ferry services has been aided by the improvement in road access to ports on both sides of the Channel, and the demand for additional facilities arises principally from the enormous growth in car-based holidays in western and southern Europe. At present British tourists bound for Europe can choose from over 120 daily crossings, although 75 per cent of all traffic is still concentrated on the traditional short-sea routes from Dover. Many of the newer services, such as those between Portsmouth, Caen and St Malo and from Plymouth to Roscoff, were initially introduced for commercial vehicles only but the continuing growth in demand from the tourist market encouraged the addition of car-carrying facilities in the 1970s. The increase in carryings on the cross-Channel routes has been accompanied by the improvement of terminal berthing facilities at Dover, Portsmouth, Calais, Ostend and Zeebrugge and by the launching of new high-capacity ferries such as the 26 000 tonne *Pride of Dover*, which can carry 650 cars and 2300 passengers and is equipped with two-level two-lane ramps to reduce loading and unloading times and thus achieve more sailings during the day. The catamaran introduced on to the Dover–Calais and Portsmouth–Cherbourg routes in 1990 has crossing times of 40 and 160 minutes respectively, although the vessel is limited to 450 passengers and 80 cars.

to less than 3 hours and attract passengers from both air and road-borne travel. In this respect British Rail and SNCF estimate that their share of the total Anglo-French passenger market could rise from its present level of 7 per cent to 25 per cent.

The extent to which tourist flows on the longer sea routes are being attracted to the tunnel depends upon the origins and destinations of travellers. Holidaymakers using ferries from Portsmouth or Plymouth and bound for western France or Spain would find little advantage in diverting to Dover and it is unlikely that the tunnel will attract much of this traffic.

Plate 5.2
The large volume of passenger traffic across the English Channel has stimulated the introduction of the hovercraft and the hydrofoil on the short sea routes linking the United Kingdom and France. The SRN hovercraft, which can accommodate 250 passengers and 30 cars, was introduced in 1969 on the longer Ramsgate–Calais service but now operates on the Dover–Boulogne crossing, which is achieved in 35 minutes at speeds of up to 110 kph.

5.4 *International transport within the European Union*

The progressive expansion of the European Union since 1958 has stimulated trade and exposed the deficiencies in many of its international road and rail links. Political changes in eastern Europe will lead to an increase in trade between this region and the Union and this trend, together with the formation of the Single Market in January 1993, has created a demand for further improvements in the inland transport infrastructure of Europe as a whole. The heaviest investments to date have been in motorway networks and railway electrification but, although many through routes have been upgraded, priority in construction has often been given to national needs, particularly in those Union states whose transport systems were severely damaged during the Second World War (Fig. 5.9). Rotterdam's rise as the principal Union port on the North Sea has been assisted by the building of motorways

Plate 5.3
Barcelona typifies the Mediterranean port which is backed by coastal highlands, making it necessary to reclaim land for port extensions. This view of Barcelona in the early phases of containerisation shows the protective breakwater and the extensive uncovered storage areas. Barcelona is also the terminal for the Balearic Island ferries.

through the Netherlands to France and Germany, and the barriers imposed by the Alps to northern European–Italian trade have been lessened with road tunnel and motorway construction. The particular implications for Germany of unification of the former federal and democratic republics and wider political changes in eastern Europe are discussed in Chapter 6.

Inland waterways continue to play an important role in international freight movements and the Rhine in particular is a key transport artery within the European Union (Box 5.4). About 40 per cent of all goods traffic between EU states, and between EU states and other trading partners, is carried by waterways, which counter the increasing challenge from road hauliers by concentrating upon bulk commodities. Within Europe the expansion of international road container trade is constrained by highway condition and capacity in certain eastern states and by customs restrictions. Much of the existing container traffic is by rail, especially in eastern Europe, but progress is being made on an integrated motorway network capable of accommodating container trucks.

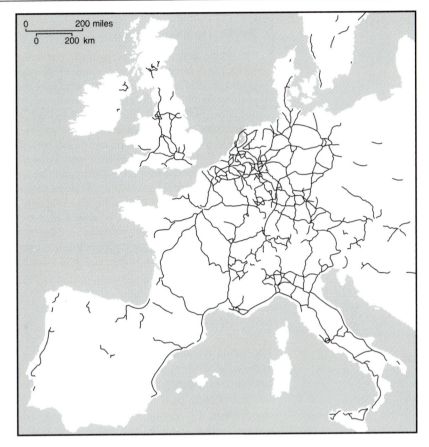

Fig. 5.9
International road transport routes: the European motorway network in 1992.

5.5 International terminals

The overall pattern of international air- and seaports and the character of individual terminals have changed considerably in response to the developments in ship and aircraft technology and cargo-handling methods discussed earlier in the chapter. As the capacity of ships and aircraft has increased there has been a concentration of traffic upon fewer routes which serve a correspondingly reduced number of international terminals equipped with more specialised facilities for handling cargoes and passengers.

Box 5.4 The Rhine waterway

The Rhine, serving river ports in the Netherlands, France, Germany and Switzerland, carries about 90 per cent of all water-borne international trade within the European Union and the annual traffic of over 250 000 tonnes is dominated by iron ore and petroleum imports via Rotterdam to the Ruhr and other destinations upstream. The navigable Meuse and Moselle rivers and a canal network extend the Rhine waterway into France, and to the east access to Germany is via canals linking up the Weser, Elbe and Oder rivers. The Rhine and Danube waterways were first linked in 1846 by a canal between Bamberg and Kelheim but this was disused by 1945. A larger canal which can accept 3300 tonne ships was opened in late 1992 with an annual capacity of 18 million tonnes, although it is unlikely that this total will be achieved for at least 30 years. With the unification of Germany and the strong prospects for increased trading between the European Union and eastern states the new Rhine–Danube canal may eventually prove to be a valuable addition to the European inland water system.

5.5.1 Seaports

The extensive literature on port geography covers locational factors, hinterland development and character, methods of classification and contemporary economic aspects. In this section an emphasis is laid upon recent trends in port trading and on how individual terminals have been re-equipped to remain competitive and, where possible, extend their services both in their maritime forelands and their hinterlands. The demise of the international passenger port is common to all nations although, as described above, seaborne traffic is still thriving on certain short routes with heavy flows of tourist traffic.

Within the freight sector the principal changes have been associated with the expansion of bulk carriage in vessels of ever-increasing capacity and the consequent necessity for seaports to adapt their berthing and docking facilities to accept these ships. Where this adaptation has proved impossible entirely new terminals have been created, often specialising in just one commodity, and the petroleum port offers the prime example of this trend.

Where the large size of tankers creates difficulties for ports with shallow approaches the single-point mooring system has often been adopted, where vessels discharge their cargoes close to shore via a pipeline, without the need to berth alongside the conventional terminal pier. In the UK the Amlwch terminal off Anglesey takes this form and accepts oil tankers formerly unloaded at Tranmere on the Mersey, with a pipeline across North Wales to the Stanlow refineries (Box 5.5). Single-point moorings are now also common in the Persian Gulf and off the coasts of Japan, Borneo, Darusallem Brunei and Nigeria.

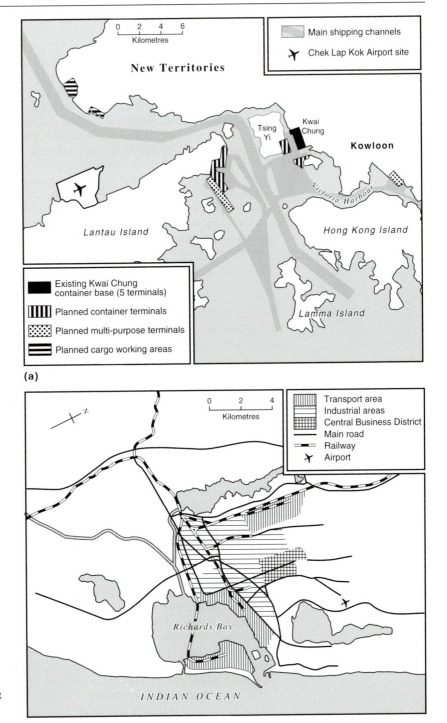

Fig. 5.10
(a) Existing and planned cargo-handling facilities in Hong Kong, 1994.
(b) Principal bulk cargo-handling areas in Richards Bay, Natal, S. Africa. *Source:* (a) Hong Kong Government Information Services; (b) based on Fig. 20.2 in Hoyle and Hilling (1984).

Box 5.5 European oil terminals

In western Europe major petroleum terminals have been built at Europoort, Marseilles, Southampton, Milford Haven and at many other ports where, following initial refining, final distribution can be carried out by pipeline, rail or coastal shipping. Europoort in particular benefits from the facilities offered by the Rhine, which has a substantial barge traffic in oil upstream to the Ruhr and Switzerland. In southern France, Fos, dating from 1968, accepts tankers up to 400 000 tonnes dead-weight, and typifies the contemporary pattern of oil-based industrialisation at a petroleum terminal, with refining and petrochemicals. Similar oil-related growth, although on a smaller scale, has occurred at Fawley, on Southampton Water, and at Milford Haven, where berths for 270 000 tonne vessels are linked by pipeline to the Llandarcy refinery. Bantry Bay, which like Milford Haven has a penetrating ria harbour with deepwater approaches, has become an important terminal for the very largest tankers in south-west Eire.

Other bulk cargoes, such as iron ore, still require dockside handling equipment but it is now possible to deal with vessels of up to 350 000 tonnes in this way. Ships with this capacity can now berth at four Japanese ports but most of the European terminals, such as Fos, Rotterdam, Dunkirk, Port Talbot and Immingham are limited to ships in the 150 000–200 000 tonne range.

The gradual diffusion of containerised traffic from the late 1960s was made possible by selective investment in ports which had the necessary deepwater approaches, sufficient land for container handling and storage and road and rail links which could accept container lorries and wagons.

Box 5.6 The Chinese port modernisation programme

The current Chinese programme for port investment is an essential part of the drive to expand foreign trade and the Seventh Five-Year Plan proposed to increase total cargo-handling capacity to 317 million tonnes by the mid-1980s. Congestion is severely hampering shipping movements at many ports and by the early 1990s China hoped to have a further 120 deepwater berths available, with many of these devoted to the increased export trade in coal. In the early 1980s China had 145 ports on her 18 000 km seaboard, but only 20 were open to international trade, and 80 per cent of this trade was concentrated upon 8 major terminals. In 1979 Shanghai became the first Chinese port to offer container facilities, followed by Tianjin in 1982, and the Yangtse port of Nantong is now also equipped. Traffic dispatched from many of the smaller coastal and river ports is sent by coastwise ship to Hong Kong for transhipment to larger ocean-going vessels, and the latter port will assume a greater significance in this role when it becomes a part of Chinese territory in 1997.　▶

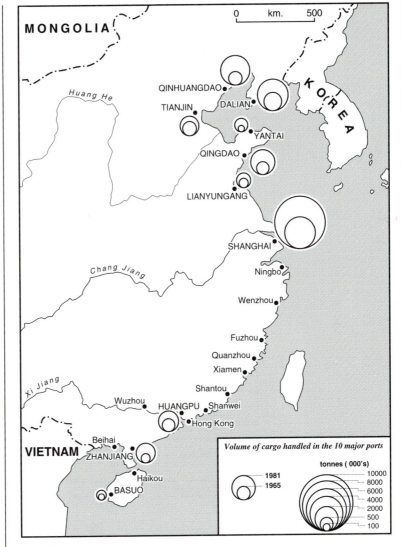

Fig. 5.11
Principal international trading ports and inland waterways in China. *Source:* based on Fig. 10.3 in Chiu and Chu (1984).

All major industrial nations now possess container terminals which have either been grafted on to existing ports, as with the Seaforth dock at Liverpool, or established at ports which had previously been unimportant trading centres (Fig. 5.10). In the UK the current success of Felixstowe, on the Suffolk coast facing the Netherlands, is also almost entirely due to its participation in the short-haul and later long-haul, container trade. Felixstowe handled 1.17 million containers (15.4 million tonnes) in 1991 and is now one of the largest container terminals in western Europe.

In the Far East Hong Kong has emerged as the leading container centre and over 40 per cent of its annual trade is now handled in this way. On a world scale it ranks second to Rotterdam in terms of container tonnage, having overtaken New York in the late 1980s, and the only constraint which hampers its continuing expansion is lack of space for additional berths and storage. The future of Hong Kong as a container port is closely linked with China, which has made substantial progress in port modernisation in the 1980s and will repossess Hong Kong in 1997 (Box 5.6 and Fig. 5.11).

The efficient working of a seaport and of its hinterland links is often a vital part of the development process in the less advanced nations. In many African states there is an urgent need for investment in this sector if international trade is to be expanded and a greater variety of specialised shipping services attracted to particular ports. Lagos, the principal port of Nigeria, typifies many of the difficulties encountered by such ports and improvements have been made in an effort to solve the chronic problems of congestion experienced in the mid-1970s. Diversion of shipping to alternative ports such as Port Harcourt, Warri and Sapele helped to some extent but Lagos remains the target for most of the current investment plans. Wharves at Apapa were extended and in 1977 the Tin Can Island scheme was completed, with an annual capacity of 3 million tonnes and container handling and roll-on/roll-off berths (Fig. 5.12).

In East Africa Mombasa and Dar es Salaam were also equipped with container berths in the 1970s, and further south in Mozambique the port of Beira is undergoing modernisation with the intention of it regaining its former role as the principal outlet for land-locked Zimbabwe and, to a lesser extent, neighbouring Malawi and Zambia. This Beira project is closely integrated with plans for the modernisation of the Harare–Beira railway, which is the port's main link with its hinterland, and when the

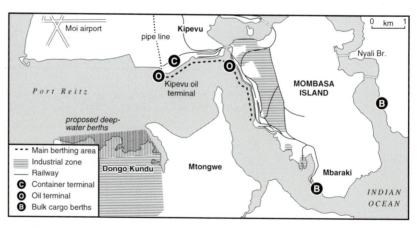

(a)

Fig. 5.12
Port installations in developing countries: (a) Mombasa, Kenya; (b) Lagos, Nigeria (facing page). *Source:* based on Hoyle (1968).

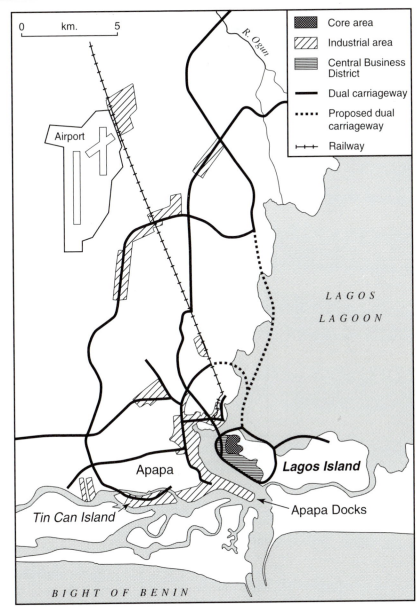

(b)

Fig. 5.12 continued

entire scheme is completed the current dependence of Zimbabwe upon South African ports as trade outlets will be substantially reduced.

5.5.2 Airports

The rapid expansion of international air passenger traffic since the 1950s has placed strong pressure upon airports, especially those in the industrial

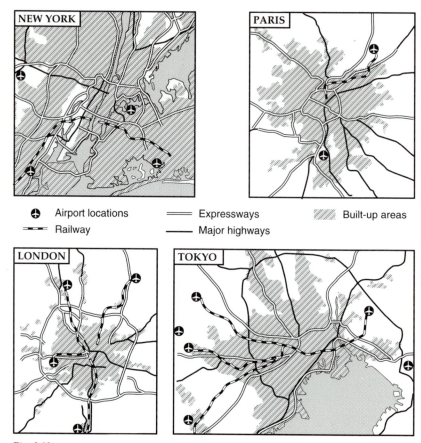

Fig. 5.13
Airports of major cities, showing principal transport links with city centres.

nations. Additional capacity has been provided by enlarging existing terminals, by upgrading regional airports to accept international flights and by building entirely new airports, but there is a continuing demand for growth in capacity.

Passenger traffic continues to form the major revenue at all airports, but a distinction must be made between domestic and international services, and between scheduled and charter operations, as these have an important influence upon the fortunes of individual terminals. In particular the overall travel time is affected by the periods spent on the land-based sections of each journey, a vital consideration which becomes of increasing significance on the shorter distance international services, as in Europe. As a result the search for the location of new airports, and decisions made in respect of the expansion of existing centres, are strongly influenced by the quality of potential or existing links between airport site and the neighbouring population concentrations which provide their markets

Table 5.7
Some measures of airport
hub performance, based
upon July 1992 flight
schedules

Airport 'hub' and airline	Average number of connections per flight	Domestic connections to European destinations within 90 minutes	Minimum connection times (mins)
Frankfurt (Lufthansa)	17	1521	45
Amsterdam (KLM)	14	236	50
Paris, CDG (Air France)	7	82	45
Madrid (Iberia)	6	246	45
London, Heathrow (British Airways)	8	611	45

Source: Dennis, personal communication (1993).

(Fig. 5.13). However, the planning of new airports and their urban links must take into account the environmental impacts, as discussed in Chapter 10.

The complex series of international agreements on air travel have endowed certain major airports with the status of 'gateways'. Such airports are entry points into a country for designated operators and are the end-points for an international route on which only one airline may provide services. Many airports occupy strategic locations on the global route network and are termed 'hubs' as they act as focal points for a large number of feeder services connecting with long-haul flights. There is often strong competition for traffic between hub airports, which must provide attractive terminal facilities for transit passengers, provide co-ordinated flight schedules with minimum waiting time between connections and, most crucially, organise convenient take-off and landing 'slots' for the major airlines. Within Europe Amsterdam and Frankfurt have emerged

Box 5.7 Airports in Germany

Frankfurt's growth as a major hub in western Europe has been based upon its ability to offer well-scheduled international connections. In 1992 passengers on over 1450 daily flights into Frankfurt from origins in Germany and other European countries could secure long-haul connections within 2 hours, but the airport is now experiencing severe congestion at peak times. This will be partly relieved by the new terminal at Munich, open in 1992, whose eventual capacity of 25 million passengers will exceed the existing total at Gatwick. There are also plans for a new international airport at Berlin to be completed by about 2005 with a capacity of 30 million. International traffic in the unified Germany will be focused upon Frankfurt and the new Munich terminal, with linking domestic services to Leipzig, Dresden, Stuttgart, Bonn and Hamburg.

Table 5.8
International air terminals
ranked by annual passenger
totals, 1990

Country	Airport	International commercial air transport movements (thousands)	International terminal passengers (millions)
UK	London, Heathrow	279	35
Germany	Frankfurt	223	22
France	Paris, Charles de Gaulle	209	21
UK	London, Gatwick	160	20
Hong Kong	Hong Kong International	106	19
Japan	New Tokyo International	111	18
USA	New York, J. F.Kennedy	109	18
Netherlands	Amsterdam, Schiphol	188	15
Singapore	Singapore, Changi	98	14
Switzerland	Zurich	153	12
Thailand	Bangkok International	81	11
Canada	Toronto, Lester Pearson	132	10

Source: Transport Statistics Great Britain 1992.

as leading hubs, but although London Heathrow is a focus of many world routes its efficiency as a hub is constrained by limited runway capacity, the tight scheduling of take-off 'slots' and the distance between its four terminals (Table 5.7 and Box 5.7).

However, despite these limitations, London Heathrow is still the world's leading airport in terms of annual totals of international passengers. Three other European airports come next in the ranking followed by Hong Kong and the New Tokyo International. Although Chicago is the world's busiest airport on the basis of total traffic much of this is domestic, and New York J. F. Kennedy airport, the leading international centre in the USA, ranks only seventh on the world scale (Table 5.8).

International air traffic growth in western Europe has created severe congestion, both in air corridors and at passenger terminals. Paris and London in particular saw the urgent need for additional airport capacity in the 1960s, but the policies adopted by the two countries to solve this problem differed considerably and offer marked contrasts in their rates of response and location strategies. In the 1960s it became clear that London's two major airports at Heathrow and Gatwick would be unable to cope adequately with the projected expansion of traffic in the period up to the year 2000. The initial choice of Stansted, 52 km from central London, as a third airport met with fierce opposition, and the subsequent Roskill commission made alternative recommendations which were, however, not implemented (Ch. 10). In 1985 Stansted was reselected and was expanded to provide an annual capacity of 8 million passengers, with a proposed capacity of 25 million when the main runway is eventually rebuilt.

In the Paris region it had become clear by the late 1960s that air

Plate 5.4
Until the opening of the rail link to central Manchester in 1993, Manchester Airport was
served by a shuttle bus service from the city centre and by a limited number of buses from
other local towns. With the expansion of international services a second terminal was
opened in 1993 and the rail connection was completed at the same time.

passenger traffic could no longer be dealt with by the airports at Le
Bourget and Orly, respectively 14 and 12 km from the city centre. In 1959
a site for an additional airport was found at Roissy, 25 km north of the
capital and construction began in 1966. This new Charles de Gaulle
airport was completed in 1972 and the Le Bourget terminal was closed
three years later. Orly has been retained and now handles 7.2 million
international passengers annually and Charles de Gaulle has 21 million,
ranking as third in the world. The Paris Airport Authority thus achieved
in 13 years what it took London about 25 years to accomplish, although
the volume of air traffic around the French capital is smaller.

The efficient handling of international air traffic requires adequate
airport–city links and most major airports are now equipped with motorway and rail connections. London Heathrow has express highway and
metro rail connections with the capital but still lacks a high-speed surface
rail link. Both Gatwick and Stansted have high-frequency rail services to
London and in northern England Manchester International Airport's
railway to the city was opened in 1993 in time to serve the new second
terminal.

These various developments in airline deregulation and privatisation

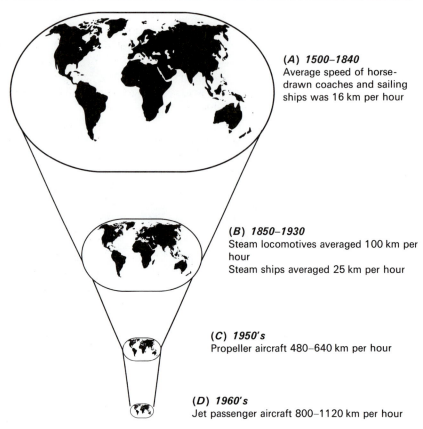

(A) *1500–1840*
Average speed of horse-drawn coaches and sailing ships was 16 km per hour

(B) *1850–1930*
Steam locomotives averaged 100 km per hour
Steam ships averaged 25 km per hour

(C) *1950's*
Propeller aircraft 480–640 km per hour

(D) *1960's*
Jet passenger aircraft 800–1120 km per hour

Fig. 5.14
The contributions in transport technology to reductions in travel time: *Source:* based on Fig. 1 in McHale (1969).

and in airport expansion have involved both government and private initiatives and local and national airport undertakings, and there is rarely any attempt to co-ordinate all aspects of air passenger transport in any one state. The steady increases in traffic which will continue into the twenty-first century can only be met by additional numbers of air movements at existing airports or by the construction of new ones, and both these alternatives have profound implications for planning and environmental issues (Ch. 10).

5.6 Concluding summary

International transport played an essential part in the European processes of industrialisation and colonisation in the nineteenth century and has assumed a new significance in the late-twentieth-century growth of transnational corporations. Advancements in shipping and aircraft technology have both stimulated and met the demands for the cheaper and more rapid movement of freight and passengers over long distances. The recent developments in global telecommunications may in future reduce the need for personal business travel but tourism will continue to provide a profitable market for international airlines.

The conquest of distance by modern forms of international transport is often described in terms of 'global shrinkage', with world maps drawn in terms of time–distance to illustrate how journey times have been reduced with technological advances (Fig. 5.14). If fibre optic and related communications systems do, as suggested above, make much of the contemporary pattern of business travel at the world scale unnecessary in the future then the 'shrinking world' will lose some of its significance to coming generations, who will be able to conduct their international linkages without leaving their own workplaces.

Further reading

Couper A D 1972 *The geography of sea transport* Hutchinson
Hoyle B S, Hilling D (eds) 1984 *Seaport systems and spatial change* Wiley
Lloyds 1990 *Lloyds maritime atlas* Lloyds
Muller G 1989 *Intermodal freight transportation* Westport (N: ENO)
Naveau J 1990 *International air transport in a changing world* Martinus Nijhoff, Dordrecht
Pearce D G 1987 *Tourism today: a geographical analysis* Longman
Pearson R, Fossey J 1983 *World deep sea container shipping* Gower
Sealy K R 1976 *Airport strategy and planning* Oxford University Press
UNCTAD 1990 *Review of maritime transport* UNCTAD

6 *National transport systems*

The transport industry – its role in national economies

When viewed at the national scale, transport systems illustrate many of the issues discussed in Chapters 2–4, especially in terms of changing markets and the ascendency of road over rail for both passengers and freight in advanced economies. In Russia and many of the eastern European states the recent liberalisation of transport policy and management is gradually bringing changes to road and rail infrastructures. Within the less developed world a continuing lack of finance severely hinders the implementation of much needed improvements in transport systems.

6.1 Introduction

Variations in the character of transport systems in individual nations are often as great as those of the economies which they serve, and there are substantial differences from country to country in the shares of passengers and freight traffic carried by road, rail, waterway and internal air services. Contemporary systems are a reflection of the extent to which governments exercise control over the transport sector by means of policies such as subsidisation, deregulation or privatisation or by state ownership. In the developing nations investment in the transport infrastructure is often determined by the policies of international aid bodies such as the World Bank.

In some states national economic development plans incorporate specific proposals for the transport sector whereas in others the individual parts of the economic structure lack any apparent co-ordination. The broad policy and planning issues which are involved are discussed in

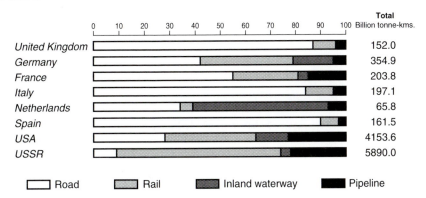

Fig. 6.1
Proportions of freight traffic carried by principal surface modes in selected countries, 1990.
Sources: International Road Federation: *World road statistics* (1992); *Transport statistics Great Britain* (1992).

Chapter 12 and in this chapter the emphasis is upon the contemporary structure of the principal means of inland transport.

The complexity of transport systems and the composition of the traffic that they carry may be measured in terms of route densities, traffic densities, network indices, and modal split of traffic. There is an obvious contrast between the intricate nature of the rail and road networks of advanced countries such as France, Germany or Japan, and the much simpler systems of developing nations typified by Zambia, Kenya or Ghana. The density of traffic on road and rail also shows a similar contrast, with daily flows of at least 150 000 vehicles on metropolitan orbital motorways such as the M25 in south-eastern England but flows of less than 1000 vehicles on many regional highways in states of southern Africa.

The allocation of traffic between various modes varies both between the more and less advanced groups of nations and within each of these categories. Differences also exist between traffic distribution in nations subject to centralised socialist planning such as China and in states where government control is much less strongly applied (Figs 6.1 and 6.2).

In order to examine and analyse these various contrasts in more detail examples will be drawn from countries which are representative of the three broad categories of (a) industrialised nations, (b) states which are or have until recently been subject to centralised economic planning and (c) the less developed world. The transport systems of Germany, the UK and the USA have been chosen to illustrate the advanced industrial nations, the continental extent of the latter contrasting with the much smaller areas of the two European states. These latter in their turn exhibit differences in terms of political geography, recent levels of investment in infrastructure, and government policy towards co-ordination within the transport sector.

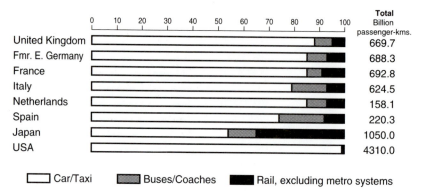

Fig. 6.2
Proportions of total passenger traffic carried by principal surface modes in selected countries, 1990. *Sources:* International Road Federation: *World road statistics* (1992); *Transport statistics Great Britain* (1992).

Poland and China, again showing great contrasts in territorial area, are selected to see to what extent transport operations and policy have been influenced by decades of centralised economic planning. Finally Nigeria, Zimbabwe, Malaysia and Bolivia illustrate the very substantial differences which can exist in the transport infrastructures of the developing nations. In each case emphasis will be laid upon the broader national and regional elements of transport networks, leaving the specific issues associated with urban and rural communications to be dealt with in more detail later (see Chs 7–9).

Although roads and railways remain the dominant modes in most countries, inland waterways still play an important part in freight carriage in many states, and domestic airways now carry a substantial proportion of inter-city and interregional passenger flows in the advanced nations.

6.2 Transport in advanced industrial nations

The USA, Germany and the UK all possess highly sophisticated transport systems, which reflect substantial levels of past and continuing investment in order to meet the demands of industry and society. However, the uses made of these communication networks for different types of traffic now differ considerably and each mode is considered in turn in terms of its significance within the transport sector.

6.2.1 Railways

During the second half of the twentieth century most railway systems in the industrialised world have been modified to meet the challenge of road transport, a challenge which in many cases has resulted in the railways becoming the minority partner in the carriage of both freight and passenger traffic (see Ch. 2). These adaptations have taken various forms, the most common being network contraction, modernisation of motive power and rolling stock, and a concentration of new investment in those sectors of the system judged to be best suited to the carriage of traffic which can still yield a profit. Substantial government aid has often been involved, and state subsidies have been applied to underwrite the losses which many systems continue to make despite remedial measures. In particular several states provide financial support for routes which are uneconomic but which have been reprieved from closure on the grounds of social necessity.

The UK and Germany, in common with all other western European nations, possess state-owned railway systems, although the former is now implementing a progressive privatisation plan (Fig. 6.3). The British network, with 16 600 km, is considerably less extensive than that of Germany with 44 000 km, and contrasts can also be drawn in terms of freight traffic allocation, with the German system carrying 24 per cent of all inland freight but the British only 10 per cent. Although the British and German railways are proportionately of less importance as freight carriers than those of France, Austria and Switzerland, both continue to play an important part within the overall transport sector.

Modernisation of the British system dates from a 1955 plan and has focused upon electrification of lines in the south-east commuter region and the trunk routes linking London with Birmingham, Liverpool, Manchester and Glasgow. With completion in 1991 of the London–Newcastle–Edinburgh scheme about 30 per cent of the total network is now electrified, compared with 40 per cent in Germany (Table 6.1) Far fewer closures have been made on the German network, which still carries about 30 per cent of all public passenger traffic (as opposed to 6 per cent in the UK). The size of the former West German national network was reduced by only 4.5 per cent to 31 500 km by the late 1970s and further closures in the 1980s reduced the overall length by another 9 per cent. During the 1980s the Deutches Bundesbahn initiated a programme of new line building to accommodate high-speed trains similar in concept to the TGV of France and those on the Italian Florence to Rome route. The 327 km between Hanover and Würzburg and the 1200 km between Mannheim and Stuttgart were completed in 1991 to allow speeds of 250 km per hour and it is hoped that these improved services will result in at least a 60 per cent increase in traffic by the year 2000 (Fig. 6.4).

The UK's experiments in high-speed travel using tilting trains on existing track were unsuccessful, and the London–Channel Tunnel line through Kent and Essex to be completed in the early twenty-first century

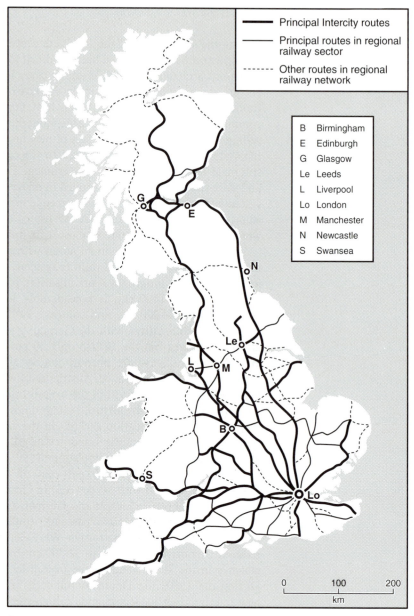

Principal Intercity routes

Principal routes in regional railway sector

Other routes in regional railway network

B	Birmingham
E	Edinburgh
G	Glasgow
Le	Leeds
L	Liverpool
Lo	London
M	Manchester
N	Newcastle
S	Swansea

0   100   200
km

Fig. 6.3
The British Rail passenger services network in 1994. Details of suburban and branch lines are excluded.

will thus be the first British example of a high-speed rail route similar to those in continental Europe.

Levels of subsidy on British railways have always been much lower than those on German and other European systems and the progressive

Table 6.1
Principal railway networks, showing percentage of route-kilometres electrified (including all countries with over 10 000 km railway in 1990)

Country	Rail km in use (thousands) 1990	Percentage of rail km electrified, 1990
USA	192.7	0.8
Former USSR	147.5	36.9
Canada	90	2.0
China	54.0	12.0
France	34.1	36.4
Germany	44.0	35.9
Japan	27.2	40.5
Italy	19.6	55.1
UK	16.9	29.0
Spain	14.3	48.2
Former Czechoslovakia	13.1	29.0
Sweden	11.3	66.4

Source: Transport Statistics Great Britain (1992); *Jane's world railways* (1992).

Plate 6.1
High voltage electrification in the United Kingdom has been confined to trunk routes to the north and north-west of London. Elsewhere, intercity services are provided by high speed diesel-hauled trains, such as this IC 125 approaching London on the main western route from Bristol, South Wales and south-west England. These trains have a limited capacity and are often followed by 'back up' services.

reductions in state support since the early 1980s paved the way for the current privatisation plans (Box 6.1). The Deutsches Bundesbahn, however, will require more state support as the former federal and democratic republics' networks are integrated and the eastern system upgraded. Rail traffic east–west across Germany is increasing and many routes are to be improved in addition to the high-speed lines described above.

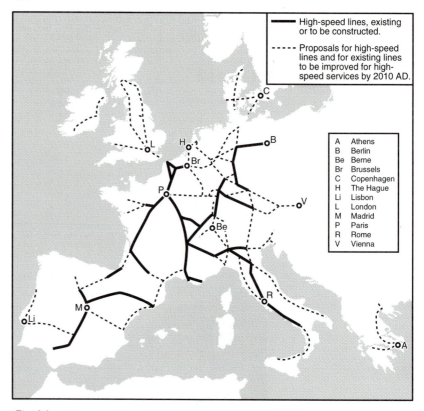

Fig. 6.4
Existing and proposed high-speed railways in western Europe.

Railways in the USA have been subject to a series of government controls, but have never passed into state ownership, and the 300 private companies which exist are dominated by about 40 principal carriers. The attraction of most inter-city passenger traffic to airlines and road services led to steeply declining revenues in the post-1950 period, and during the 1970s several of the major rail companies became bankrupt, notably the Penn Central Railroad and two Midwestern undertakings serving Chicago. Federal aid was provided to establish the Consolidated Rail Corporation (CONRAIL) in 1976 from elements of the former Penn Central and other failed companies in the north-east USA and this new grouping initially received government subsidies to assist its operations. The federal government was also involved in the formation in 1971 of the National Railway Passenger Corporation (AMTRAK) to provide passenger services between selected cities over privately owned railways (see Ch. 2).

American rail freight traffic, using train-loads of up to 15 000 tonnes, has been maintained with the introduction of containers and flat wagons

Box 6.1 British railway privatisation proposals

The 1963 Beeching Report on the reshaping of British Railways was an attempt to combat ever-rising deficits caused mainly by growing competition from road freight and passenger transport. There was a gross imbalance in the use of the network, with the concentration of lucrative traffic on very small sections of the system and with one-half of the railways earning insufficient revenue even to meet essential track maintenance costs. A drastic programme of passenger service reductions and track closures was introduced together with a selective programme of investment in the major inter-city routes, the London commuter network and trunk lines with heavy bulk freight traffic, these being judged as being best able to attract and retain revenue-earning traffic. Some threatened passenger services were supported with government subsidies and since the 1970s the size of the network has been stabilised and few major cutbacks have taken place. During the 1980s a series of corporate plans were published under which the state subsidies for passenger services were progressively reduced and by 1989 the InterCity sector, which accounts for about 40 per cent of all rail passenger traffic, had recorded a profit. In early 1993 government proposals for restoring British Rail to the private sector were published, with plans to retain a smaller state-owned system known as Railtrack which now controls most of the track, signalling and stations. Seven routes, which together generate about one-third of all passenger revenue, will be franchised. They include the trunk lines from London to Edinburgh, South Wales and south-west England, most railways in Scotland, the London–Southend commuter line, the London–Gatwick Airport shuttle and other commuter routes in southern England. Many of these trunk routes are profitable, but the Scottish and some London commuter systems are dependent upon subsidy and require substantial track and rolling stock improvements.

The transfer of freight from road to rail will also be encouraged by state contributions towards lines carrying freight and the cost of freight rolling stock.

Government claims that the private sector will provide more efficient management and hence better services are countered by supporters of continuing state control, who argue that uneconomic lines will be closed, facilities for family and other concessionary travel withdrawn and that through long-distance services may be under threat.

carrying road trailers (the 'piggy back' system) for long-distance hauls. The percentage of all freight tonne-kilometres accounted for by rail is now 35, compared with 56 per cent in 1950, but the rail-borne proportion of national passenger traffic has fallen heavily over the same period from 47 to 4 per cent. By the late 1980s, however, over 190 000 km of track were still carrying traffic, and the percentage fall in length between 1967 and 1990 was only 9 per cent compared with a decrease of 30 per cent in the UK.

Table 6.2
Countries ranked by
numbers of cars/taxis per
head of population in 1990

Country	Cars/taxis per 1000 persons	
	1980	1990
USA	548	648
Germany	331	437
Sweden	347	421
France	357	417
UK	277	374
Netherlands	322	370
Spain	202	308
Japan	204	283
Italy	219	228
Czechosovakia (former)	141	207
Yugoslavia (former)	109	147

Source: Transport statistics Great Britain (1992).

6.2.2 Roads

The road transport sectors in advanced nations have received massive
investments since 1950 in terms of infrastructure and vehicle building
(Table 6.2). Expansion of motorway networks, increases in rates of car
usage and ownership and in the haulage of freight by road transport are
common to all these countries, although government policy on road
finance and, more recently on environmental issues, and the extent to
which traffic allocation between road and rail is state-controlled vary
considerably (see Chs 10 and 12).

The UK, Germany and the USA all show very clearly the dominance
of road transport for freight and passengers, and in all these nations there
have been substantial improvements in road networks, with a particular
emphasis upon the building of limited access high-speed motorways
(Table 6.3). The rates at which these new roads have been constructed,
and other highways improved, have rarely kept pace with traffic increases
however, and the UK in particular experiences severe congestion problems both within urban areas and on major inter-city routes (see Ch. 7).

As a carrier of freight the British road network is of relatively greater
significance than those of Germany or the USA, where rail and waterways
still account jointly for over 40 per cent of all freight tonne-kilometres.
By 1990 roads in the UK were carrying 87 per cent of all freight lifted,
compared with 42 per cent in the former West Germany and about 28
per cent in the USA. Inland waterways, and particularly the Rhine in
Germany (see Ch. 5), still carry about one-fifth of all freight and in the
USA long-haul rail services are important.

Table 6.3
Motorways and principal
road networks in 1990

Country	Motorway network (thousand km)	Principal road network (thousand km)	Motorways as percentage of principal network
USA	84.5	650.4	13.0
Germany	10.7	45.0	23.7
Canada	7.5	69.2	10.8
France	7.1	35.2	20.2
Italy	6.2	51.9	11.9
Japan	4.7	50.9	9.2
UK	3.3	15.4	20.8
Spain	2.4	20.6	11.6
Netherlands	2.1	4.2	50.0
South Africa	1.8	52.5	3.4
Belgium	1.6	14.5	11.0
South Korea	1.5	13.8	10.9
Switzerland	1.5	19.9	7.5
Austria	1.5	11.7	12.8
Venezuela	1.2	36.3	3.3

Source: International Road Federation: *World road statistics* (1992).

Plate 6.2
High capacity trucks and trailers on a Californian freeway near Mt Shasta. The completion
of the US Interstate Highway network has facilitated the rapid expansion of the trucking
industry and the capture by road haulage of much long-distance freight traffic from the
railways.

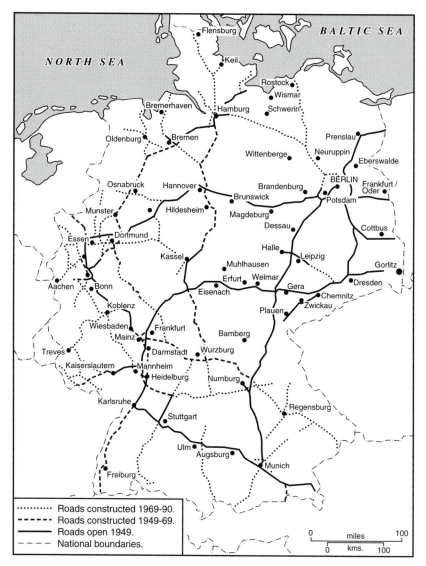

Fig. 6.5
The motorway network of Germany, 1992. *Source:* based on Fig. 7.3 in Hoyle and Knowles (1992).

The introduction of motorways specially designed to cater for the growing road traffic after the First World War came in the early 1920s, when the first Italian *autostrade* were built between Milan, Como and the Swiss border. The system was extended to serve the industrial areas of the Po valley, and in the 1930s Germany followed the Italian example by initiating construction of a nationwide system of *autobahnen* linking Berlin with the Ruhr, Hamburg and Bavaria (Fig. 6.5).

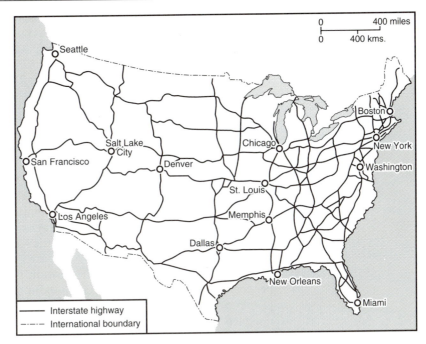

Fig. 6.6
The Interstate highway network in the USA.

The first long-distance motorways in the USA date from the 1940s, but in the UK a high-speed road programme was not begun until 1958, with the opening of the first section of the Lancashire M6 motorway at Preston. Building has continued through the later twentieth century, although the relative importance of motorways within national trunk road systems varies considerably.

The USA, with about 960 000 km, has the world's most extensive system of limited access highways, accounting for 15 per cent of the total road length. The inter-state highway network was the largest civil engineering programme in American history, and is a combination of state and federal toll roads and motorways providing a nationwide network which is at its densest in the north-eastern industrial states and in California (Fig. 6.6).

The 10 700 km of *autobahnen* in Germany is the largest network in Europe, the pre-war system inherited by the former federal republic having been rebuilt and remodelled to (a) focus upon the Ruhr and Hamburg and (b) provide more direct north–south links between the Ruhr, Frankfurt and Bavaria. With the development of increased trade with Union partners, who were themselves also building motorways, this federal system was extended to connect with the French, Belgian and Dutch roads. Road planning was also influenced to some extent by the need to improve the designated European International Road network which

Table 6.4
United Kingdom: traffic
distribution on different road
categories by type of vehicle
in 1991, showing percentages
of total vehicle-kilometres

Road type	Cars/taxis	Buses and coaches	Goods vehicles	All vehicles
Motorway	22.6	17.1	37.4	23.7
Major roads in built-up areas	31.8	47.2	17.7	30.9
Other major roads	45.6	35.7	44.9	45.4

Source: Transport statistics Great Britain (1992).

extends from the UK to the Russian border. Following German unification in 1990 there are plans to improve road links between eastern cities such as Leipzig, Dresden and Chemnitz and the Ruhr, Frankfurt and Munich, and the eastern motorway system as a whole requires substantial upgrading to bring it up to the standard and capacity of the western autobahns.

The 3300 km of motorway now open in the UK account for only 1 per cent of the total network but carry about 16 per cent of all road traffic (Table 6.4 and Fig. 6.7). As in Germany the British system has been designed to link up major urban centres and to relieve the more seriously overloaded sections of the conventional road network. With the reductions in new road construction programmes made in the early 1980s it appeared likely that the orbital London motorway and the Oxford–Birmingham motorway would be the last elements in the programme. However, in 1988 a new 10-year road investment plan of £12 billion was announced to increase the capacity of many existing motorways which are now seriously congested and to build a limited number of new roads. This motorway-widening programme has begun but has aroused fierce opposition from many environmental interests.

There are indications in the roads policies of both Germany and the UK that future investment will be less heavily concentrated upon motorways and that more funds will be diverted to improving existing routes. In the UK it is proposed that a relief route to the north of Birmingham will be built as a toll road with private capital, but progress will depend upon available funding. The government has also suggested that any other major new highways should be financed by the private sector, thus introducing the concept of toll roads to a nation where usage of almost all roads has traditionally been free. Germany has also considered plans to introduce tolls on its *autobahnen* as a first move towards privatisation, with some of the proceeds directed towards railway subsidies in an effort to persuade more freight on to rail.

Within the transport sector as a whole many industrial nations are now allocating higher proportions of funds to railways and public transport in general than in previous years, and it is likely that the era of rapid expansion of motorways is at an end. Pressure groups advocating

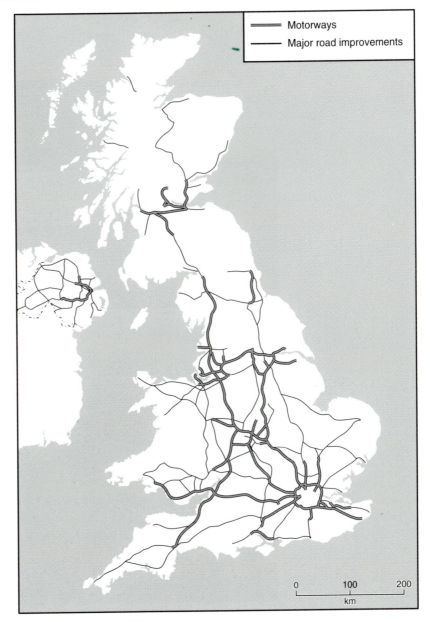

Fig. 6.7
The British motorway network in 1994.

greater investment in public transport and less in roads are increasing in strength and effectiveness, and governments are beginning to take more note of environmental issues when implementing new road schemes (see Chs 10 and 12).

(a)

(b)

Plate 6.3
(a) Motorway construction and trunk road improvements have allowed the introduction
in the UK of the National Express intercity passenger coach network, with frequent services
and reliable scheduling (although this is now being undermined by congestion). Over shorter
distances, as between London and Birmingham, the coach can compete effectively with rail
and air transport. (b) In contrast the intercity bus services in Zimbabwe, mainly used by
urban workers returning to their families in rural areas, are slow, unreliable and invariably
overcrowded. There is a marked discrepancy between the good quality roads and the old
and indifferently maintained vehicles using them.

6.2.3 Air transport

Internal air services are increasing their share of inter-urban passenger service traffic in advanced countries over distances exceeding 250 km and in areas where road and railway facilities are either inadequate or absent, as in parts of Canada (Box 6.2). The densest and most heavily used network of domestic air services has been created in the USA and now carries about 18 per cent of all inter-urban travellers. Traffic has increased steadily since 1950 and services are provided by private airlines whose operations are overseen by the Civil Aeronautics Board. Long-haul flights connect eastern seaboard cities such as Boston, New York and Philadelphia with the major Pacific coastal cities and there is a complex network of flights connecting major inland centres. The growing adoption of the 'hub and spoke' system since deregulation in 1978 has led to improvements in many services (Box 12.1).

Air traffic has also increased within the countries of the European Union, with many more flights between provincial centres in France, Italy and the UK. With unification of the former German federal and democratic republics the Lufthansa company has become the national carrier, and more frequent flights between cities in east and west are being introduced as part of the efforts to integrate the transport system, with a particular emphasis upon upgrading airports serving the east.

Centres such as London, Frankfurt and Zurich act not only as national focal points on domestic networks but as hubs for the international European air route system (Fig. 5.8). The relatively short distances

---

Box 6.2 Changes in the Canadian domestic air transport system

Domestic air services date from the formation in 1937 of the government-controlled Trans Canada Airlines (now Air Canada) which had exclusive rights over many trunk routes until 1959. Canadian Pacific Airlines (CP Air) began operations in the early 1940s as a privately-owned competitor and these two large undertakings, flying the transcontinental routes, were supplemented by a small group of regional airlines, each providing services at a provincial level. By the 1960s Air Canada, exploiting its state monopoly, controlled 50 per cent of the domestic market, followed by CP Air and the smaller provincial airlines. In 1979 the government relaxed restrictions on competition with Air Canada over its trans-continental routes, a move which followed deregulation of US domestic airlines in the previous year. The National Transportation Act of 1987 sanctioned the privatisation of Air Canada and its major competitor CP Air was then bought by the US company Pacific Western Airlines and renamed Canada Airlines International. During the 1990s all these Canadian airlines have faced increasing deficits and it is possible that US airlines may be allowed to operate domestic flights within Canada.

between most cities within European states means that rail can often compete effectively with air over many domestic routes, and in many respects air services within Europe are better understood as an international system rather than several national undertakings. Most of the hubs are linked by two national airlines on an agreed basis, and it has been recommended that many of the difficulties faced by the various national air traffic control systems could be solved by establishing a unified European operation which could double the capacity of existing corridors.

6.2.4 Inland waterways

The opportunities to exploit the relative cheapness of water transport for freight are restricted to those countries with major rivers penetrating the interior, and with important riverine industrial concentrations making use of bulk raw materials. Within western Europe the importance of the Rhine and its tributaries has already been described (see Ch. 5) and in North America the Great Lakes–St Lawrence Seaway system is a vital transport artery for industry in both Canada and the USA. Of the ten leading ports by tonnage in Canada six are located on the Seaway, with Montreal, Hamilton and Quebec City acting as general cargo centres and Thunder Bay, Sept-Iles and Port Cartier being terminals for the mineral and grain trade. Exports of grain from Canada to Europe can exceed 45 billion tonne-miles each year and a large proportion of this trade is shipped out through the Seaway.

Within the USA the economic impact of the 1993 floods in the Mississippi–Missouri basin was less a consequence of the loss of crops in the inundated areas and more a result of the disruption of the Mississippi navigation and the failure to export grain that had survived in states upstream of the flooded areas.

6.3 Transport in planned socialist economies

Efforts to develop the economy through the medium of fixed-length plans are characteristic of many socialist nations, and the transport sector usually assumes an important place in these expansion programmes. Russia (formerly the major part of the USSR) Poland and China have been selected to illustrate this section as each of these states has been faced with particular transport problems which have been approached in varying ways and with different levels of success. All three lagged behind the Western industrial nations in respect of rail construction, and modern Poland in particular had to remodel its pre-war network to meet the requirements of the post-1945 state which was established within newly defined national boundaries.

Heavy industry, with an emphasis upon mining and iron and steel production, was initially seen by these three states as the major way forward in the drive to achieve higher levels of economic development, and railway systems received massive investments in order to support these industries. Restricted consumer purchasing power has resulted in very low levels of private car ownership, so that the impetus for improvement of road systems has been very much less than in the USA or the UK. As a result the proportions of freight and passengers carried by railway systems in this group of nations is generally very much higher than in other states, and the level of new railway construction has been unequalled elsewhere in the world in the twentieth century.

Since 1990, however, Russia, Poland and the other east European states have begun to abandon many of their central planning approaches and to adopt more relaxed attitudes towards the operation of their transport sectors, especially in terms of the shares of passengers and freight carried by road and rail. It is too early to examine the results of this changing attitude within their transport systems but many new investments are planned for both road and railway networks.

6.3.1 Railways

China has relied very heavily upon its railways to serve existing and recently established centres of mining and heavy industry and to consolidate political control over parts of its territory previously lacking any form of modern communication. In the four decades since the communist state was established in 1949 the network has been extended from its original nucleus in the south-east coastal area to provide rail access to all the Chinese regional and provincial capitals except Lhasa.

During this 40-year period almost 60 per cent of the present 52 000 km system was built, a rate of growth which no other modern state has ever achieved (Box. 6.3 and Fig. 6.8).

Polish progress in the rail sector cannot approach that of China, but the scale of construction and modernisation since the Second World War has been more ambitious than that of many other European nations. Railways within the present territory of Poland were originally built in the nineteenth and early twentieth centuries by the three occupying powers of Russia, Austria-Hungary and Prussia, but with the formation of the first modern independent Polish state in 1918 attempts were made to create a system suited to the needs of the newly defined nation. The most important advance was the opening of a line from the Upper Silesian

Box 6.3 Chinese railway growth

War damage sustained during the conflict with Japan in the 1940s resulted in only 17 000 km of the 22 900 km network being in action at the time the Chinese communist state was formed. In 1953, when the First Five-Year Plan was introduced, 50 per cent of China's territory, mainly in the west, had less than 2 per cent of its railways. Many of the railway developments proposed in the four Five-Year Plans between 1953 and 1975 were delayed or substantially altered as a result of economic crises, the withdrawal of USSR technical support and the Cultural Revolution. Despite these disruptions many new trunk lines were completed, and from 1949 to 1970 the rate of new building and upgrading of existing track averaged 1000 km per year. Remote western centres such as Urumchi and Lanchow were connected to the system and many additional routes were provided between Beijing and the southern provinces of Yunnan and Guanxi Zhuang. Many of the pre-war lines, such as those linking Beijing with Canton and Tientsin with Shanghai, were converted to double track and the more heavily trafficked freight routes in the east were electrified to aid the expansion of the iron and steel industry. The proportion of freight carried by rail has risen from 35 per cent in 1949 to about 47 per cent in the late 1980s, but on a tonne-kilometre basis the railway's share of all inland freight is about 72 per cent.

Although the Chinese railway system is seen primarily as a major economic asset it also provides a specialist tourist attraction with its surviving steam-hauled trains. As other states such as India continue to phase out steam locomotives the Chinese have the potential to expand tourism on their railways.

coalfield to the coal port of Gdynia on the Baltic, but further improvements were delayed by the Second World War and by 1945 the devastated Polish railway network was regarded as being at least 40 years behind those of other European industrial nations in terms of its equipment and layout.

State transport plans concentrated upon reconstruction of the pre-war system and new building to facilitate east–west trade with the former German Democratic Republic and the USSR. Further improvements to links between Silesia, Warsaw and the Baltic ports have also been made with electrification of the Gliwice–Jarocin–Gdynia and Upper Silesia–Wrocław–Szczecin routes. Polish membership of the former Comecon organisation resulted in the building of a mixed standard/USSR gauge line conveying Russian iron ore to the Katowice steel mill. By the mid-1980s the Polish railways were carrying 77 per cent of all freight and about 50 per cent of all passenger traffic and 30 per cent of the network had been electrified (Fig. 6.9).

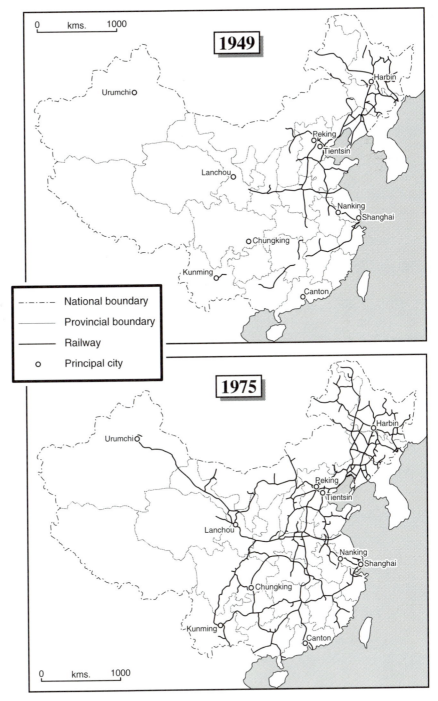

Fig. 6.8
The expansion of the railway network of China, 1949–75. *Source:* based on Figs 9 and 14
in Leung (1980).

Fig. 6.9
Planned and projected transport schemes in central and eastern Europe. *Source:* based on Fig. 1 in Hall (1993).

6.3.2 Roads

Both China and Poland have recently devoted increased attention to road improvement programmes, particularly to encourage more use of motor transport for short-distance carriage of freight and passengers in order to relieve growing congestion on the railways. Although road network densities and the vehicle-to-population ratios are still far lower than in Western industrialised nations, both states are increasing vehicle production and building lengths of motorway-standard highway in urbanised areas.

The Chinese transport authorities have announced plans to extend the national road network from the 1980 total of 890 000 km to 2 090 000 km by the year 2000, a massive 57 per cent increase, but it is difficult to qualify this objective with any accuracy as road standards vary widely throughout the state. However, there have been extensive road building and upgrading of existing routes in peripheral regions, with sealing of former earth and gravel tracks. By the mid-1980s it was estimated that about 93 per cent of the Chinese national territory had access to motorable roads, and a limited 1000 km motorway programme has begun, concentrating upon the eastern coastal areas. The Beijing–Tanggu road has been completed and others are under construction between Guangzhou and Zhuhai, Shenyang and Dalian and Shanghai and Nanjing.

These various schemes are designed to help realise the aim of increasing threefold the amount of freight carried by road between 1980 and the year 2000, and about 90 per cent of the existing Chinese road network has been either built or substantially improved since 1956, an achievement which in its scale approaches that in the railway sector. It is also planned to relieve the chronic congestion on passenger trains by diverting more traffic over shorter distances to bus transport, and by the mid-1980s about 75 per cent of all public journeys were on the roads, leaving the railways to concentrate more upon long-haul freight and passenger transport.

The shortage of suitable road trucks in China has meant that rail transport, where available, has had to be used for freight over distances of less than 100 km. The latest economic plan aims to use road transport for all freight hauls below this distance and the majority of passenger trips of less than 200 km. Until recently most trucks were limited to a 4-tonne capacity but larger vehicles are now being introduced in the efforts to secure a more efficient balance between rail and road. Many short-distance trips made by public transport or car in the Western world are carried out on bicycles in China and the significance of this mode must not be overlooked.

Although the Polish rail system is denser than that of the UK or France the sealed road network is much more sparse than that of other European nations. Vehicle ownership, at 57 cars per 1000 persons is also much lower than that in the UK and Germany. During the post-war period Polish road planning has had much the same objectives as the railway sector, with new building concentrated upon improving inter-urban connections in the newly defined state. Until 1970 there was little increase in the road vehicle fleet, as the railways received priority, but in the last two decades domestic car and truck production has increased in line with an expanded highway construction programme. The sealed road network increased from 96 000 km in 1946 to 115 000 km by 1990 and a limited plan for the building of motorways was made in 1972. A continuous motorway now exists between Berlin, the Upper Silesian industrial area, Warsaw and Gdansk, and by the late 1980s Poland possessed 5000 km of road to international standards, a higher total than any other east European state. With the expansion of trade with the EU a greater proportion of Polish exports now travel by road, assisted by the formation of a state road haulage organisation for foreign trade.

These recent developments in both China and Poland indicate that road transport is becoming of much more importance within their economies, although it is unlikely that either country will ever embark upon motorway construction on the scale of Western nations. China in particular, with its very long hauls, will continue to rely upon railways for much of its freight and passenger transport.

6.3.3 Air transport

Distances between major urban and industrial centres in China, and in Russia and its neighbours in the Commonwealth of Independent States (CIS), are as great as in the USA, but the lower living standards in these

---

Box 6.4 Inland waterways in China

China's waterway system is founded upon the three great rivers of the Yangtse, Huang Ho and the Hsi Chiang, supplemented by a network of canals. Shipping can make use of 30 000 km of waterway, although vessels larger than 8000 tonnes are restricted to only 8000 km of the system. The Yangtse, which is navigable for 10 000 tonne vessels 1100 km upstream to Nanjing, and for smaller craft a further 1800 km to Wushan, carries one-half of China's water-borne freight. When the dam and reservoir being constructed in the Three Gorges section of the Yangtse below Wushan are completed in about 2010 it will enable 10 000 tonne craft to penetrate as far upriver as Chungking; it is estimated that the flow of freight on this section of the waterway will then be increased fivefold from the present 10 million tonnes.

Although the Huang Ho is the second largest river it is limited in its capacity, but the channel is being deepened to allow 1500 tonne craft upstream to Lanchou. If inland and coastal shipping is included within the total freight haulage pattern for China then this mode currently accounts for about 18 per cent of all tonnage lifted. If many of the shallow sections can be deepened and lengths of formerly navigable river restored to the network then it is possible that some of the freight at present carried by rail could be transferred to the waterways, thus alleviating congestion on rail routes.

countries have severely restricted any development of a flourishing internal market for air transport. Domestic flights in Russia are also often disrupted by winter snow, icing and blizzards in Siberia. The division of the former USSR airline Aeroflot, once one of the largest fleets in the world, into several smaller units each operating within a CIS member country has resulted in severe problems associated with maintenance and fuelling, and scheduling standards have fallen in many areas.

6.3.4 Inland waterways

Both Russia and China owe much to the facilities offered by navigable rivers in the development of their heavy industries (Box 6.4). The former USSR made extensive use of the Volga and Don waterways for the assembly of raw materials in the iron and steel industries and certain of the Siberian rivers carry freight traffic.

6.4 Transport in the developing nations

The general significance of transport in the development of colonial and subsequently independent territories has been discussed in Chapter 4. Problems associated with transport operations in developing countries are now examined in more detail, with reference to two former African colonies, a former South-east Asian colony and a land-locked South American nation which secured its independence in the nineteenth century but still experiences severe problems in expanding its economy.

In Nigeria, Zimbabwe and Malaysia the larger part of the railway systems were built during their periods of British colonial occupation, and Bolivia, which became independent from Spain in 1825, established its first railways in the early twentieth century. Road transport now carries a substantial share of freight and passenger traffic in these countries and their railways often require state subsidies in order to survive.

6.4.1 Railways

In Zimbabwe, Nigeria and Malaysia railways were first constructed to promote colonial agricultural and mining enterprises and to provide a convenient route to the sea for exports and imports. In Nigeria the Niger river system, which was available for navigation throughout the colony during the summer season, partially influenced the routeing of the two railways penetrating inland from Lagos and Port Harcourt, but in the

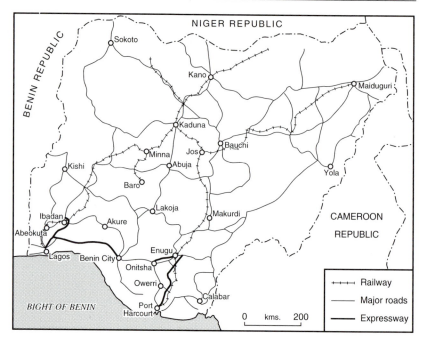

Fig. 6.10
Principal roads and railways of Nigeria.

other states inland waterways have played very little part in transportation. Interchange between rail and river took place at Jebba and Baro on the Niger and Makurdi on the Benue but since the 1960s river trade has steadily declined. The Maiduguri extension of 1964 from Bauchi into the cotton belt of north-east Nigeria was one of the last large-scale railway schemes of the mid twentieth century, and although the Nigerian system is more comprehensive than other West African networks it has now lost its former significance as a major carrier of freight. In common with other African systems, the Nigerian railways were often unable to meet the seasonal demand for the transfer of perishable agricultural produce from farm to port and gradually their role has been taken over by road transport (Fig. 6.10).

The Zimbabwean railway system was initially an extension northwards beyond the Limpopo of the South African network, Bulawayo and Harare being reached in 1897 and 1902 respectively. A route was also extended inland from the Portuguese colonial port of Beira in Mozambique to Harare, and further lines were built beyond Bulawayo to the Hwange coalfield and across the Zambezi to the Zambian and Congo copper belts; the latter were then linked westwards with the Angolan port of Lobito. Zimbabwe, in contrast to Nigeria, thus came to occupy a nodal position on the railway network of southern Africa, with connections to adjacent colonial territories (Box 6.5 and Fig. 6.11). Additional links to the coast

Plate 6.4
The single-track railway between Hwange and Bulawayo carries over 10 000 tonnes of freight each day, mainly local traffic from the Hwange coalfield in liner trains destined for the Redcliff steelworks or for the thermal power station in Harare.

were provided with the Limpopo valley line to Moputu (1955) and the trans-Limpopo railway via Messina and the South African system to Durban (1974).

Malaysia's distinctive peninsular character aided the initial exploration of the interior by river from ports on the western coast, and the same direction was followed by railways after 1885. Rail construction was stimulated with the introduction of rubber plantations in the 1890s and the various isolated coastal lines were connected in 1918 with completion of the spine railway from Singapore to Kuala Lumpur and northwards to Bangkok. Few lines were built in the less productive eastern areas of the peninsula apart from a second route to the Thai border. The port of Singapore, at the southern end of the Malaysian system, thus became the terminus of a tenuous network of colonial railways extending through Malaysia and Thailand to Cambodia.

Railway construction in Bolivia was severely hampered by the Andean

Box 6.5 Road transport in Zimbabwe

Zimbabwe inherited a comprehensive and well-maintained road system on achieving independence in 1980, and since that date the government's objective of developing the rural areas has depended largely upon an expansion of the road network. The inter-urban system of tarred roads carries well-organised trunk freight services, and a programme of rural road improvement is extending sealed gravel routes into the more remote areas of communal farming land. The 1986 National Development Plan proposed the creation of a state-owned transport corporation as a means of improving goods carriage and personal accessibility in these rural areas, but at present progress is severely hampered by a shortage of foreign currency to finance purchase of new trucks, buses and spare parts. ▶

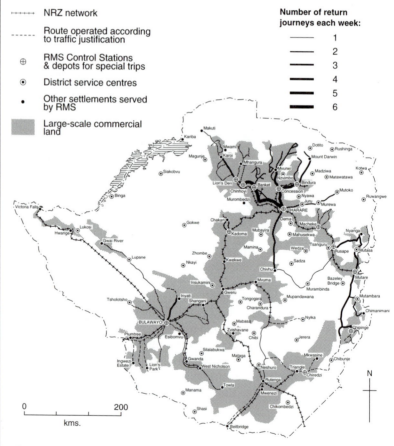

Fig. 6.11
The Road Motor Services (RMS) network in Zimbabwe, showing frequency of services, the distribution of large-scale commercial farming land and the National Railways of Zimbabwe (NRZ) network. *Source:* based on Fig. 1 in Turton (1991a).

The commercial marketing of maize and groundnuts by African communal farmers depends largely upon small-scale hauliers who are often unable to meet the demand for transport during the harvest periods. The railway-owned Road Motor Services (RMS) was established in the colonial period to provide scheduled freight transport for the white commercial farming areas and contract haulage for grains and minerals is also available. Although the RMS carries only 5 per cent of all road freight in Zimbabwe it does perform a valuable function in collecting crops from some communal areas where private hauliers are reluctant to operate because of poor road conditions.

Inter-urban bus services carry heavy loads of urban workers returning to their homes in communal areas at regular intervals, but journeys are often delayed and disrupted because of unreliable vehicles. Until the initiation of the World Bank's Economic Structural Adjustment Programme in 1992 essential repairs and maintenance were hampered by the lack of spares, especially tyres, which required scarce foreign currency, and the road passenger transport sector as a whole urgently needs improving.

mountains, with heights of over 4000 m, and by the poorly drained areas of the Oriente lowlands, and two physically separate networks were built (Fig. 6.12). A rudimentary western system links the Andean mining centres with the Chilean ports of Arica and Antofagasta and, via the Lake Titicaca transit, with Mollendo on the Peruvian coast. La Paz is also linked eastwards with Buenos Aires by a route completed in 1925 but with many severe gradients. Santa Cruz, to the east of the Andes, has rail connections with Brazil and Argentina and in the 1950s it was proposed to construct a 500 km line between Santa Cruz and Cochabamba to link the two Bolivian networks. This plan was abandoned and the connection was eventually made by road, but current plans to improve the transcontinental rail route between Antofagasta and Santos, on the Brazilian coast, involve the upgrading of lines in Bolivia and providing improved road–rail transhipment for the Santa Cruz road section. Projected improvements to the Matarani–Buenos Aires railway involve the modernisation of lines in southern Bolivia and a new Lake Titicaca train ferry.

There are marked differences between these four states in respect of the contemporary links with railways in neighbouring territories. The west coast of Africa was colonised by the UK, France and Germany and each colony tended to be developed as an individual unit, with self-contained railway systems and few intercolonial links. In contrast, the Zimbabwean network was a northern extension of the South African system and was also connected to the railways in Mozambique and Zambia and hence those of Zaïre and Angola, thus forming an integral element of a system of subcontinental extent. The Economic Community of West African States and the Southern African Development Community both see an important future role for their railway systems in promoting more intra-regional trade between member states, and the Zimbabwean network in particular could carry increased traffic in transit from Zambia to Beira on the Mozambique coast.

6.4.2 Roads

Since independence many African and South-east Asian states have paid increasing attention to upgrading and extending their road networks and investment in this sector has often exceeded that in all other areas of transport. Almost all roads were initially built by colonial authorities with primarily economic motives, but now the emphasis has shifted towards providing better roads in rural areas as a part of plans to improve access to welfare facilities and to involve more farmers in commercial agriculture.

In Nigeria the main objectives of the roads programme have been to provide a higher percentage of all-weather sealed roads, to build additional links to reduce journey lengths and to construct new highways designed for high-capacity freight vehicles. By the early 1980s 100 000 km of road were in use, of which 30 per cent were sealed, and during each of the four national plans produced in the 1962–85 period the road system's share of total transport investment never fell below 54 per cent. Roads now carry the major share of both exports and imports from Lagos and Port Harcourt and the southern states are gradually being equipped with expressways, such as that linking Lagos with Ibadan.

In Malaysia road building and improvement projects are an essential component of national plans for rural development, especially in the more remote eastern states of Terengganu and Kelantan. New trunk roads, such as those between Kuala Lumpur and Seremban and Karak, link the western and eastern coasts and a greatly extended network of rural roads has been completed to promote commercial farming in the formerly isolated eastern peninsular region. Over 80 per cent of all pensinsular roads are now sealed and Malaysia is now the leading nation in South-east Asia in terms of the quality of its roads.

In Bolivia the physical difficulties associated with high altitude in the Altiplano and poor drainage in the Oriente which restricted railway building have also severely constrained road improvement plans. There are only 4000 km of roads suitable for motor transport and only about one-half of this network is of an all-weather character. Most of the trunk roads have been built since the 1950s and many form part of overseas-financed programmes to establish international routes within South America. For example, the Pan-American Highway system incorporates a route from southern Peru to Argentina via La Paz and Potosi but the Bolivian section still requires substantial improvement. The planned inter-oceanic highway from Arica, on the Pacific coast of Chile, to Santos in Brazil will also pass through La Paz and Corumba and thus benefit Bolivia. Both the Cochabamba–Santa Cruz and the La Paz–Beni valley

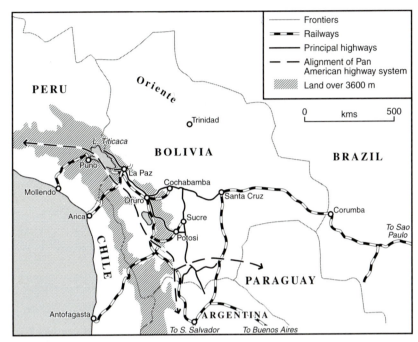

Fig. 6.12
Principal roads and railways in Bolivia.

roads were initially planned as extensions to the railway system but were ultimately opened as highways to assist in the integration of the Andean and eastern lowland regions (Fig. 6.12).

6.4.3 Air transport

Domestic air traffic in these countries is generally limited to business and official travel and to tourists. Malaysia has a growing tourist trade focusing upon Kuala Lumpur, and Zimbabwe is following the example of East African states by encouraging more visitors to safari parks at Hwange and to the Victoria Falls. The increasing numbers of international tourists arriving in Zimbabwe will, however, require improvements in domestic air transport between Harare and the major resorts.

In Bolivia air transport is essential in areas such as the Oriente where few reliable surface links exist. La Paz and Santa Cruz have flights to most urban centres but travel is mainly for business and official purposes and tourism is very limited.

6.4.4 Inland waterways

Inland waterways have played only a limited part in the transport sector of developing economies. Although the Amazon, Orinoco, Niger, Congo and Zambezi rivers flow through several of the developing independent states of South America and Africa their use as waterways has been restricted by fluctuating regimes, falls and rapids and the absence of suitable sites for deepwater ports at their deltas. Until the 1960s the Niger was used for the transport of cotton, groundnuts and palm products to the coast but this traffic has now been transferred to the roads, although the Nigerian government has proposals to improve navigation and revive trade.

6.5 Concluding summary

The second half of the twentieth century has seen substantial changes in the transport systems of countries at all stages of economic development. Within the advanced industrial states road transport now carries the dominant share of freight and passenger traffic, but in recent years there has been a resurgence of interest in the potential that rail can still offer for the high-speed movement of passengers and freight, free from the costly congestion which is an ever-growing problem on both urban and inter-urban roads.

States such as China, the countries of the CIS and those of eastern Europe still exercise strong control over the distribution of traffic between inland modes of transport, and their railway systems have been upgraded and continue to play a leading role in the economy. Road transport in these states is, however, increasingly being recognised as an efficient means of carriage over shorter distances, and within eastern Europe in particular rising private car ownership is creating demands for additional road capacity.

Many of the developing countries have introduced programmes designed to upgrade transport facilities in rural agricultural areas in order to improve social and economic life. These schemes depend heavily upon international financial aid but are essential if the aims of decreasing the reliance of a large part of the rural population upon subsistence agriculture and increasing their accessibility to basic services are to be achieved.

Many of the issues discussed in a national context in this chapter are also examined in subsequent sections dealing with transport policies and planning (Ch. 12), environmental aspects of existing and planned transport operations (Ch. 10) and the question of equity between the various categories of transport users (Ch. 11). The particular difficulties encountered in urban and rural areas are discussed in Chapters 7 and 9 respectively.

Further reading

Button K (ed) 1991 *Airline deregulation: international experiences* D Fulton Publishing
Dunn J A 1981 *Miles to go: European and American transportation policies* Mass. Inst. Technology Press, Cambridge, Mass.
Hoyle B S (ed) 1973 *Transport and development* Macmillan
Leung C K 1980 *China: railway patterns and national goals* Department of Geography Research Paper 165, University of Chicago
Owen W 1989 *Transportation and world development* Johns Hopkins
Starkie D 1982 *The motorway age: road and traffic policies in post-war Britain* Pergamon

Part Three

Transport in urban and rural areas

7 Urban transport problems

What is the 'urban transport problem'?

Los Angeles is the quintessentially car-orientated city, yet even in the 1930s there were complaints of stagnation and delay. Since then huge expenditure has made it the most mobile city in the world – for twenty hours a day – but unfortunately half of the trips occur in the other four hours and are congested and polluting. The urban transport problem is clearly very deep seated indeed if it cannot be solved in this rich and sophisticated city.

The first task of this chapter is to shed some light on the complex relationships which govern urban trip-making, such as journey purpose, distance travelled, personal circumstances, the transport systems available and city size. The principal modes available in cities of the developed and developing world are described and an attempt is made to define the urban transport problem and its economic basis. The point is emphasized that the problem is world-wide and worsening as city populations grow, car ownership rises and public transport faces deepening financial crises.

7.1 Introduction: urban areas and transport

Travel can be an adventure, undertaken for the fun of it, though in cities it is hardly ever so, for throughout the world urban transport systems are strained to breaking point. People struggle to get to work on time and get their shopping back home; they fear for their children's lives on the roads and for their own security on a lonely station at night; they are deafened by the traffic noise and sickened – literally – by the fumes; and everywhere, in rich countries and poor, they are maddened and frustrated by the perpetual congestion, overcrowding and delay.

The significance of urban transport difficulties stems from the fact that around half of the world's population lives in urban areas, a proportion that is rising very quickly. The number of very large cities, in which transport problems are especially acute, is growing particularly rapidly. The rate of urbanisation is fastest in the developing world where the resources to pay to overcome such problems are scarce. For example, in 1950 there were 5 cities with more than 5 million residents, but by the year 2000 there will be over 50, some 40 of which will be in the developing world.

If the problems seem acute at present they pall beside the potential social, economic and environmental difficulties in store. The giant city of the future will not only have to try to cope with many more people wishing to make trips but also with the likelihood that the number of trips per person will rise as the population becomes more youthful and changing cultural attitudes lead to more work trips by women. As city areas expand so will these trips become much longer.

City growth and transport go hand in hand, for transport is part of the city, not just an addition to it (Ch. 4). For future growth to be accommodated methods need to be developed to plan transport in a logical way. These methods go to make up the 'urban transport planning process' that is examined in Chapter 8. However, a precondition of planning is an understanding of what the existing problems are and what is causing them. It is the aim of this chapter to shed some light on these issues.

7.2 *Activities in urban areas*

7.2.1 *The movement of people*

We begin by trying to develop an understanding of why and how people move in the contemporary city. Chapter 2 presented a discussion of personal trip generation and how it relates to mobility and accessibility in general terms. Here we sharpen the focus to concentrate more particularly on the basic principles of urban personal movement. The first of these is the co-ordination of activities whereby some activities can only take place if people congregate at the same place at the same time. Factory shifts, market days, school lessons and religious celebrations are examples of this. The frequency and regularity of travel to activities such as school and work produce the familiar daily and weekly ebb and flow of urban movement (Fig. 7.1). It is, of course, this need for co-ordination that produces the most intractable problems of city transport, with high concentrations on certain routes at certain times of the day producing the most serious congestion.

A second fundamental principle underlying the patterns of urban movement is that of distance decay, whereby we attempt to minimise the

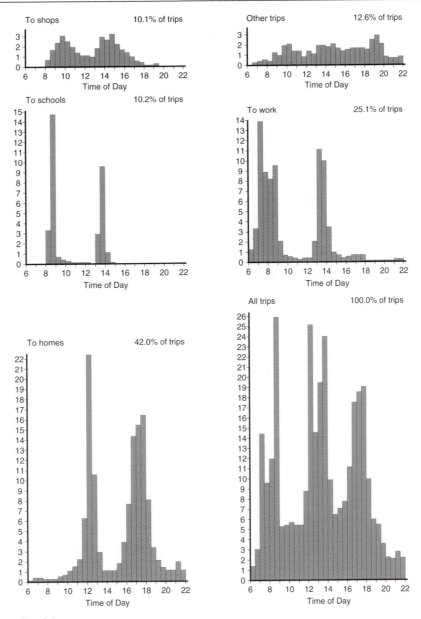

Fig. 7.1
The 'peaking' of urban travel: the half-hourly distribution of person trips in Northampton, 1964. *Source:* Fig. 45 in Taylor (1968).

cost or inconvenience of travel for a given purpose. Other things being equal, we use local shops when we can and define catchment areas for schools to allow safe, short-distance movement for schoolchildren. For the journey to work, we trade off where we would like to live against the

Plate 7.1
Traffic congestion is a common experience in almost every city: Mainz, Germany.

cost and length of the commuting journey, with different people's perceptions of the values of these factors and their differential ability to pay for them producing widely differing commuting behaviour and thus a complex mosaic of residential areas. The combination of this principle with that of co-ordination provides the main explanation for urban movement patterns, with trip numbers on any route being a function of the trip-generating attraction of the destination and the frictional drag caused by the distance from the origin.

The third principle is the conditioning of individual travel behaviour by personal circumstances which dictate the need and ability to engage in particular activities. For example, we know that there are very real differences in the travel patterns of men and women, related to elements of the personal environment such as disposable income and availability of personal transport. As Chapter 11 will show, responsibility for children, particularly taking them to and from school, frequently leaves blocks of time too small for other activities to be completed and thus limits women's ability to undertake discretionary trips. Many such personal circumstances change as stages of the life cycle are passed, each being characterized not only by different activities, such as school trips by the young and work trips by older individuals, but also by varying volumes of trips, with the high mobility of the young contrasting with the restricted length and frequency of journeys by the elderly.

The fourth basic principle that influences personal movement is the existence of economies of scale in provision of transport facilities. Where

Table 7.1
A typology of characteristic residentially based intra-urban personal movements

Geographical description			Destination and modes in order of incidence					
			Daily journeys		Weekly journeys		Infrequent journeys	
'Field' description	Maximum range	Range population	Destination	Mode	Destination	Mode	Destination	Mode
Neighbourhood (small town) journeys	1 km	5 000	Convenience-goods shops, primary schools, relatives, clubs, pubs work	Walk Car Bike	Food shops, clubs, pubs, places of worship	Walk Car Bike	Clincs and doctors, parks	Walk Car Bike
Suburban (medium town) journeys	5 km	50 000	Secondary schools, work	Bus Car Bike Walk	Food shops, personal business, friends or relatives, sports	Car Bus Bike Walk	Durable goods shops, personal business, friends or relatives, entertainment	Car Bus Bike Walk
Urban (city) journeys	15 km	500 000	Work, higher education	Car Bus Rail			Specialist shops, entertainment	Car Bus Rail
Metropolitan or conurbation journeys	75 km	10 000 000	Work	Rail Car			Specialist shops, entertainment	Rail Car

Source: Daniels and Warnes (1980).

the density of potential users justifies new public investment in transport, some journeys will be made easier by the provision of higher-speed or higher-capacity services and this will have knock-on effects on people's travel behaviour, particularly their destination and modal choices. A hierarchy of roads and a predominantly radial pattern of public transport routes result. In turn this will be modified by the fifth principle, the increasing differentiation of activities, land uses and movements with increasing urban size. The large urban area will contain a greater diversity of trip lengths and patterns than the small town, including some very long distance commuting. The greater volume of movement produces a more complex public transport system and encourages more investment, which in turn concentrates movement along certain corridors and by certain modes, providing strong contrasts with the lower levels of movement more characteristic of smaller towns.

These principles underlie the hypothetical scheme of movements provided by Table 7.1, which attempts to show for each distance category and frequency group the purposes and modal characteristics of journeys.

Four journey-distance categories are suggested, with marked differences between the frequency, modal and purpose characteristics of the trips which are most common within each distance range. The classes of journey accumulate with increasing city size, so that the largest cities will have all four types of movement but the small town only the first.

7.2.2 *The movement of goods*

Goods traffic does not display the severe peaking that is so problematical for passenger movement, so it will not be dealt with explicitly as part of the urban transport problem in the remainder of this chapter. Yet it is by no means unimportant: anything from a tenth to a quarter of urban road traffic is made up of goods vehicles, which are essential to the efficient functioning of the urban economy. They too demand space for movement, delivery and parking and, moreover, the greater bulk and weight of trucks and vans may amplify their perceived impact on the urban scene, not least in terms of noise, air pollution, visual intrusiveness and traffic danger.

Broadly speaking, urban commodity movements can be divided into those that facilitate economic activity on the one hand and residential functions on the other. In the former case there are flows of fuels, components and raw materials for processing, and the resultant outflows, which may include waste products and semi-finished goods destined for other plants. However, the most important category of movement here is the final distribution of products to wholesalers and retailers, an activity that employs large numbers of people and vehicles and focuses mainly on the city core.

In developed countries fuels, food and household goods are delivered to residences using modes as diverse as oil tankers, vans and electric milk delivery 'floats', with newspapers and mail being delivered on foot or by bicycle. These short-distance, regularly scheduled movements account for a large part of urban freight traffic, some 41 per cent of urban commercial vehicle movements in a sample of eleven American cities, for instance. In developing countries the need for daily food deliveries may generate as many vehicle movements as does the need to move people, and such movements may be sharply focused in time and space, as on markets and central area retail districts. The outflow of these materials after processing by the household, particularly in the form of sewage and refuse, provides some of the most intractable urban management problems.

There is in addition a whole category of goods movement that is largely unseen, by virtue of being concealed in pipes and cables beneath the city surface. There is a flow of water, gas and electricity, as well as increasingly significant networks of cables for television and other forms of information delivery and exchange. Whether they are carrying tangible commodities or not, these networks play a vital role in the transport system of the city.

Table 7.2
Vehicles in selected cities and countries

City	Year	Bicycles (thousands)	Motor vehicles (thousands)	Population (thousands)	Bicycles per 1000 residents	Motor vehicles per 1000 residents
China	1988	300 000	1 200	1 104 000	272	1
Beijing	1982	3 773	··	9 231	410	··
Shanghai	1988	5 600	200	12 400	445	12
India	1985	45 000	1 500	765 000	59	2
Bombay	1981	984	90	8 200	120	11
New Delhi	1981	945	313	5 800	163	54
Madras	1979	272	64	4 000	68	16
Indonesia	1985	··	··	164 050	100	··
Jakarta	1985	··	··	7 600	35	··
Surabaya	1976	200	144	2 300	··	··
Bangladesh	1982	1 500	250	92 585	··	··
Korea, Rep. of	1982	6 000	··	39 000	154	··
Thailand	1982	2 500	400	49 000	51	53
Thailand	1988	··	6 300	54 960	··	116
Malaysia	1982	2 500	900	14 000	179	64
Japan	1988	60 000	30 700	122 000	492	252
Netherlands	1985	11 000	4 900	14 000	786	350
United States	1988	103 000	139 000	245 000	420	567

·· Indicates data not available.
Source: Replogle (1992).

7.3 Modal choice

There are persistent and two-way relationships between the purpose of a journey undertaken, its frequency, timing, length, characteristic participant and the choice of the mode to use. Obviously popping round to the local store to get some vegetables for dinner is different in almost every way from the long-distance journey to work, and it follows that the most suitable mode for one will not necessarily be the best for the other. Moreover, the mode used may not be ideal because access to particular modes is frequently limited by income; Table 7.2 shows that though we might mentally tend to associate the USA with the car and India with the bicycle, in reality the greater prosperity of the USA means that American bicycle ownership per capita is seven times greater than that in India. The following section will examine the principal modes of urban travel, bearing in mind that these are critical matters for the urban transport planner. Individual decisions on the selection of the preferred mode for a trip, when multiplied by hundreds or thousands, produce extremely complex movement patterns and have profound implications for the density of traffic on the network, the provision of infrastructure and the quality and convenience of the journey.

7.3.1 *Walking*

Walking is the most important form of urban transport. Nearly everyone walks and children and women walk most. Some types of journeys are more likely to be made on foot; though more than one in three of all journeys people make in urban and rural UK are made on foot, for education journeys the figure is 60 per cent. Car availability is also a strong influence on the frequency and length of walking trips, with car owners making fewer and shorter journeys. However, most trips are of necessity multimodal and walking is a component of almost all of them, whether it is to the parked car or to and from public transport stops and stations.

In developing countries walking dominates urban transport for the poor, who walk most often and furthest. In India, for example, the lower-income groups depend upon walking for almost 60 per cent of all urban journeys. Many of these walkers do not have a modal choice as such, for their poverty denies them the opportunity to use anything else but their feet. Even when buses are available they are frequently full, so that even many of the not-so-poor cannot actually use them. But reluctant walkers are still travellers in the urban system and planning must recognise that walking is, and will remain, a perfectly valid form of transport for most people. It is one that is entirely appropriate for many types of urban trips and it is often the most efficient, both for the walker and for the urban transport system as a whole.

7.3.2 *Non-motorised vehicles*

In industrialised countries the pedal cycle is the principal vehicle in this category, but it rarely accounts for more than 10 per cent of the modal split. However, there has been a resurgence of cycling in recent years and many countries, notably the Netherlands, demonstrate that the bicycle may make a very significant contribution to urban transport when proper facilities are provided and a pro-cycling ethos is established. Its key role in transport planning for cities is taken up in Chapter 8.

However, it is in developing countries that non-motorised vehicles assume dominance as a means of mobility, though there is much variability between countries in the particular vehicle used. In China the cycle predominates, with ownership levels as high as 460 per 1000 people in some cities and 80 per cent of all trips being made by cycle. In Indian cities essential trips (work and education) are made primarily by non-

Table 7.3
Percentage share of total trips
by non-motorised travel for
major trip purposes in
Indian cities

| | Trip purposes | | | |
City	Work (%)	Education (%)	Others (%)	Total (%)
Delhi	75	6	19	100
Bangalore	39	30	31	100
Madras	51	45	4	100

Source: Rao and Sharma (1990).

Plate 7.2
Bicycles dominate transport in China, although space allocated to motorised movement is
often greater.

motorised modes (Table 7.3) with cycles predominating in metropolitan
areas but smaller cities depending more on the cycle rickshaws which
operate as two-person taxis over much of South-east Asia. Bangladesh,
for example, has a fleet of 8400 buses, but has 14 000 auto-rickshaws and
700 000 cycle rickshaws. It also uses 160 000 bullock carts for personal
transport and in other countries horses, camels and donkeys are used.
The role of water transport should not be ignored either, especially in
South-east Asia: again Bangladesh with its 300 000 boats is a good
example.

7.3.3 Private cars

In the industrial world the car is now the leading mode for all categories
of journey. Its popularity is explicable in terms of its flexibility, personal

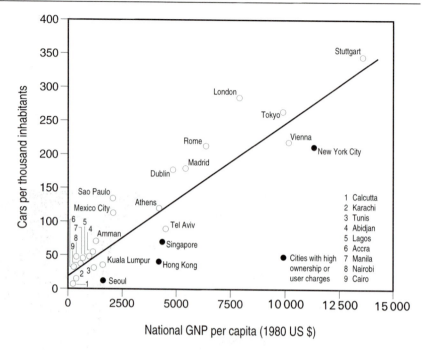

Fig. 7.2
Income and car ownership. *Source:* Dimitriou (1990).

convenience and the status that ownership confers. Though ownership levels are static in some inner urban areas, generally speaking the number of households with one, two or more vehicles continues to grow throughout industrial societies, with one car for every 1.7 people in the USA and every 2.9 in Europe (excluding Portugal and Greece). There are evident transport advantages of car ownership and these are augmented by powerful social and economic pressures that have made the car a symbol of wealth, choice and success.

The contrast with cities of the developing world is marked and is principally a function of income (Fig. 7.2). Though ownership rates may be growing quickly this is from a low base, so that cities such as Calcutta and Nairobi still have fewer than 50 cars for every 1000 inhabitants. John Adams (1981) has calculated that the 59 poorest countries of the world, containing over 60 per cent of its population, together own less cars than the American city of Los Angeles. The potential for future growth in the number of cars is thus enormous, although there are strong environmental and social reasons why current growth rates cannot continue for much longer as Chapters 10 and 11 will show. For the time being, though, the headlong dash towards global mass car ownership continues unabated.

7.3.4 Public transport

Bus-based systems

Buses come in many shapes and sizes, including mini- and midi-buses, standard or articulated single-deckers, and variants on the double-decker which can carry up to 150 people. Because they all benefit from the free use of existing roads they are cheap and flexible to operate as compared with rail-based public transport. They are of particular importance in developing countries, where they are intensively used, with the largest Indian cities, for example, having up to 40 per cent of trips made by bus. However, many vehicles are old and unreliable and although fares are low, they are still too high for many people.

In contrast to the problems of reliability and cost that afflict bus systems in developing countries, the biggest task facing bus operation in advanced nations is to escape car-caused congestion. Pulling out back into the stream of traffic causes additional delay, as does the need (on one-person-operated buses) to wait at the kerbside whilst tickets are sold to boarding passengers. These delays prompted the infamous remark from one English bus company that its buses would be better able to keep to the published timetable if they did not have to keep stopping for passengers! One solution is to use prepaid tickets or travel cards whilst another is to return to the use of conductors, a labour-intensive solution but one favoured by passengers vulnerable to assault, such as women and elderly people.

Buses are less efficient users of energy than urban rail systems and emit more pollution, though their record on these criteria is much superior to other forms of motorised road transport (Ch. 10). The flexibility and low cost of bus operation and maintenance, together with the prior existence of the necessary roadway, mean that buses are likely to remain firm favourites in cities in both industrialised and developing societies. The problems lie in clearing the road space of cars so that the advantages of the bus can be properly exploited.

Rail-based systems

These systems can run at average operating speeds of up to 60 km per hour, carrying from 15 000 to 60 000 passengers per hour, and are designed to serve high-density corridors in order to justify the high costs of construction. There are four principal types of system and these are described in Table 7.4.

Rapid transit systems are self-contained urban railways, known as *U-bahn* in Germany and as 'metros' elsewhere, after the Paris system begun in 1900. Typically, such systems are underground in city centres and offer up to 30–40 trains per hour, to which people walk and on which they pay a flat or zonal fare. New rail development can only be justified if there is a high potential ridership, so metros are typical of cities of more than a million people, though by no means do all such cities have them.

Table 7.4
Typical characteristics of urban rail systems

	Streetcars	Light rail	Surburban rail	Metro
Urban size				
Population	200 000–500 000	100 000–1 million	Over 500 000	Over 1 million
CBD employment	Over 20 000	Over 20 000	Over 40 000	Over 80 000
Route characteristics				
Route length from CBD	Under 10 km	Under 20 km	Under 40 km	Under 24 km
Track	On street	Over 40% segregated	Segregated	Segregated
CBD access	Surface	Surface	Surface to CBD edge	Underground
Station spacing in suburbs	350 m	1 km	1–3 km	2 km
Station spacing in CBD	250 m	300 m	—	500 m–1 km
Maximum gradients (%)	10	8	3	3–4
Minimum radius (m)	15–25	25	200	300
Engineering	Minimal	Light	Medium	Heavy
Rolling stock				
Carriage weight (tonnes)	16	Under 20	46	33
Number of carriages	1 or 2	2 or 4	Up to 12	Up to 8
Carriage capacity	50 seats 75 standing	40 seats 60 standing	60 seats 120 standing	50 seats 150 standing
Carriage access	Step	Step or platform	Platform	Platform
Performance				
Power current	DC 500–750 V	DC 600–750 V	DC 600–1.5 kV or AC 25 kV	DC 750 V
Power supply	Overhead	Overhead	Overhead or 3rd rail	3rd rail
Average speed (km per hour)	10–20	30–40	45–60	30–40
Maximum speed (km per hour)	50–70	80	120	80
Typical peak headway (minutes)	2	4	3	2–5
Maximum hourly passengers	15 000	20 000	60 000	30 000

Source: Knowles and Fairweather (1991).

There are nearly 80 systems operational world-wide including 27 in Europe, 17 in Asia, 17 in the former Soviet Union, 12 in North America, 7 in Latin America and 1 in Africa.

Suburban railways (*S-bahn*) are the routes of main line surface railways on which a service is offered to meet local needs. As Table 7.4 shows, *S-bahn* speeds are higher than metros and stations wider-spaced. Trains have high capacities and new investment is expensive due to the nature of the infrastructure. Nevertheless, building new track and stations in order to achieve segregation from long-distance services or to provide new links is often good value, as seen in Hamburg and Glasgow. There are systems in all 5 continents, in over 100 cities in more than 100 countries.

Light rail transit (LRT) systems are developments from the electric street- or tramcar and today tend to serve densely populated areas of cities within short walking distances of stops. The term LRT is sometimes confused with 'light rapid transit' which includes not only light rail transit but also guided busways and monorails. LRT proper fills the gap between streetcars and rail rapid transport, being much more segregated from other traffic than streetcars and yet demanding lower standards of engineering than heavier rail systems. Trains of up to three or four cars may achieve headways of 90 seconds, average speeds of 20–40 km per hour and capacities of 15 000 passengers per hour in each direction. Light rail has relatively low costs (compared to metros), enabling extensive systems to be built and integrated with heavy rail and buses. The harmonious integration with pedestrians in shopping areas and the superior image of modern LRT relative to buses is a spur to investment in LRT by cities seeking an attractive image. There were only around 100 systems in operation at the start of the 1990s, including 20 in Germany and 16 in the USA, but there were others being built or planned in hundreds of cities around the world.

Electric streetcars or 'trams' developed from horse-buses in the 1880s and spread to most North American and European cities. They run in city streets, negotiating tight curves and accelerating quickly from the frequent stops. They are subject to car-caused congestion, so that some European cities are now upgrading them to LRTs by enclosing the tracks, granting priority at lights and running them through pedestrianised areas and tunnels in city centres. In smaller cities streetcars are usually the only rail-based public transport, but in larger ones they complement metro or *S-bahn* systems. About 250 systems exist world-wide, including over 150 in the former communist bloc.

There are variants and hybrids of all of these systems, such as the trolley-buses of Seattle and Lyons, powering buses through overhead wires, and the double-decker trams of Hong Kong. Increasingly systems are becoming automated, rather like airport 'people movers'. Automated light rail, with no drivers or station staff, was first tried in a traditional city centre in Lille, France. This system, known as 'VAL', opened in 1983, and consists of driverless trains on rubber wheels, operating a 90-second service at peak hours and carrying over 3 million passengers annually. Its success has encouraged the introduction of other lines such as those in Vancouver and in London's Docklands.

Taxis and 'informal' modes
Taxis are small vehicles that usually carry one passenger over short distances, often in city centres where parking shortages make the use of private cars difficult. They have high unit costs, especially of labour and fuel, and thus tend to be used principally by those on higher incomes and for business travel. Their contribution to the modal split in Western cities is typically below 1 per cent, though they may have particular importance

Table 7.5
Proportion of trips by car,
bus and informal sector
public transport in selected
cities

City and survey date	Percentage of trips by			
(a) Motorised modes	Car	Bus	Informal sector and other	
Bangkok (1970)	29	59	12	
Hong Kong (1977)	30	39	31	
Jakarta (1972)	29	49	22	
Karachi (1971)	16	63	21	
Kingston (1978)	36	49	15	
Manila (1974)	29	22	49	
(b) Bus and informal/other				
Delhi (1979)	—	78	22	
Hyderabad (1979)	—	49	51	
Bangalore (1979)	—	48	52	
Kampur (1980)	—	6	94	
Jaipur (1979)	—	18	82	
Agra (1979)	—	13	87	
Baroda (1979)	—	45	55	
Cheng Mai (1976)	—	7	93	
(c) All modes	Car and motorcycle	Bus	Minibus	Other
Kuala Lumpur (1978)	50	19	11	20

Source: White (1990).

at certain times (late at night for example) and for certain groups such as women and the disabled, to whom they offer the advantages of security and door-to-door convenience.

Where taxis and other small vehicles can be shared by more than one passenger the costs to the individual may be lowered. Partly for this reason – and partly because of the unreliability of conventional bus operations – many cities in developing countries are largely dependent upon such 'informal' bus and taxi services which, together with various forms of non-motorised modes, provide what is known as paratransit (Box 7.1). In Manila, for example, nearly 50 per cent of motorised trips are via informal modes, most of them on the estimated 17 000 jeeps converted into minibuses ('jeepneys') (Table 7.5). In Africa informal transport is also extremely important, taking, for example, a greater share of passengers in Lusaka and Lagos than formal buses. Examples of different styles of vehicles include the *matutus* of Nairobi (mostly pick-ups and vans) and the mammy-wagons (trucks) of Lagos.

The usefulness of such modes cannot be questioned, but these small-capacity vehicles are less efficient than conventional buses and pass substantial congestion costs to the community, because of the obstruction to other traffic caused by frequent intermediate stops in the nearside lane. They also have a wretched safety record. For example, Zimbabwe's 'emergency taxis' are 16.5 years old on average; lights are missing, brakes are unreliable, doors fail to shut, there are holes in the bodywork and mechanics make highly creative modifications to the vehicles in order to

Box 7.1 Paratransit in the developing world

(a)

(b)

(c)

(d)

Plate 7.3
Paratransit in the developing world: (a) Malucca, Malaysia – rickshaws and bicycles; (b) Sonora, Mexico – truck passenger transport; (c) Harare, Zimbabwe – communal transport; (d) Singapore – motorised and non-motorised tri-shaws and scheduled buses.

keep them on the road. For the would-be traveller, the decision to use an overloaded bus or an unsafe taxi can hardly be called a modal choice: it is a dilemma.

There is also an 'informal' sector of a kind in developed countries in the form of 'car-pools' and 'van-pools', whereby existing drivers share their vehicles. This is the second most important means of commuting in American cities, where it achieves better use of the existing private vehicle fleet in circumstances where public transport is often rudimentary in coverage.

7.4 *The urban transport problem*

7.4.1 *What is the urban transport problem?*

Because transport is a means to an end rather than an end in itself, it is much more relevant to ask what kind of city people want than what kind of transport system they would prefer. One should ask: can most employees get to work easily, cheaply and quickly? Are children's journeys to school free from traffic danger? Can shopping be done without traffic noise and pollution? In virtually every city in the world, the answer to these questions is no, no and no. For ordinary people going about their everyday lives, *that* is the urban transport problem.

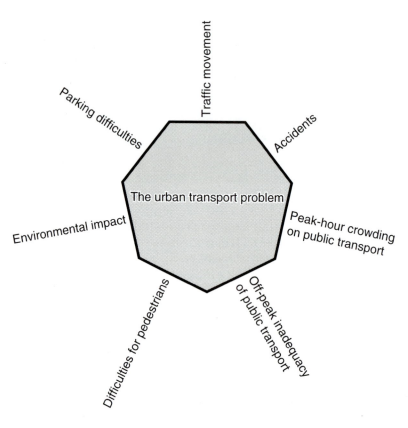

Fig. 7.3
The seven facets of the urban transport problem. *Source:* Thomson (1977).

Michael Thomson's classic book *Great cities and their traffic* (1977) provides a useful breakdown of the ways in which most people are dissatisfied with the transport systems of their cities and his scheme is followed here (see Fig. 7.3). The issues are interrelated: the urban transport problem can thus be treated as a seven-sided problem.

7.4.2 Facets of the urban transport problem

Traffic movement and congestion

Congestion may be defined as waiting for other people to be served and is particularly prevalent in service trades like transport where it is not economic to provide sufficient capacity to meet the highest levels of demand. 'Vehicle congestion' is the delay imposed on one vehicle by another, while 'person congestion' characterises public transport subject to severe demand fluctuations through time.

Congestion occurs in all developed cities no matter what their transport provision and is now a long-term and apparently immutable fact of life: average traffic speeds in London, for example, have scarcely changed in the past 80 years and in central districts fell from 20.7 km per hour in 1972 to 17.6 km per hour in 1990. It appears that peak-hour traffic speeds reach an equilibrium of about 16 km per hour; if traffic moves faster than this more people are tempted to travel by car, thereby increasing congestion and forcing speeds down again. In cities in developing countries congestion is worse still, with few motorised vehicles per capita, but narrow streets, multiple street uses and high population densities.

The effects of such congestion are to delay goods movement, frustrate passengers and clog streets with stationary traffic. When traffic is at 98 per cent of its maximum on a road, journey times are seven times longer than in light traffic conditions. These delays apply to all vehicles, of course, but buses are particularly susceptible. Figure 7.4 shows that at 10 km per hour a fully occupied car requires double the space per person as a full bus (6.2 and 3.1 m² respectively). In reality this comparison is even less favourable to the car: for example, 70 per cent of cars entering central London in the morning peak in 1981 had no passengers while the average bus occupancy was 40. So the more efficient bus is slowed by the less efficient car and because congestion increases operating costs of buses much more than it does for cars, the bus user has to pay more for the journey. It is thus said that the car driver 'passes on' time and financial costs to non-car commuters.

Public transport crowding

The 'person congestion' occurring inside public transport vehicles at such peak times adds insult to injury, sometimes literally. A very high proportion of the day's journeys are made under conditions of peak-hour loading, during which there will be lengthy queues at stops, crowding at

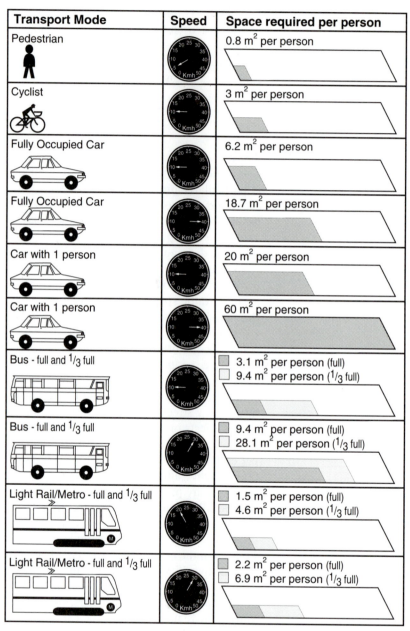

Fig. 7.4
Consumption of space by different modes of transport, occupancy and speed. *Source:* Navarro et al (1985).

terminals, stairways and ticket offices, and excessively long periods of hot and claustrophobic travel jammed in overcrowded vehicles (Fig. 7.5). In Japan, 'packers' are employed on station platforms to ensure that passengers are forced inside the metro trains so that the automatic doors can close properly. Throughout the world, conditions are difficult on good days, intolerable on bad ones and in some cities in developing countries almost unbelievable every day. Images of passengers hanging on to the outside of trains in India are familiar enough. Quite what conditions are like inside can only be guessed at.

Off-peak inadequacy of public transport

If public transport operators provide sufficient vehicles to meet peak-hour demand there will be insufficient patronage off-peak to keep them economically employed. If on the other hand they tailor fleet size to the off-peak demand, the vehicles would be so overwhelmed during the peak that the service would most likely break down.

This disparity of vehicle use is the nub of the urban transport problem for public transport operators. Many now have to maintain sufficient vehicles, plant and labour merely to provide a peak-hour service, which is a hopelessly uneconomic use of resources. Often the only way of cutting costs is by reducing off-peak services, but this in turn drives away remaining patronage and encourages further car use. This 'off-peak problem' does not, however, afflict operators in developing countries. There rapidly growing urban populations with low car ownership levels provide sufficient off-peak demand to keep vehicle occupancy rates high throughout the day.

Difficulties for pedestrians

Pedestrians form the largest category of traffic accident victims. Attempts to increase their safety have usually failed to deal with the source of the problem (i.e. traffic speed and volume) and instead have concentrated on restricting movement on foot. Needless to say this worsens the pedestrian's environment, making large areas 'off-limits' and forcing walkers to use footbridges and underpasses which are inadequately cleaned or policed. Additionally there is obstruction by parked cars and the increasing pollution of the urban environment, with traffic noise and exhaust fumes affecting most directly those on foot.

At a larger scale, there is the problem of access to facilities and activities in the city. The replacement of small-scale and localised facilities such as shops and clinics by large-scale superstores and hospitals serving larger catchment areas has put many urban activities beyond the reach of the pedestrian. These greater distances between residences and needed facilities can only be covered by those with motorised transport. Whereas the lack of safe facilities may be the biggest problem for the walker in developing countries, in advanced countries it is the growing inability to reach *anything* on foot, irrespective of the quality of the walking environment.

UN JOUR COMME LES AUTRES

Fig. 7.5
A day just like the rest: the commuting rat-race as seen by an unknown cartoonist. *Source:* unknown.

Environmental impact and accidents

These form two separate 'sides' to Thomson's seven-sided problem. However, because the effect of these issues is felt much more widely than just in built-up areas and because both have become key elements in any modern consideration of transport impacts on society, detailed discussion will be left to Chapters 10 and 11. Setting them aside at this juncture, however, does not mean that they are unimportant. Environmental problems in particular have become of central concern; indeed, one could make a strong case for arguing that they are the most critical issues facing cities of the late twentieth century.

Parking difficulties

Many car drivers stuck in city traffic jams are not actually trying to go anywhere: they are just looking for a place to park. For them the parking problem *is* the urban transport problem: earning enough to buy a car is one thing but being smart enough to find somewhere to park it is quite another.

However, it is not just the motorist that suffers. Cities are disfigured by ugly multi-storey parking garages and cityscapes are turned into seas of metal as vehicles are crammed on to every square metre of ground. Public transport is slowed by clogged streets and movement on foot in anything like a straight line becomes impossible. Providing more parking

Plate 7.4
In many Western cities finding space to park the growing number of cars has become a problem and motorists respond, as here in Milan, by leaving them in the road, on the verges and the footways.

cannot solve the problem, for in large cities of the industrialised world there are far more cars owned by a city's residents than could ever be accommodated in its centre. If London's million-plus daily commuters all brought their cars with them, the city centre would have to be demolished to make space for them. The supply of parking thus has to be limited – by manipulating prices, availability, length of stay and so forth – in order to control car use in the city. Parking is thus different from the other elements of the urban transport problem in that in many cities parking difficulties are deliberately maintained in order to relieve the other symptoms.

7.4.3 What is causing the problem?

The economic roots of the problem

Cars have not caused the urban transport problem. Ancient Rome was strangled by traffic congestion and blighted by pollution; then the horse was to blame. Of course the car has particular characteristics that exacerbate the problem, in particular its speed and exhaust emissions, but many of the seven sides of the urban transport problem would be present whatever the vehicle, provided that the same cost structures were in place as today. Why is this so?

The reason is that transport is different from other items of consumer expenditure. As society becomes more affluent, so more people can afford items such as televisions until they become commonplace in virtually every household. Provided that we can produce enough electricity to power them, resources to make them and enough programmes to watch on them, the number of televisions in a city can be unlimited. In short, everyone who wants a television can have one without imposing any cost on anyone else.

However, in contrast, all transport expenditure imposes a cost on everyone else and some things are required that cannot be bought by any individual at any realistic price. A new car owner may want to buy a parking space – but in a crowded urban area with lots of other car owners, there may be no parking space available unless it is taken away from someone else. (In Tokyo, you cannot register a new car unless you can show the authorities that you have got somewhere to park it.) You may be willing to pay a fuel tax so that the authorities can provide you with road space on which to drive your car, but as others do the same the traffic clogs and the clear road disappears. As a pedestrian you may want freedom from traffic danger, but it is not available at any price.

Of course, it is technically feasible to solve many of these problems but not economically realistic. For example fast, reliable, comfortable, frequent, safe and free public transport could be provided throughout the city, but what city could afford it? All roads and car-parks could be put underground, but who could pay the fantastic price that this would cost? What is more, we are now coming to realise that all forms of movement have a price and that some are so demanding of the world's resources of fuel and clean air that they are not sustainable in the long term – that is to say that they are beyond any price that the world can pay.

The root of the problem is thus not the nature of the vehicle; rather it is economic, in that the price system fails to keep a balance between the demand for transport facilities and the cost of supplying them. There are four ways in which this is so. Firstly, the use of urban roads is effectively free. The cost to the community of cars using a particular kilometre of urban road in the peak hours is extremely high in terms of congestion passed on to the other users, environmental costs and the costs of providing new roads, yet it is no more costly to the driver than any kilometre of road anywhere else. Raising the money to pay for such things is not the problem; rather it is one of justifying the expenditure of huge sums of money to provide effectively free movement for users.

Secondly, although most cities now charge for parking space, the income is much lower than could be yielded if the land were rented for alternative potential uses, such as for offices or shops. Parking is thus artificially cheap as motorists are subsidised by the community and there is, as a result, more car-driving than there would be under free-market conditions. Thirdly, logical pricing of public transport would demand that passengers pay more for a ride in the peak hours, when demand is high, than they should in the off-peak period when it is low, but rarely is this done. Instead, fares are equalised throughout the day, or indeed may be discounted for peak-period passengers who are prepared to buy a monthly ticket. That this perverse practice has become commonplace is shown by the way that the 'commutation' of fares to lower levels for daily passengers on American suburban railways gave us the term 'commuter'.

Lastly, the polluter does not pay for the environmental impacts inflicted on others, such as noise or loss of amenity. Making the polluter pay would be fairer and also more efficient, forcing the traveller to decide whether the high cost of the journey was justified by the personal benefits gained. These issues will be discussed at greater length in Chapter 12.

The dynamic nature of the system

Many of the pricing distortions discussed above have been present for a very long time and their accumulated effects have been widespread. For example, the fact that travel by car is artificially cheap (in the sense that the congestion, environmental, accident and social costs of car use are paid by society rather than by individual drivers) has encouraged people to live on the edge of the city or beyond and commute to the centre by car. As many others have made the same decision the roads have become congested, leading to demands for new roads. Thus most new urban transport investment has gone into roads in an attempt to cure congestion, but has simply provided new road space to be used free of charge by newly generated trips. In this way new investment follows old, even if the original investment was misguided. The effect of pricing deficiencies thus lasts for a long time, becoming enshrined in the structure of the city which in turn perpetuates the same inefficiencies.

Little investment has remained for public transport, which has had travel times increased and schedules disrupted by the congestion from private cars and seen passengers seduced away by the comfort and convenience of inexpensive private travel (Fig. 7.6). As Thomson (1977: 65) summarises: 'As more and more people have deserted public transport in favour of private transport the pressure on governments both to build new highways and to prevent public transport from going bankrupt have been too great, and cities have sunk helplessly into the condition of chronic congestion and collapsing transport services . . .'.

Constrained choices for individuals

The pricing distortions in the urban transport market produce an opposite effect to that in classical free markets, for the collective result of actions taken by individuals in their own best interests is to produce the worst possible outcome for everyone. We can illustrate this by way of a simple example.

As more people commute by car so does congestion increase for everyone, including bus passengers. Thus all journeys become slower, but at any particular mix of car and bus commuting the bus is always slower than the car, because it has to stop to pick up passengers and get back out into the traffic. If only 10 per cent of a city's commuters use a car, the journey times for bus and car users, door to door and including walking, waiting and parking times, might be as set out in Fig. 7.7. In year one the bus passenger takes 53 minutes for the trip, but could do it in 35 minutes by car, and in greater comfort too. Self-interest and logic demand that she switch to commuting by car. However, others will reach the same conclusion, so the proportion of car commuters will continue to rise year by year and the traffic will get slower and slower.

The key point to grasp is that this process is not controllable by any individual and will continue long after it is in everyone's interest to stop it. When 60 per cent of all commuters are travelling by car it takes our original bus passenger either 108 minutes by bus or 84 by car, both of them much longer than the 53 minutes it took when she originally travelled by bus. She might like to turn the clock back to that earlier situation, but she cannot: she can only select from the choice available to her now. The choice she would really like – to travel by car when everyone else travels by bus – implies control over other people's behaviour that she does not have.

The process is similar to what happens at a football match, when some fans stand up forcing those behind to follow suit. No individual fan can change anything, for the choice that he faces is to stand and see, or to sit and not see. The sensible thing would be for everyone to sit down again simultaneously, but that would require either co-operative action among

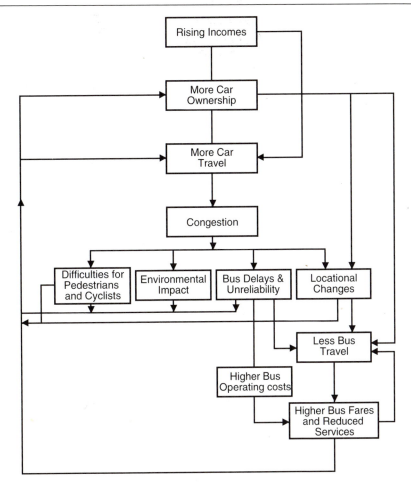

Fig. 7.6
The rise of the car and the fall of the bus: vicious circles in urban transport. *Source:* Goodwin (1969).

the fans or some central direction. The urban transport equivalent of the standing football fan is the commuter travelling at 16 km per hour. When millions of commuters make logical, self-interested, economic decisions in conditions where the pricing system does not allocate real costs to users the result is much worse traffic conditions for them all. There is thus a powerful argument for intervention to control the system in the public interest, for the *laissez-faire* city seems doomed to chaos and congestion.

7.4.4 *The urban transport problem in the developing world*

The transport problem in developing countries is frequently much worse than in cities of the developed world. Much of the problem stems from a

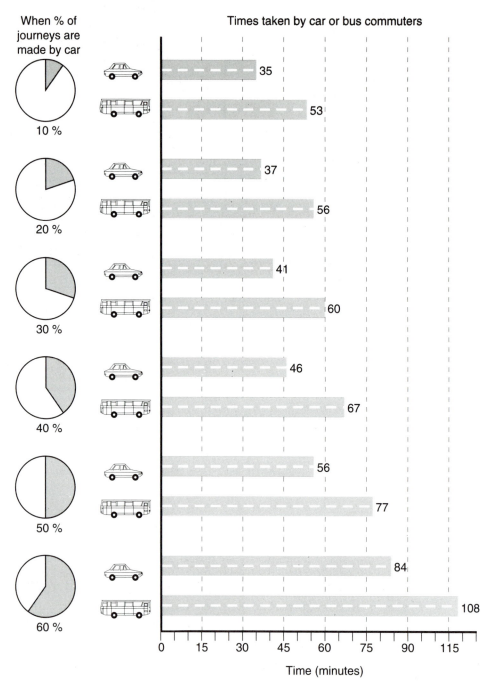

Fig. 7.7
Hypothetical times of commuter journeys by car and bus. *Source:* Plowden (1980).

Plate 7.5
One of the major problems facing many bus services in developing countries, is that of insufficient serviceable, reliable and safe vehicles: here passengers at Mbare Musika bus terminus in Harare, Zimbabwe wait patiently for overcrowded, overloaded and overdue buses.

lack of money. The urban poor cannot afford to pay much for transport but constitute a high proportion of the population, sometimes 50 per cent or more, and they frequently have the longest journeys to work from peripherally located shanty towns. As Box 7.2 shows, they often pay the highest costs (in time and as a percentage of income) to get less mobility than other groups.

The lack of funds is also apparent when it comes to the operation of public transport systems, despite the fact that many cities spend between 15 and 25 per cent of their budgets on transport. There are low levels of

---

Box 7.2 'You'll always be late': getting to work in Harare

Zimbabwe's capital city illustrates the problems that many cities in LDCs have with public transport. The bulk of the workforce lives in high-density suburbs on the periphery of the city and is almost entirely dependent on public transport, with 34 per cent of work trips by bus and 38 per cent by informal modes, principally emergency taxis. This article shows what it is like on a day-to-day basis.

Source: The Southern African Economist, August/September 1989.

'You'll always be late'

Thomas Mapfumo, Zimbabwe's most popular musician, belts out the words: '*Vanoshanda muZimbabwe, vanoshanda!*' (They work in Zimbabwe, they work!) and Zimbabweans do indeed work hard. But what a hassle to get there and back! Mabasa is a worker who lives 25 km away from Harare where he works. Here is a typical start to his day. ▶

Fig. 7.8
A humorous view of travelling in Harare's emergency taxis: exaggerated, but not by much.

It's four in the morning and Mabasa is shocked awake by the alarm clock. Mechanically he switches the light on, rubs his sleepy eyes and stumbles to the bathroom. Five o'clock finds him at the bus stop. Already there is a queue and he is at the back. It's winter and cold, but there is no sign of a bus. The men stick their hands in their pockets and hunch their shoulders against the pre-dawn chill. The women pull their jerseys tighter around their bodies. There is no knowing how long one might have to stand here. Mabasa shuffles from one foot to the other, his eyes glued on where the bus should be coming from. But no such luck. He glances at his watch. Half past five.

At long last a bus does arrive. But already it is packed to bursting. Maybe next time, Mabasa consoles himself. But the same thing happens again and again, bus after bus passing overloaded at alarmingly long intervals. Time is moving on.

After what seems like a whole day of waiting, a bus with a few empty seats stops. Though there is an inspector who is supposed to keep things in order, there is a scramble to get in and the inspector is pushed aside. Arms fly out, shoulders heave and shoes stamp. In the melee someone loses his wallet. No matter. This is the way things are. Not only today or tomorrow or the day after, but apparently for ever. To make matters worse it is pay day and the wives are going along too, following after their husbands to make sure the pay does not lose its way home or get ambushed by the beerhall. Once again Mabasa is unlucky in the scramble to climb aboard.

He looks at his watch again. Time is now flying. Yet here he still is, stuck 25 km from town, and it's 7.30 a.m.

An idea strikes him. Why not catch an emergency taxi – one of those jalopies about whose unroadworthiness there is rarely any doubt, which are grudgingly licensed to ferry passengers. Such a vehicle happens and stops. The bus queue is suddenly abandoned and everyone dashes for the taxi, even though there is room for only a few. Once again they don't include Mabasa, who rushes back to the queue now re-forming itself.

Along the road, gleaming and not so gleaming vehicles rush headlong townwards, some with only two or three persons and others with only the driver, while the bus queue casts an envious eye. And now comes the winter drizzle to add to the commuters' mistery. Mabasa grits his teeth. Yesterday, ▶

the day before and the day before that he had been late. The boss had been furious. That kind of thing can't go on forever. Too many people out job-hunting on the streets.

Then there is a rumble in the distance and a bus looms up. Mabasa jostles his way to it, powered by desperation, and hangs on to the hand rails. He sighs with relief. Today, thank God, he might just be in time. The bus chugs along. But will it get there? In snorts along and then, halfway into town, the engine stutters, coughs and staggers to a jerky stop. Frustrated faces look about them. There is no anger, no cursing. It happens too often for that to be worthwhile. Promises of a better bus service have remained just that – promises. As for Mabasa, he has a single thought: late again! What about the job this time, the rent, the school fees, the family?

Dry of mouth, hard of eye, he stands at the roadside, his arm mechanically thumbing down passing cars. But then, the spirits be praised, someone who knows him screeches to a halt. With a broad smile and a thousand thanks, Mabasa steps into his friend's car. He is tired and sweating, but happy. His friend drives him right to the doorstep of his workplace.

As he closes the car door behind him he sighs a mighty sigh of relief. One punctual arrival at work successfully achieved. But at the back of his mind lurks the perennial question: when will the powers that be recognise his plight, which is that of thousands like him, and do something about it?

provision of conventional transport in comparison to the industrialised countries. Fares are low, state policies towards subsidies are inconsistent and raising fares is politically impossible. Many city transport undertakings recognise that the poor cannot afford to travel, so fares are pitched at levels which do not cover costs. The attempts made to meet mobility needs are thus seen as a 'welfare service', but it does involve a substantial loss on the part of the operators of the system. As a result, there is lack of funds for new projects and finance for maintaining old and unreliable vehicles, so that services cannot keep pace with growth in demand. It is not unusual for over half the bus fleet to be off the road at any one time due to shortage of spares or skilled labour. Those services that are operated are seriously affected by congestion, because of the inadequate road networks, narrow streets, bad traffic management and mix of street uses.

Even though car ownership levels are low, they are rising very rapidly in some cases and ownership is heavily concentrated in capital cities: Bangkok, for example, has 10 per cent of Thailand's population but 70 per cent of its cars. Every day, half of Nigeria's cars attempt to crowd on to six-millionths of the country's area, downtown Lagos. This crush of humanity in and on so many different modes of transport, the use of the street for trades and professions and the shortage of off-street parking spaces, produce congestion levels that reduce typical speeds to barely half those in the city centres of industrialised countries. When the poor quality of the travelling environment is taken into consideration – especially the frightening accident levels and the widespread pollution – it can readily be seen that the urban transport problem is at its most intractable in the cities of the developing world.

7.5 *Concluding summary*

It is clear that the urban transport problem is universal. The specific manifestations of the problem may be different from place to place, but the underlying economic causes are the same. Broadly speaking, they revolve around the fact that all new transport infrastructure generates new traffic and this is especially true in the case of roads because they are free to their users. New roads generate faster trips, longer trips, more trips by car and higher car ownership, all of which adds up to more traffic. They also undermine the use of public transport.

The problems are getting worse as urban populations grow and become more affluent. It is clear that doing nothing is not a solution, but is part of the problem, because left to itself, the urban transport problem tends to get worse. If there is no intervention, eventually all roads will become congested and although this will spread to larger areas of the city and last for longer periods of the day, the city will not 'grind to a halt'. Instead, traffic speeds will fall until significant numbers of travellers desert their cars or the area, maintaining an equilibrium whereby the city is afflicted by high levels of congestion and pollution and travellers experience extreme discomfort and delay. It is, however, possible to raise the equilibrium level so that congestion is limited, pollution minimised and the majority of people can travel in safety and relative comfort, but this requires intervention by the urban transport planner. That is the subject to which we turn in Chapter 8.

Further reading

Adams J 1981 *Transport planning: vision and practice* Routledge
Daniels P W, Warnes A M 1980 *Movement in cities* Methuen
Dimitriou H T (ed) 1990 *Transport planning for Third World cities* Routledge
Hanson S (ed) 1986 *The geography of urban transportation* Guilford Press, New York
Jane's 1993 *Jane's urban transport systems 1993/4* Jane's Information Group
Plowden S 1980 *Taming traffic* André Deutsch
Thomson J M 1977 *Great cities and their traffic* Penguin

8 Urban transport planning: the search for solutions

> **How do we solve urban transport problems**?
>
> Chapter 7 has shown that the urban transport problem is widespread, deeply entrenched and has been with us for a very long time. It also demonstrated that left to itself the problem gets worse. So what can be done and what has been done so far? This chapter describes the planning methods in use and evaluates them in cities in the developed and developing world. Reasons for their lack of success are outlined and an alternative approach discussed, based on encouraging the use of public transport and the non-motorised 'green' modes and on limiting traffic rather than attempting to cater for it.

8.1 Introduction

Physical planning is about how much of what gets puts where. In the transport sense, it concerns the location of traffic generators, the distribution of population and provision of transport facilities. One's view of how these things should best be arranged will depend very much on one's professional outlook, training and political perspective. The civil engineer will see the solutions in terms of building, the ecologist will focus on conservation and energy use, the architect will give priority to the aesthetics of the urban scene, the politician will be concerned with public opinion and the power of vested interest groups, the transport operator will want the most efficient use of the vehicle fleet and the resident will be worried about the safety of children and the health of the neighbourhood.

As the urban transport problem has grown and mutated so have valuable insights and ideals emerged from planners and philosophers and been incorporated in plans of the time. For example, Ebenezer Howard's garden cities of the early twentieth century were planned as settlements in which everyone could walk to work, while the USA, with its higher levels of car ownership, was planning spatial arrangements that protected residents from the car by the 1920s. Clarence Stein and Clarence Wright's new town at Radburn, New Jersey, pioneered the idea of groups of houses segregated from surrounding traffic arteries and connected internally by walking routes, an arrangment that became very popular in the post-war UK's New Towns and influenced the ideas of Colin Buchanan that are discussed in section 8.7. Other planners have encouraged the use of cars in towns, notably Le Corbusier with his notion of 'cities in the sky' – high-rise developments connected by high-speed motorways – which was so influential in the 1960s and 1970s in the UK. The American architect Frank Lloyd Wright also wanted space for living, but envisaged it spread horizontally, with housing densities of two to the hectare, universal car ownership and a new city which would encompass the entire country, what he called the 'Broadacre City of tomorrow'. More recently philosophies have turned towards conservation, higher densities and a resumption of urban living in compact communities connected by patterns of non-motorised movement, with a rallying cry provided by Jane Jacobs' book *The death and life of great American cities* in 1962.

Of course, none of these ideas has taken root universally, but it is easy to see their influence in the existing built environment on which modern planning has to work. As philosophies change, particularly with regard to the role of the car in urban life, so must the planner adapt and synthesise in the task of ensuring that essential movement can be accommodated whilst the quality of urban life is improved for all city dwellers. The methods that have been developed to meet this challenge will be the focus of the first part of this chapter.

8.2 The urban transport planning process

The general framework of the urban transport planning (UTP) process is derived from the pioneer urban transport studies in Detroit (1953–56) and Chicago (1959–62). The approach recognises the influence of transport and accessibility on shaping the structure of urban areas by adopting a systems approach. If we remember the familiar definition of a system as a 'set of objects together with the relationships between the objects', then we can think of land uses (such as shops, pubs and offices) and transport facilities (such as roads, bus stations and airports) as the 'objects' in our transport 'system'. The relationship between them is of course traffic. How any object behaves has an impact on other objects, so that the amount of traffic is determined by the level of land-use activity and the physical characteristics of transport facilities.

To understand systems requires systems analysis, a method which allows complex and dynamic interactions to be understood in broad outline and thus provides a useful framework for planning, designing and managing large-scale systems. The basic components of this analytical approach are:

1. *Definition* – what problem is the plan intended to solve?
2. *Projection* – how will the situation develop if the problem continues?
3. *Constraints* – what are the limits of finance, time, etc. within which planning must take place?
4. *Options* – what are the alternatives and their pros and cons?
5. *Formulation* – what are the main alternative plans, i.e. packages of available options within the prevailing constraints?
6. *Testing* – how would each of the alternative plans work out in practice?
7. *Evaluation* – which plan gives greatest value (within the constraints) in terms of solving the problems already defined?

After these analytical stages the results and proposals – whether for city districts or entire conurbations – would be fed back into the political process for appraisal and for appropriation of necessary resources. At each stage the initial analysis may be modified or even rejected; the process is thus a learning one, with the apparent problem and the initial objectives seen as tentative until the real nature of the problem is agreed upon and new ideas for its solution are developed.

8.3 A framework for the urban transport planning process

The basic stages of the process, as Fig. 8.1 shows, are the pre-analysis, technical analysis and the post-analysis phases.

8.3.1 The pre-analysis phase

Once the problems have been identified it is possible to define the desired future state of the urban area. The goals are broad statements (e.g. 'provide a safe, energy-efficient transport system') and these are then operationalised by a set of more specific objectives against which the performance of the alternative courses of action could be evaluated. These might include 'reduce the number of traffic accidents' or 'reduce the consumption of fossil fuels per person'.

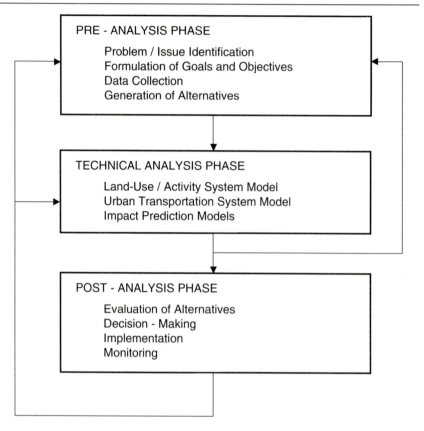

Fig. 8.1
A general representation of the urban transport planning process. *Source:* Pas (1986).

Once the goals are established, data need to be collected in order to prepare land use, transport and travel inventories of the study area. The availability of good quality, extensive and up-to-date data is an essential precondition for the preparation of an urban transport plan. Accordingly, there will need to be an inventory of the existing transport system and the present distribution of land uses; a description of current travel patterns; and data on such matters as population growth, economic activity, employment, income levels, car ownership, housing and preferred travel modes.

8.3.2 The technical analysis phase

The technical phase involves predicting the traffic flow on the links of a specified network. Forecasting techniques are used for this which are generically known as the urban transport planning modelling system. A major input is the future distribution of houses, employment, shops and other land uses in the urban area and these are predicted by models relating to land uses, population trends, employment and income levels and the performance of the urban economy and by references to urban physical planning proposals.

In many ways the urban transportation modelling system is the most well-developed part of the UTP process (Fig. 8.2). It consists of a set of submodels which tackle the problem in four stages:

1. whether to make a trip (trip generation);
2. where to go (trip distribution);
3. which mode of transport to use (modal split);
4. which route to use (traffic assignment).

The objective of the trip generation model is the estimation of the number of trips that originate and end in each traffic zone, using information such as number of households, family structure, household income and car ownership. A technique known as category analysis is used, whereby it is assumed that households with the same characteristics will have the same level of trip-making, so that if the number of households with particular characteristics in a particular zone is known, it is possible to estimate how many trips will emanate from that zone. In turn, the number of trips that end in a zone is estimated from a trip attraction model, which relates attractions to a zone to the intensity of activities within that zone. Land-use types are used to describe activities and are modelled from variables such as the number of workers, households and educational establishments.

At the end of the generation stage there should be estimates of the total number of trips generated by, and attracted to, the various zones of the study area as shown in the row and column totals in Fig. 8.2(a). The object of the trip distribution stage is to estimate the trips between all pairs of zones, i.e. to fill in all the blank cells. A number of types of model have been developed to do this, many of the gravity model type, in which trips to a given zone are assumed to be made in direct proportion to its relative attractiveness and in some indirect proportion to the distance separating it from other zones. The basic form is

$$T_{ij} = \frac{KG_i A_j}{F(D_{ij})}$$

where T_{ij} are the future trips between zones i and j, G_i the total number of trips generated in zone i, A_j the total number of trips attracted to zone j, D_{ij} the separation between i and j, and K and F are constants.

The result will be the ability to assign trips as shown in the example in Fig. 8.2(b) where we can now see that of the 60 trips *generated* by zone one, 10 go to zone two, 30 to zone three and 20 to zone four. Similarly, of the 100 trips *attracted* to zone one, 20 come from zone two and 40 each from zones three and four.

Variants of this model are widely used for different planning purposes

Attraction zones / Generation zones	1	2	3	4	Total Generations
1					60
2					40
3					80
4					120
Total attractions	100	60	80	60	300

(a) The results of the generation stage.

Attraction zones / Generation zones	1	2	3	4	Total Generations
1		10	30	20	60
2	20		10	10	40.
3	40	10		30	80
4	40	40	40		120
Total attractions	100	60	80	60	300

(b) The results of the distribution stage.

Fig. 8.2
The results of the generation (a) and distribution (b) stages in the UTP process. *Source:* Bell et al (1983).

ranging from large urban freeway projects to estimating the attractive effect of new shopping centres. They are easy to understand, simple to calibrate and are robust enough to cope with changes in land use and travel costs.

The last two stages are the forecasting of modal split and the assignment of traffic. Using the estimated total trips by car-owning and non-car-owning households, and the estimated inter-zone costs by private and public transport, the likely modal split on all journeys between all pairs of zones can be predicted. All that remains is to forecast which routes will be taken. This is done separately for public and private transport.

The characteristics of the transport network can be studied to determine the least-cost routes between all origin–destination pairs. All trips are then assigned to these cheapest routes subject to capacity constraints. For public transport a least-cost network is identified, using time, distance, fares, walking- and waiting-time plus an interchange penalty, and this allows trips to be assigned to bus, train and other public transport modes.

8.3.3 The post-analysis phase

The outcome of the technical analysis phase that we have just described is a set of predictions of the likely impacts of various plans. The final stage involves the evaluation of the alternatives, using the appraisal techniques described in Chapter 10, followed by implementation of the decision and then monitoring of the performance of the system. Once a decision has been taken, the system changes have to be programmed, to match the predicted flow of funds. The performance of the system is monitored to allow the identification of short-term, low-cost, fine-tuning measures which will help to keep the new system on track.

8.3.4 Summary

In summary we can see that the UTP process has four principal characteristics – quantification, comprehensiveness, systems thinking and a scientific approach. Modelling travel demand depends on quantification and this derives from the ability of the planner to measure inputs such as vehicle flow and speed and from the need to cost recommendations. In theory the process is comprehensive in that it does not support one mode more than any other but in practice, as we shall see, things have been rather different. The UTP's need to handle complex interrelationships and analyse large amounts of data has led to a formal approach based on systems thinking and this implies the use of a rigid set of procedures which are scientific in their approach to problem solving. However, it is this inflexibility which has emerged as one of the major weaknesses in the UTP system.

8.4 Has the urban transport planning process 'worked'?

8.4.1 Introduction

The UTP process just described has become a standard approach to the planning of transport in cities all over the world. It has, for example, been employed in every major urban area of the USA and the UK and in cities as diverse as Athens, Calcutta, Lagos and São Paulo. It could not be expected that it could resolve all of the transport problems which confront these cities, but it is reasonable to ask how successful it has been. Of course there is no easy answer, if only because we have no way of knowing how things would have turned out otherwise. Nevertheless, an examination of the experience of cities in the USA and in developing countries may shed some light on the difficulties of applying the UTP methodology in different economic and cultural circumstances and in cities of widely different physical form.

8.4.2 Urban transport planning in the United States

Detroit is an interesting case study because it is the home of the world's automobile industry and because it pioneered the application of modern land-use and transport planning techniques. The latest traffic and land-use study (TALUS) reported in 1969 with a plan for 1990 envisaging that a further 558 km of freeway would be needed to cope with an anticipated trebling of traffic and a growth in average trip length from 9.5 to 15.4 km. As Table 8.1 shows, public transport's share of the modal split had fallen by 1980 to 4.1 per cent from 13.8 per cent in 1960, compared to a rise for private motorised transport from 80 to 93 per cent.

The main problems concern the carless in such a city, for most of Detroit's population is now too dispersed to be properly served by any conceivable system of public transport. The streetcar system, one of the world's largest, has disappeared and the buses leave large parts of the city almost completely inaccessible. We should note also the environmental strains caused by the development of a wholly car-based transport system. Between 1960 and 1980 the urbanised area grew by 42 per cent, average work trip length by 35 per cent and fuel consumption by 11 per cent. To put this in perspective, gasoline use per capita in Detroit was 67 000 megajoules in 1980, compared to 40 000 in Melbourne, 10 000 in Amsterdam and 5000 in Hong Kong. It is very clear that lower urban density is a critical factor in increasing car dependence and energy consumption.

Detroit's problems are by no means unique. Robert Cervero and Peter Hall began an article on urban traffic conditions in American cities with the words 'As America enters the 1990s, traffic congestion seems to have reached a new phase, a new dimension. ... Within a few short years, traffic

Table 8.1
Detroit: transport and land
use indicators, 1960–80

	1960	1980
Population (millions)	3.7	4.0
Inner city population (millions)	1.67	1.20
Urbanised area (thousand hectares)	190	270
Jobs in the inner city (thousands)	744	497
Jobs in the outer area (thousands)	570	1186
Average work trip length (km)	10.2	13.8
Modal share, public transport (%)	13.8	4.1
Modal share, automobiles (%)	79.8	93.1
Average auto occupancy	1.5	14
Average auto speed (km per hour)	37.9	44.4

Source: Newman and Kenworthy (1989).

congestion has eclipsed every other concern – be it crime, unemployment, or air pollution – as America's number one urban menace' (Cervero and Hall, 1989: 176).

The cost of this delay is very great, estimated to be $67 per capita every year in Los Angeles. There are costs too in terms of increased stress, declining worker productivity and deteriorating quality of life. As a result more and more communities are passing no-growth or slow-growth measures, but these eventually backfire, because they force new commercial developments away from established residential areas, often pushing them to the metropolitan fringe where they cause more sprawl, aggravate job–housing imbalances and thus generate more traffic and congestion.

One example is the Bishop Ranch Office Park some 50 km east of downtown San Francisco. This huge development covers 250 ha, has 370 000 m² of floor space and includes one so-called 'horizontal sky-scraper' which is only three storeys high but is nearly 0.8 km long and is surrounded by a parking area which provides more than one space per employee. Many employees used to work in San Francisco's central business district (CBD) to which they commuted on the BART (Bay Area Rapid Transport) rail system, but most have now switched to travelling alone by car. Thus mass transit use amongst employees has fallen from 58 to 3 per cent; there has been a threefold increase in the number of car-kilometres, with a comparable increase in fuel consumption and exhaust emissions. Cervero and Hall (1989: 178) conclude that 'increasingly, the "desire-line" maps of today look like thousands of pick-up sticks dropped on the floor – trips flow from everywhere to everywhere. This has given rise to a serious mismatch problem – the mismatch between the geography of commuting and the geometry of highway networks'.

In the USA then, we have seen the demise of public transport and growing car use, with several social, economic and environmental consequences. From this evidence, the UTP process can hardly be said to have been a success, an outcome for which there are numerous explanations.

With hindsight we can now see that the process does not comprise an objective, value-free technical process, because even the mathematical models in their selection of inputs and the methods of measurement used implicitly contain subjective judgements and reflect particular perspectives on human behaviour. For example, the 1950s and 1960s, when the large urban plans were begun by Detroit and Chicago, were a period of growth and optimism during which it was assumed that car ownership would and should spread. Land-use patterns were planned with this in mind and once the land uses were in place they could not be served effectively by public transport, so a huge increase in road building became necessary to serve the new developments.

There have also been technical flaws such as when simplified networks have been drawn up which necessarily only included the major roads. Much of the predicted traffic would be loaded on to these roads (since alternatives were not in the model) resulting in apparent undercapacity to cope with traffic and a resultant case for more road building. Such technical difficulties are soluble but in many countries it is the political agendas of individual government agencies and the technical advice which they give to policy-makers on transport and land use which is at the heart of the failure of the UTP to deliver improved transport conditions. Successive rounds of road building, as in the UK for example, have led to the creation of very powerful government road-building bureaucracies with a strong vested interest in ensuring that only roads-based solutions are put in front of politicians at the final decision stage. As has been stressed before, the whole process is driven by the goals and objectives which are adopted at the outset and in the past these have principally been to try to keep supply and demand in equilibrium through providing capacity to meet unlimited demand. It is now evident that the grand plans have failed to achieve this, with congestion worsening and the side-effects of unbridled mobility becoming some of the principal environmental and social concerns of our age.

8.5 The application of the UTP process to cities in developing countries

Having seen the problems posed by the application of the UTP process in cities in developed countries, it should come as no surprise that the difficulties have proved even greater in cities in developing countries. The process has led to the implantation of Western solutions like urban expressways, underground railways and computerised traffic signals in cities where they were not likely to be effective. Understandably, the process has come in for considerable criticism, with one observer commenting that 'we have a series of excessively complicated and expensive models using unsubstantiated and biased techniques to provide information of dubious accuracy for answering the wrong questions' (Atkins, 1977: 58).

The root of the problem lies in the fact that many of the assumptions that planners have employed are just not relevant for the conditions encountered in the cities of the developing world. Harry Dimitriou (1990a) has compiled the following list of such assumptions:

1. the belief that the urban transport problem is basically one of how to overcome motorised traffic congestion, despite the fact that the overwhelming majority of households in developing countries are not vehicle-owning;
2. the premise that increasing vehicle ownership levels are inevitable, which then becomes self-fulfilling;
3. the idea that informal public transport is not worth detailed study, so that transport studies thus exclude important means of mobility and employment for the urban poor;
4. the belief that improving the operational efficiency of existing transport systems will produce the greatest benefits, which aggravates equity issues in view of the fact that transport systems have often been planned and managed in favour of selective sections of the community;
5. the premise that variables affecting transport are predictable, whereas volatile circumstances in developing countries can jeopardise forecasts for as little as 10 years ahead;
6. the idea that urban transport problems are essentially the same world-wide, a belief contradicted by the very different evolutionary development of transport problems in cities in developing countries.

These misconceptions can be at least partly explained by the fact that many of the planners employed to solve developing countries problems come from – and have been trained in – the Western cities where the UTP was first devised and applied.

8.6 A new direction for planning: the needs of developing countries

If present approaches are flawed how should they be modified? One essential step seems to be to acknowledge better the contextual importance of transport planning. In the past the UTP process has had much to say about traffic engineering, transport operations and travel behaviour, but has said much less about the impact of transport improvement on urban communities, land use and wider urban development policies. The planner must be aware of what efforts are being made to provide the city with basic needs of urban living, to enhance its economic productivity and to improve its distribution of opportunities, and must see transport as an instrument in the achievement of these goals. It demands greater understanding of the role of informal transport systems, an objective appreciation of past transport planning practices and greater attention to the practical restraints of finance, manpower, energy, and institutional arrangements.

This 'developmental' approach to urban transport planning has considerable potential for improving the assessment and implementation of plans and policies. By integrating the process with urban development it is likely to generate a quite different set of investment priorities from the often inappropriate ones seen in the past, concentrating in the future on buses and the non-motorised technologies on which the great majority currently rely and which can be best enhanced to meet their basic needs. This approach has been adopted in Madras where, as Box 8.1 and Fig. 8.3 show, the most important task has been identified as the need to

---

Box 8.1 The need for street management in cities in developing countries: the example of Madras

As the text shows, many grand plans for cities in the developing world – such as the studies of Bombay (1955) and Calcutta (1967) – have been based on the principle of easing traffic congestion for the benefit of improved car travel. They recommended that when the roads become congested they should be widened, other street activities be removed, pedestrians should be confined behind barriers and slow-moving vehicles banned. However, this approach is inconsistent with the realities of the urban transport situation. Generally speaking, vehicular traffic includes large numbers of animal-powered vehicles, bicycles, and depending on the city, taxis, mini-buses and rickshaws. The dominant means of access to facilities in the street is on foot but a characteristic of many cities is the neglected state of the pavements, which are often narrow and rough and blocked by other street activities that involve the informal sector, such as crafts, selling, entertainment and even living. As a result the pedestrian is often forced to walk in the roadway along an intermediate zone which is shared with other non-motorised traffic.

More recently, however, it has been recognised that plans based on Western assumptions are not the best use of scarce resources, interrupt the operation of a vital informal sector and are socially unjust. A new approach was adopted in the development plan for Madras in 1980 which noted the significance of the street for the informal sector. Though most of its recommendations related to rail and bus services, the study also emphasizes the importance of managing the street for the benefit of all its users. Three observations that it made were:

1. that as the majority of trips are made by pedestrians and cyclists, facilities for them need to be radically improved;
2. that the efficient use of the existing road network should be improved, principally through integrated policies of street management involving not only the highway authorities, public transport agencies and the police, but also bodies responsible for social services, housing, education and commerce;

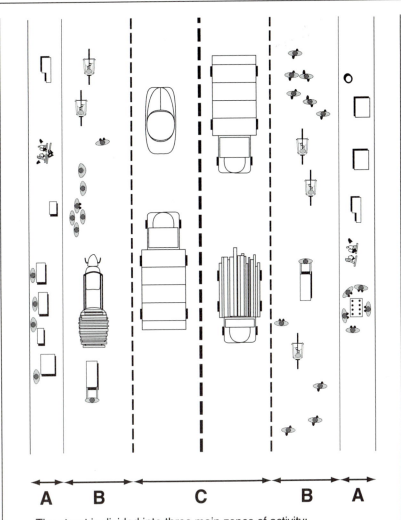

A B C B A

The street is divided into three main zones of activity:
A Narrow sidewalks obstructed by piles of rubbish, street vendors, missing manhole covers, excavation or municipal equipment.
B Inner lanes used mainly by cars, bicycles, pedestrians, parked vehicles and groups of people waiting for buses. Muddy or dusty surface is evidence of lack of use by motor traffic.
C Outer lanes used by motor traffic; cars, buses, lorries, taxis, auto-rickshaws.

Fig. 8.3
Typical street use in Madras. *Source:* Proudlove and Turner (1990).

3. that the improvement of pavements is essential in order to discourage people from walking in the roadway. This, said the study, 'is the single most important measure to which transport policies should be addressed'.

manage street space for the benefit of non-motorised transport and in particular to improve conditions on the pavements so that people do not spill out into the roadway and thus exacerbate the congestion there.

8.7 A new direction for planning in industrialised countries: environmental traffic management

8.7.1 Introduction

The apparently irresistible forces of rising populations and car ownership in cities seem likely to intensify the transport problems in circumstances where the UTP process is failing to solve the existing ones. What then can be done? We may begin our search for an answer by returning to one of the classic contributions to the urban transport debate. There can be few students of transport who are unaware of the Buchanan Report, published in 1963 as *Traffic in towns*. The ideas contained in it were not new, yet it succeeded in capturing the public imagination, popularising an intellectual foundation for the development of policy to cope with the problem of urban traffic.

What the report said in essence was that if environmental standards are set for an area, then a limit is automatically set to the amount of traffic that it can accommodate. This capacity can be increased, however, without reducing the quality of the environment, but only at a price. Growth of car traffic and preservation of environmental standards were seen to be compatible if society was prepared to spend the money. It all depended on the relative value placed on environment, accessibility and cost.

The report recommended the use of a hierarchy of urban roads, with through-traffic diverted away from 'environmental areas', on to primary arteries. An analogy was drawn with a large hospital, where the quiet of areas like wards and theatres is preserved by diverting traffic along major corridors. However, in most established cities such corridors frequently have other functions such as shopping and living which would be undermined by heavy flows of through-traffic. In reality the distinction between 'corridors' and 'rooms' is not clear, for the city is actually more like an open plan office, where spaces between the desks carry long- and short-distance traffic as well as much interaction between those sitting nearby. Funnelling increasing volumes of through-traffic along this multi-purpose space creates enormous problems of disruption. The disruption could be restricted by putting primary arteries underground but few cities could afford this on any scale. In actuality it seems that Buchanan's goal of finding civilised ways of accommodating growing volumes of traffic is not achievable, because the economic cost is too high.

However, the concept of environmental areas could still be applicable if there were a general policy of traffic restraint under which traffic volumes are reduced throughout the urban area. Under these circumstances any extra traffic decanted on to the boundary roads would be offset by the overall reduction in traffic that would result from general traffic restraint. The reform implied in such city-wide traffic restraint or 'environmental traffic management' (ETM) is a radical one, combining infrastructural works with spatial, economic, legal, psychological and educational approaches into a coherent policy. As authors such as Plowden, Roberts, Tolley and Whitelegg have argued, it is a new framework for urban transport planning, one that recognises that relationships exist between speed, access, environment and the quality of urban life. How it might be achieved is described in this section through an examination of its five strands: promoting the bicycle, encouraging walking, practising traffic-reducing town planning, promoting public transport and restraining car traffic.

8.7.2 *Promoting the bicycle*

The benefits of cycling have long been recognised. The bicycle is cheap to buy and run and is in urban areas often the quickest door-to-door mode (Fig. 8.4). It is a benign form of transport, being noiseless, non-polluting, energy- and space-efficient and non-threatening to most other road users. A pro-cycling city would promote fitness among cyclists and health among non-cyclists. Cycling is thus a way of providing mobility which is cheap for the individual and for society. Table 8.2 gives indicative annual cost savings which could be achieved in the UK for various levels of cycle use.

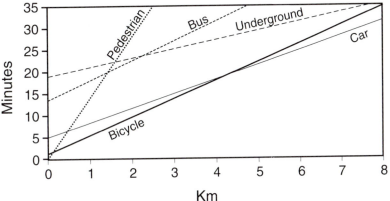

Fig. 8.4
Travel time in urban areas from door to door. *Source:* Bracher (1988).

Table 8.2
Estimated annual cost
savings from a shift to
cycling (£m)

	Level of cycle use			
	20%	30%	40%	50%
Road costs				
Congestion	725.0	1248.8	1811.7	2515.1
Roads				
Capital	38.7	66.6	96.7	134.2
Cleaning	—	—	—	—
Maintenance	—	—	—	—
Accidents				
Pedestrian	132.5	228.2	331.1	459.7
Cyclist	—	—	—	—
Motorist	58.6	103.1	153.4	220.3
Policing	—	—	—	—
Pollution				
Air pollution	94.0	161.9	234.9	326.0
Global warming	23.9	41.2	59.7	82.9
Noise	22.3	38.3	55.6	77.2
Health benefits				
Heart disease	22.6	42.3	63.0	82.3
Working days lost				
General	108.0	208.0	308.0	408.0
CHD	86.4	165.6	244.8	320.4
Total savings	1311.9	2304.1	3358.9	4626.1

CHD = coronary heart disease.
Source: Earth Resources Research (1993).

    Advocates of ETM frequently cast envious glances at the Netherlands, where cycle planning is set in the context of national planning for sustainability, described in Chapter 12. The *Master Plan Bicycle*, which aims to increase bicycle-kilometres by at least 30 per cent between 1986 and 2010, not only tackles the traditional concerns of cycle infrastructure and road safety, but also addresses issues of mobility and modal choice; how to encourage businesses to improve the role of the bicycle in commuting; reducing bicycle theft and increasing parking quantity and quality; improving the combination of cycling and public transport; and promoting consideration of the bicycle amongst influential decision-makers. These 'pull' measures are part of a national transport strategy of discouraging car use which 'pushes' motorists towards use of the bicycle.

    Of course the Netherlands has ideal conditions for cycling – it is flat and everything is close together – so that it may be thought that this is a solution for the Dutch alone and one which cannot apply elsewhere. It is true that bicycle use in developed countries in the post-war period has rarely been significant outside the countries of mainland north-west Europe, but the growing realisation that continued motorisation is not sustainable is focusing attention on the bicycle as the best alternative to

Plate 8.1
Simple facilities, when coupled with a pro-bicycle planning philosophy, can encourage considerable levels of cycling even in car dominated cities, as seen here in Montreal, Canada.

the car over short distances. The potential for substitution is very great: in the UK more than 60 per cent of car trips are less than 8 km in length and 60 per cent of trips by all modes are shorter than 5 km. In Germany – the centre of the European car industry and by no means a flat country – the bicycle's share of the modal split is now up to 11 per cent nationally and is over 20 per cent in many cities. The promotion of the health benefits of cycling by the medical profession helps to explain the resurgence of cycle use in the most unpromising circumstances. In the USA between 1983 and 1990 the number of adults who regularly use a bicycle has risen from 10 to 25 per cent, though it is true that many more are used for leisure than for daily transport purposes.

The main drawback to cycle use is the danger from motorised modes. One approach is to build segregated cycle routes, as in the Dutch city of Delft, which is a model of cycle planning. (Box 8.2 and Fig. 8.5). If, however, promoting cycling is part of a package of measures as the ETM approach advocates, then segregated cycle facilities become unnecessary, for the reduced speed and dominance of motorised traffic would turn the road system into one which is safe for all modes.

8.7.3 Encouraging walking

Walking is the most important mode of transport in cities, yet frequently data on it are not collected and many planners do not think of it as a form of transport. As a result of this neglect, facilities provided specifically for walking are often either absent or badly maintained (see Box 8.1 for an example from a developing country) and pedestrians form the largest single category of road user deaths. There are social, medical, environmental and economic reasons for promoting walking, for it is an equitable, healthy, non-polluting and inexpensive form of transport. Moreover, 'foot cities' tend to be pleasurable places in which to live, with access to facilities within walking distance frequently cited as a key indicator of neighbourhood quality of life.

The ETM approach would argue that walking should be promoted, not just permitted. In the first instance this would be by encouraging official recognition, that it plays a vital role in everyday life and deserves consideration alongside other modes as a serious form of transport. Secondly, ETM requires that the walking environment be improved to one which is clean, visually attractive, comfortable, convenient, protected

---

Box 8.2 The demonstration cycle network in Delft

The Delft Cycle Plan of 1979 proposed the construction of a city-wide network of cycle facilities, to be arranged as a functional hierarchy, with each of three levels in the network being specifically designed to meet particular needs of cycle traffic. The 'city level' network comprises a rectangular grid of segregated tracks some 400–600 m apart, connecting the major destinations. The 'district' network has a finer mesh (200–300 m) of less sophisticated facilities and the 'sub-district' level is of short-distance and simple facilities designed to route cyclists from their homes up to the higher levels in the system.

The thinking behind the design was that its coherence would have a greater value than the sum of its individual parts, particularly so in the case of door-to-door journey times. In other words, time-saving is seen as a more important factor in encouraging cycling than safety gains. It was hoped that it would demonstrate that the construction of whole networks would be able to arrest the ongoing modal transfer from bicycle to car. In the event, the network has been more successful than expected, with cycle speeds rising by 15 per cent, safety levels improved, attitudes towards cycling becoming more positive and a modal switch of 3 per cent from the car to the bicycle achieved. It is expected that the long-term results will be more significant still and it is thought that the results are not specific to Delft but could be achieved in other medium-sized cities.

The recommendations are clear:

- the construction of a hierarchical network of bicycle facilities does promote cycling;
- expensive facilities are not necessary, for small-scale facilities which are part of a master plan are more effective, especially when they improve the continuity of a route and cut down travel time;
- public participation and publicity are essential: it is not enough to build cycle tracks, it is also vital to market them.   ►

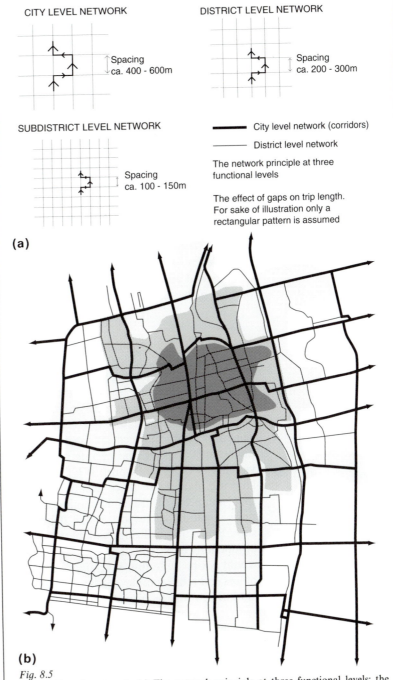

CITY LEVEL NETWORK

Spacing
ca. 400 - 600m

DISTRICT LEVEL NETWORK

Spacing
ca. 200 - 300m

SUBDISTRICT LEVEL NETWORK

Spacing
ca. 100 - 150m

City level network (corridors)

District level network

The network principle at three
functional levels

The effect of gaps on trip length.
For sake of illustration only a
rectangular pattern is assumed

(a)

(b)

Fig. 8.5
The Delft cycle network. (a) The network principle at three functional levels: the
effect of gaps on trip length. (b) The network pattern. *Source:* Ministry of Transport
and Public Works (1987).

from the worst effects of the weather, personally secure and free from conflict with vehicles. This would not be done by creating pedestrian islands in the traffic-orientated city, but instead would involve the selective diversion of vehicular flows away from routes of high pedestrian demand, together with the modification of the urban fabric by pedestrian-friendly measures. Continuous and close-mesh pedestrian networks would be required, with better same-level crossings, access to public transport stops and positive environmental treatment. These would be designed to meet the needs of the traffic-vulnerable – the disabled, elderly and children – and would have to connect residential areas with pedestrianised town centres and provide safe routes to school.

Walking and cycling show how ETM must be a package of measures, for encouraging the 'green' modes without restraining the principal 'red' one, the car, would increase traffic danger. Conversely, the transformation of cities to places where one can stroll, shop and play in comfort and security would be a crucial stage in persuading motorists to abandon their cars. Clearly, walking and cycling will not increase unless motorised traffic is restrained and activities placed so that they are within reach of people on foot or bicycles. The next section will show how this might be done.

8.7.4 *Practising traffic-reducing town planning*

Virtually all cities in industrialised countries have experienced reductions in density since car use became widespread, so that journeys have become longer as distance between necessary functions has increased. The resultant dispersed and segregated town is in part consequence and in part generator of unrestrained car use and is one that by its spatial separation discourages walking and cycling and by its low density undermines the viability of public transport. If ETM is to work it would be essential for land-use planners to place urban functions and activities in such a way that the need to travel is minimised. This would require conscious reurbanization, achieved through deliberately reducing the pressures for suburbanisation, utilising available city land and discouraging segregation of functions by mixing land uses where possible. Large traffic generators would not be built, thus reversing the trend towards ever larger hospitals, offices and retail complexes. Careful planning would ensure that particular land uses would be placed at appropriate points on the transport system, so that for example new university campuses would be sited close to public transport interchanges and away from motorway junctions, thus discouraging commuting by car.

By the manipulation of land uses in this way, so may accessibility be accentuated and trips shortened. This would be an essential achievement if ETM is to work, because shorter trips would encourage a transfer to the green modes and would produce less deleterious side-effects from the trips that remain motorised, since these would now be fewer and shorter.

Whatever motorised traffic remained after implementation of these approaches could then be restrained using the measures described in section 8.7.6.

8.7.5 Promoting public transport

If ETM aims to shift trips away from cars, then attractive alternatives are required. Cycling and walking may be appropriate for the shorter distances, but transferring longer trips requires that a good quality public transport system is in place to ensure that the city can function efficiently. This means that:

1. fares need to be low enough for poor people to be able to afford them;
2. there must be sufficient vehicles for a frequent service to be run throughout the day;
3. routes must reflect the dominant desire lines of the travelling public and there should be extensive spatial coverage of the city so that no one is very far from a public transport stop;
4. speeds of buses need to be raised relative to cars by freeing them from congestion;
5. it is not enough to provide public transport: it also has to be co-ordinated. Multimodal tickets may be one essential ingredient of a functional urban transport system, but the key item is the integration of services by the provision of connections between modes (Box 8.3).

Box 8.3 The transport interchange

In an ideal network interchanging is avoided, but this is difficult in large cities where passengers invariably use more than one mode in reaching their destination. Interchanges may vary from the simple roadside bus stop to the grandeur of the world's great railway stations, but good interchanges all have key features in common: they are simple, safe, comfortable and uncrowded, providing the passenger with short walking distances and waiting times, clear information display and an attractive environment. Plate 8.2 shows some examples.

  One special kind of interchange is 'park and ride', which keeps cars out of cities by encouraging drivers to park at suburban stations that offer fast public transport services directly to the city centre. This is of particular value to those in low-density suburbs of large cities where walking distances to the station are excessive: London's suburban rail stations, for example, provide some 150 000 spaces. Park and ride is also becoming more significant in provincial UK cities such as Chester, Oxford and Hull, where buses take drivers into the CBD. Most of these are provided specifically for shoppers at weekends or at peak retailing times like the Christmas season. Bike and ride is becoming more significant too, with 50 000 Berliners parking their bicycles at the city's railway stations every weekday. In the Netherlands 35 per cent of all trips from home to the railway station are made on the bicycle, so there is a clear need to ensure that secure and convenient cycle parking is available. ▶

Plate 8.2
An essential ingredient of the promotion of public transport is making connections between different services and different modes convenient and attractive at well designed transport interchanges. Some examples would be (a) bus/rail (now LRT) (Altrincham, Cheshire); (b) bicycle/metro (Rotterdam, Holland); (c) car/rail/air (Maglev track connecting Birmingham International railway station with the nearby Birmingham International airport); (d) off-vehicle multi-modal ticketing is essential to efficient interchanging (Nantes, France).

Promoting public transport in this way is unremarkable. However, what makes the ETM approach more radical is the use of public transport as a way of restraining the car. For example, from this perspective, building separate busways and putting rapid transit systems underground both free up surface space for use by the car and are thus seen as counter-productive. Instead, it is argued that public transport must be provided with protected rights of way that simultaneously speed public transport and reduce the amount of circulation space available for cars by, for example, enclosing street tramlines with kerbs so that cars cannot drive down them; designing traffic light systems that give automatic priority to public transport; and putting public transport stops in the roadway, which simultaneously slows car traffic to the speed of the bus and speeds buses by avoiding the need to have to wait to pull back out

Plate 8.3
An example of promoting public transport whilst simultaneously restraining car use in
Nantes, France. The removal of the asphalt from the street centre track restricts cars to the
space left over and prevents them from delaying trams.

into the traffic. The integrated and interdependent nature of the ETM
proposals is well seen in this notion of simultaneous promotion of public
transport and the restraint of car traffic.

8.7.6 Restraining the car

The fifth and last strand of ETM has as its goal the restriction of the car's
demand for space and the suppression of its speed. It is based on the view
that many streets now have as much traffic as they can carry in a technical
sense, but much more than they can accommodate without environmental
degradation. Under ETM the environmental limit, rather than technical
capacity, would become the controlling force. The critical step is seen as
reducing the space set aside for the car's use, whether for its circulation
or parking, for any additional space would increase car volume or speed
or both and thus counteract the goals of restraint. ETM's four main
space-reduction strategies are outlined below.

Restrictions on road capacity and traffic speeds
The principal approach here is traffic calming, the attempt to achieve
calm, safe and environmentally improved conditions on streets. Beginning
around 1970 attempts were made in the Dutch *Woonerven* to calm streets
by legal means (taking priority away from motorised vehicles), infrastructural measures (humps, chicanes, ramps, gateways, pinch-points, etc.)

and psychological signals, whereby the impression was given to drivers that they were in a street where motor traffic was admitted, but on the terms set by residents, playing children and other 'soft' traffic.

Woonerven were almost totally successful in reducing the speed and dominance of the automobile, but they were, in view of the need to completely rebuild the street, very expensive. Moreover, the earlier schemes were for single streets in residential areas, but much motorised traffic responded by diverting to adjoining non-treated streets. Consequently, later schemes adopted an area-wide approach, with a blanket area-wide 30 km per hour (20 miles per hour) speed limit made self-enforcing by calming measures such as throttled junctions and speed tables. Planting is not as profuse as in *Woonerven*, but there are still conscious attempts to achieve environmental as well as safety improvements (Box 8.4).

Such area-wide traffic restraint schemes are now commonplace: there are virtually no towns in the Netherlands without such areas and in the former West Germany there were reckoned to be some 2000 schemes by 1988. They have produced great benefits in housing areas (Table 8.3); speeds, noise and pollution levels are lower and accidents have been reduced to one-third of their previous levels on average. They are so popular that many local authorities cannot keep up with requests from residents to calm their streets. Calming measures are progressively being extended to larger and larger areas, including main roads too. The French city of Bordeaux has adopted a city-wide classification of its streets, one-quarter of which are for motor traffic with a 50 km per hour speed limit, another quarter being 30 km per hour streets restricted to buses and delivery vehicles and the remaining half will be relandscaped and dedicated to pedestrians and cyclists. This is effectively city-wide calming and the outcome will be watched with great interest.

Though the benefits of traffic calming are clear, it is important to realise the limitations of this approach. As a set of physical measures located to manage the existing volume of traffic it can dramatically improve the quality of the immediate environment, especially in residential areas. It does not, however, diminish the total volume of motorised movement and so its role in ETM is again as part of a package of measures.

Regulating traffic access to a link or area
A complete ban on motor vehicles – pedestrianisation – is the most extreme form of traffic restraint though it is not often promoted in this way. Instead it is usually seen as a device to remove traffic from historic or retailing areas in order to create a more pleasant environment for the pedestrian. The area involved is usually not large, though if a ring road is in place and public transport access is good, extensive systems of pedestrianised streets may develop, well seen in Munich (Fig. 8.6). Pedestrianisation is popular and usually increases trade – world-wide only 2 per cent of schemes have had a negative effect on trade – so that they

are now part of the modern city's competitive strategy to win business from other towns and from out-of-town retail developments.

Even where full pedestrianisation is in force it is always necesary to make exemptions for some types of vehicle. Of course it must always be possible to gain access for emergency vehicles and many pedestrian zones permit cycle and public transport access, usually with speed restrictions. Other users who are felt to have a need of a car or who have more 'right' to be in the area also have to be accommodated and these might include disabled drivers, doctors, residents, etc. Delivery vehicles are a particular problem. In newly built areas, shops can be serviced from the rear away from pedestrian shoppers, but this is often impossible in older centres. In many cases delivery vehicles are let in for limited periods, say from 07.00 to 09.00, before the zone becomes a pedestrian one for the rest of the day. This creates major difficulties for suppliers when many towns operate the same system, since all their trucks will only be required for a concentrated part of the day – the freight delivery version of public transport's peak-hour problem. One solution might be for deliveries to be transhipped on the edge of town into small, environmentally friendly vehicles for onward transmission to the shop, though this may result in a competitive advantage accruing to out-of-town retail sites close to motorway junctions for which transhipment would not be necessary.

Another way of restricting access by certain vehicles is based on some characteristics of the vehicle, e.g. the licence plate. Both Athens and Lagos have attempted to restrict cars by allowing them to enter the centre only on alternate days based on whether the licence plate number is odd or

Box 8.4 Some examples of traffic calming in practice

(a)   (b)

Plate 8.4
Area-wide traffic-calming aims to achieve calm, safe and environmentally-improved conditions on streets. It can take many forms as these photographs show: (a) slowing motorised traffic at 'speed tables' (or 'raised crossings') and road narrowings; (b) closing roads, greening them with planting and giving them over to play and other uses for local residents (Berlin, Germany); (c), (d) reducing space for the car by 'necking down' junctions and extending the footway and cycle tracks (Buxtehude, Germany, before and after); (e), (f) restricting circulation space for motorised traffic but not for cyclists, by one-way regulations (Erlangen, Germany) and bollards (Delft, Holland). *(continued)*

(c)

(d)

(e)

(f)

Plate 8.4 continued

even. The Athens scheme achieved a reduction in vehicle-kilometres of 15 per cent and of vehicle-hours of 22 per cent. There are likely to be problems of enforcement and evasion with such schemes as in Lagos, for example, where many affluent families equipped themselves with one car

Table 8.3
Summary of applications and effects of traffic-calming measures

	Speed reduction rating	Space reallocation for other uses	Visual enhancement of street scene	Suitability			
				L	C	M	T
Speed reducing measures							
Vertical shifts in the carriageway	A	×	—	★	★	+	○
Lateral shifts in the carriageway	B	√	—	★	★	+	○
Carriageway constrictions	B	√	√	★	★	+	+
Roundabouts	B	×	×	+	+	+	+
Small corner radii	B	√	—	★	★	★	○
Priority management	B	×	×	+	+	○	○
Road markings	C	×	×	○	○	+	★
Electronic enforcement	C	×	×	○	+	+	+
Supporting environmental and safety measures							
Optical width	C	×	√	★	★	★	+
Narrow carriageways	C	√	√	★	★	★	+
Occasional strips	C	√	√	○	+	★	+
Surface changes – type/colour/location	C	×	√	★	★	+	○
Entrances and gateways	C	×	√	★	★	+	+
Central islands	C	√	√	○	+	★	+
Shared surfaces	C	√	√	★	○	○	○
Footway extensions	C	√	√	★	★	★	+
Planting/greenery	C	×	√	★	★	★	★
Street furniture and lighting	C	×	√	★	★	★	★
Regulations	C	×	×	+	+	★	★

Speed reduction rating:
A = guarantees 85 percentile traffic speeds below desired maximum.
B = reduces speeds but does not guarantee 85 percentile level.
C = serves as a reminder or encouragement to drive slowly and calmly.

Suitability: (for different street/road classifications)
L = Local streets   √ Positive effect   ★ Suitable
C = Collector streets   × Negative effect   + Possible
M = Mixed priority streets   — Neutral   ○ Not recommended
T = Traffic priority roads
Source: Devon County Council (1991).

with an even number and another with an odd one. In Athens, where doctor's cars were exempt, one survey found that 18 per cent of cars entering the area were displaying doctors' permits!

Charging for the use of roads on a link or area basis
If the cause of the urban transport problem is seen as economic distortion, it follows that one approach is to make motorists pay the full external costs of their travel. This stance views congestion as a market failure which can be addressed by increasing the price for the use of the road space which is in short supply. The cost of noise, accidents, pollution and time wasted by congestion would set the price and the market would allocate space to the users who most need it at that time and place.

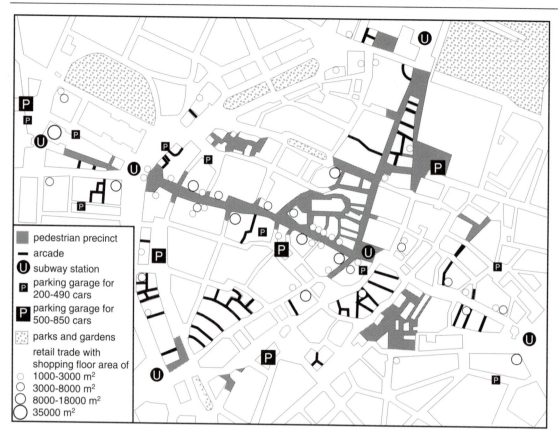

Fig. 8.6
Munich: pedestrian precincts, traffic arrangements and retail trade. *Source:* Monheim (1990).

Assuming that revenue from road pricing is spent on upgrading public transport and that some motorists would be priced out of city centres at peak times, the modal split would in theory settle at the point where the attraction of faster (but more expensive) private motoring just balanced the perceived benefits of cheap, comfortable, speedy and reliable public transport. For all travellers, the equilibrium would be at a more acceptable level than at present, which is at the point where the repulsion of expensive, uncomfortable and unreliable public transport just matches the unattractiveness of using a car on congested roads with no guarantee of a parking place at the end of the trip.

Without doubt, road pricing systems are becoming technically more feasible. Electronic road pricing (ERP) schemes involve fitting cars with 'electronic licence plates' which can be interrogated by roadside devices. A central computer would then invoice drivers for the use of roads, with the payment graduated to reflect the high cost of using city-centre streets at peak periods. Applicablity is, however, more problematical, with governments concerned to harmonise systems between cities and countries, civil liberties groups watchful of the possible implications and individual city governments concerned to ensure that the economic health of their centres is not endangered. Such a system was applied as a pilot scheme in Hong Kong in the 1980s and proved to be thoroughly workable. It was, however, abandoned for political reasons, not least the concern that data that tracked a driver's movements at all hours of day and night might be misused by politicians (or misunderstood by suspicious spouses!). ERP may have a role to play in some circumstances, but it is a partial solution at best. For example, residents suffering from noise, accidents and pollution from a busy road will not be any better off just because motorists are paying more. It is true that if congestion eases, pollution may fall – but only if the newly cleared road space does not generate more trips. Faster-flowing traffic will mean more noise and greater danger, aiding the rich commuter at the expense of everyone else. If prices are set at levels that deter car commuting and only some areas are priced, trips will be reorientated to new destinations and congestion will burst out somewhere else. There is a view that ERP is thus a convenient device for raising revenue masquerading as a comprehensive solution.

Parking restrictions
As we have seen, it is not possible to provide sufficient space for all who might like to drive and park in the central areas of large towns. Parking thus must be restricted and this is usually done by banning all-day parking by commuters or making it prohibitively expensive. Restrictions are less severe-off peak, so that shoppers and other short-term visitors who benefit the economy of the centre are not deterred. Separate arrangements must be made for local residents, perhaps through permits or reserved parking.

City authorities can thus control public car-parking places, but many other spaces are privately owned by businesses and reserved for particular employees. The effect of this is to perpetuate commuting to work by car. The future provision of such space can be limited through planning permission for new developments, as is done in London, but controlling the use of existing private spaces raises problematical issues of rights and freedoms that many countries are reluctant to confront.

Overall, parking restrictions have the advantage of being simple to administer, flexible in application and easily understood by the public. Their Achilles' heel is enforcement, for motorists are adept at parking where and when they should not and evading fines once caught. Fines in many cities are so low that being caught once or twice a week works out cheaper than paying the parking charge. Indeed, in London in 1982 a survey showed that illegal parkers outnumbered legal ones and only 60 per cent of the fines were ever paid. Parking controls have to be stringent and be enforced if they are to make any significant contribution to reducing congestion in the city. If such practical difficulties could be overcome, the potential contribution of parking restrictions to ETM

Plate 8.5
Installation of roadside kerbstones is a simple and inexpensive way of cutting down car commuting by restricting parking space, as seen here in central Amsterdam.

would be very great indeed: very few motorists would set out on a trip in the certain knowledge that there would be nowhere to park at the end!

Summary: combinations of restraint measures
A number of cities now exhibit ETM policies, including Freiburg, Hong Kong, Zurich and Singapore. The last has adopted a long-term co-ordinated package of traffic restraint measures, comprising both land-use and transport policies in order to cope with traffic congestion. The main elements are:

1. restraint on car ownership through high import duties and first registration charge and annual taxes proportional to engine size;
2. the area licensing scheme (ALS) introduced in 1975 which requires cars to have a prepurchased licence to enter the city centre in the morning peak. The number of cars entering the licensed area fell by 70 per cent immediately and, despite growth of the city population by over 30 per cent since, it is still 20 per cent below pre-scheme levels. Car modal share has fallen from 46 per cent in 1974 to 22 per cent in 1988 whilst that of public transport is up from 46 to 63 per cent;
3. extensive use of traffic management measures, including on-street parking bans, computerised traffic signal control and pedestrianised streets;
4. improvements in public transport services, especially through the

Plate 8.6
Environmental traffic management can produce city centres that are car-free but permit access to various forms of public transport and non-motorised modes (Freiburg, Germany).

operation of faster and more frequent bus services and the construction of the mass transit railway system;
5. the construction of new high-capacity roads outside the CBD linking major foci;
6. new land-use developments planned in conjunction with the public transport services.

Overall, the Singapore scheme is simple but very effective. The ALS has had a major influence for a very long period and it is expected that an electronic scheme will be introduced in the future. Much of the success is due to the fact that Singapore as a city state can raise car ownership costs in order to restrain car traffic, but this seems likely to be eased in the future by permitting lower registration costs for leisure-use cars, which could only be operated overnight during the week and after 1500 hours on Saturdays.

ETM advocates would argue that although the approach in these cities does not contain all of the ideal elements, it has shown that the suppression of motorised traffic is possible. They demonstrate that inessential motorised trips, which in the past have not been discouraged by traffic policies which have been favourable to the car, under ETM can be deterred by traffic strategies that make car journeys less attractive. The central idea is that ETM reduces the space available for the car so that traffic volume shrinks to fit it. In simple terms, building roads generates traffic, removing them degenerates traffic. Co-ordinated land-use and transport planning should shorten trips, so there is less traffic, as traffic = vehicles × distance travelled. If trip length reduces, more trips can be made by walking, cycling or using public transport, so that a higher proportion of the reduced overall amount of traffic can now be by ecologically sound modes. If ETM can be achieved it promises to produce much more civilised cities than the traditional transport planning approaches have done.

8.8 Concluding summary

We have seen that there are serious problems with the application of the existing UTP processes to cities throughout the world. Political pressures have meant that to date the process has favoured roads and suburban sprawl to the detriment of other transport modes and more compact forms of development, but there is no inherent reason why the planner should not be able in the future to adapt the models already available to produce more pro-urban, less car-based plans. These may, for example, attempt to minimise travel, maximise accessibility, revitalise central cities, reduce environmental impacts and lessen the social burdens of transport decisions that currently fall hardest on the disadvantaged.

Two last points ought to be made. First, though new technology may contribute to improving the efficiency of urban transport, we cannot rely on it to solve the urban transport problem. We have had 100 years of futuristic visions of new technologies triumphing over city congestion problems, yet the mess that we are in is worse than ever and the basic technology – the motor car – is still the same. Though new technology may aid system operation, highway management, public information systems and may be particularly important in the fields of automatic public transport vehicles, road pricing and telecommuting, existing modes are likely to continue well into the twenty-first century. Such gradualism is to a degree dictated by the pattern of existing investment, since it is almost always easier to upgrade existing systems as short-term solutions than to introduce new ones, which frequently require expensive new vehicles and rights of way.

Secondly, the role of land-use planning needs to be stressed. All cities have a dynamic relationship between their structure – that is their size, density, shape and distribution of land uses – and their transport systems. As structures change, so new trips are generated and new transport needs emerge. Transport investment takes place to meet these needs and itself alters the relative accessibility of places. The result is changes in land value and thus in land use – which starts the whole cycle off again by generating new transport demands.

To date policies which have attempted to 'build the problem away' by providing new capacity to meet unlimited demand have been undermined because new roads have promoted new development along their corridors and at their terminals. Such new development has generated new traffic which has overloaded the roads almost as soon as they are completed. The response to this has been to plan yet more new roads as a way of alleviating the new congestion, justifying them to the community in terms of time and fuel savings from reducing traffic jams. The process quickly becomes a vicious circle of congestion, road building, sprawl, congestion and more road building.

The way out of this impasse lies in the realisation that just as the promotion of new transport infrastructure changes land uses, so can the planning of land uses influence the demand for transport. Rather than the transport planner providing facilities to attempt to meet the demand, it is vital that the land-use planner places facilities in locations where they will minimise the need for movement. It is this rational location of facilities that is the environmentally sound, minimalist solution, demonstrating that non-transport approaches are every bit as significant in the achievement of civilised cities as planning for transport.

Further reading

Bruton M J 1985 *Introduction to transportation planning*, 3rd edition, Hutchinson
Dimitriou H T (ed) 1990 *Transport planning for Third World cities* Routledge
Friends of the Earth 1992 *Less traffic, better towns* FoE
Jacobs J 1962 *The death and life of great American cities* Jonathan Cape
Newman P W G, Kenworthy J R 1989 *Cities and automobile dependence: a sourcebook* Gower
Plowden S 1980 *Taming traffic* André Deutsch
Roberts J 1989 *User-friendly cities* Rees Jeffreys Discussion Paper 5, Transport Studies Unit, Oxford

Roberts J et al (eds) 1992 *Travel sickness: the need for a sustainable transport policy for Britain* Lawrence and Wishart

Tolley R S 1990 *Calming traffic in residential areas* Brefi Press

Tolley R S (ed) 1990 *The greening of urban transport: planning for walking and cycling in western cities* Belhaven

Whitelegg J (ed) 1992 *Traffic congestion: is there a way out?* Leading Edge

9 Rural transport problems, policies and plans

> **Isolation, inaccessibility and structural changes in rural areas**
>
> Although the transport problems of the world's cities usually form a major focus of attention for urban planners, the populations living in rural areas of the advanced and less developed nations also face severe difficulties in meeting their needs for movement. This chapter looks at these problems by considering rural transport deprivation at various levels of economic and social development and assessing the effectiveness of the policies and plans that have been proposed to improve levels of mobility and accessibility for people in more remote and isolated communities.

9.1 Introduction

In both the industrialised world and developing countries rural transport is a vital component of the system of services necessary for the continuing existence of dispersed settlement in less densely populated areas. Remoteness, isolation and inaccessibility are key characteristics of many of the world's rural regions and the economic and social deprivation which these areas suffer is often due in large measure to inadequate transportation services.

Several methods of defining rurality have been proposed based upon population density, settlement pattern, economic structure and aspects of remoteness and accessibility. In highly urbanised industrial states a distinction can usually be drawn between those rural areas which lie within urban hinterlands and the more remote districts where distance severely constrains travel to work into a major employment centre. Levels of commuting, combined with indices such as population density and the proportion of the labour force in agricultural and related occupations, have been used in official definitions of rurality in the UK and a more refined definition incorporates sixteen socio-economic variables to identify four main categories of rural area. Within the UK about 85 per cent of the total area, containing 12–15 per cent of the population, may be classed as rural, with the largest tracts being located in south-west England, central and North Wales, East Anglia, northern England and most of Scotland apart from the central lowlands. In terms of remoteness in England and Wales 2.7 million people in settlements of fewer than 5000 persons are at distances greater than 16 km from towns with populations of at least 20 000. However, remoteness is a relative concept and these small settlements in the UK cannot be considered as remote at the global scale.

Within much larger territorial units such as Australia or the USA the degree of remoteness, coupled with low population density, provides a convenient guide to the identification of rural areas. About 36 per cent of the USA is made up of areas where the population density is less than 14 per km^2 and where rural settlement is more than 200 km from cities of over 250 000. This area contains only 2.2 per cent of the total population and within it are tracts where over 1 million persons are distributed at densities of less than 0.5 per km^2. In Australia the identification of sparsely settled regions presents few difficulties, since the entire continent outside of the five metropolitan city areas can be classified as rural, with densities varying from less than one person per km^2 to less than one per 64 km^2. Rurality in these extensive continental areas has a totally different connotation from that in the UK since much of the Australian interior is totally uninhabited, and transport and other services are almost non-existent.

In most developing countries at least two-thirds of the population can still be classified as rural, although densities vary considerably according to levels of economic activity. In Nigeria, for example, rural densities of 400 per km^2 are recorded in the south-east region whereas the drier interior savannahs support densities of only 20–30 per km^2. If rural isolation is interpreted in terms of the absence of motorable roads then 196 million village dwellers in India come into this category and a survey of sixteen provinces in Indonesia indicated that 30 per cent of all villages had no link to the road network.

Many states, however, occupy an intermediate position between the two extremes of industrialised and underdeveloped, and within these states the levels of isolation and remoteness in rural areas can vary substantially. Nations in the Middle East and in Latin America, for example, contain some rural areas where personal mobility is relatively high but in other regions inaccessibility is still a severe problem.

9.2 Mobility and accessibility in rural areas

9.2.1 Mobility problems

Terms such as isolation, remoteness and inaccessibility must be used with caution, since the means of overcoming these constraints to communication vary widely with national living standards. In Australia the extensive use of air transport and radio links brings the most remote farmstead into easy communication with essential services, whereas in much of Southeast Asia even the bicycle is beyond the means of many villagers.

The 'minimum socially acceptable' levels of mobility identified in an advanced world context by Stanley (1975) would be irrelevant in most developing rural societies, given the great differences in overall living standards. Distinctions between the demands for, and provision of, rural transport in the developed and less advanced worlds must therefore be considered in terms of their contrasting economic and social backgrounds.

A general review of accessibility and mobility has already been presented in Chapter 2, and within rural societies these two issues are strongly linked to the dispersed nature of the population and with the trend for vital services to become more concentrated into larger settlements rather than being widely distributed throughout the area. In the industrialised world levels of mobility and access are closely related to the availability of public transport and to private car ownership, whereas in the developing countries the quality of the infrastructure and levels of personal and household income are often the most important factors. For example, in many African rural societies there is a regular acceptance of journeys on foot over 15 or 20 km to vital services such as schools and clinics, distances which would not be tolerated by pedestrians in the Western world. In these developing societies problems of accessibility are experienced not only in terms of linkages between settlements but also within agricultural holdings in terms of daily trips between home and fields (Box 9.1).

Agricultural improvement schemes in developing states which involve changes in settlement, land use and crop yields can also produce negative effects by creating a need for longer trips by farm workers to producing areas. Increased time spent on walking means that the period available for farming activities is decreased and the effectiveness of farm labour will be even further reduced. When subsistence farmers perceive distances between house and the more remote fields as being unduly long and time-consuming the consequence can be the abandonment of those tree crops which are the most difficult to carry as headloads. This then allows the cultivators to concentrate their efforts upon the more accessible food crops, especially where the headloading of produce to market or a collecting depot is a necessary part of the farming pattern.

The need for regular supplies of drinking water and firewood in rural areas can also involve long trips, which in parts of East Africa can take

Box 9.1 Trip-making in rural Africa

Studies in sample Kenyan villages found that 72 per cent of all trips were carried out on foot, with 80 per cent being directly associated with the daily farming routine. One-third of all journeys made in connection with the farm or the family household involve head- or back-loading and the majority of these are made daily by women and children, particularly for water collection and ranging in length between 2 and 12 km. Marketing trips are made once or several times a week, depending upon the type of crop and the market schedule, but journeys to larger towns for goods are usually only once a month. The carriage of seed and fertiliser from stores back to the farm is less frequent and can vary between 5 and 25 km in length.

In parts of rural Borno state in Nigeria each person made an average of 3400 trips per year, but 68 per cent of all households surveyed had no pack animals although bicycles were often available. Only 15 per cent of households owned motor vehicles of some kind and in many cases the wealthier families would have the use of both animal-drawn and mechanised transport.

Table 9.1
Rural transport trips by length and loads transported in selected developing countries

	Kenya	Malaysia	India	Bangladesh	Western Samoa	Republic of Korea
Typical distance of transport	90% of trips <7 km	75% of trips <7 km	90% of trips <5 km	Most trips <12 km	Most trips <5 km	Most trips 10 km
Average on-farm distance	0.8 km	1 km	1.5 km			
Average off-farm distance		10 km	8.3 km			
Loads transported	70% of trips <25 kg			Most trips <50 kg	Most trips <80 kg	30–80 kg

Source: Barwell et al (1985).

up to 4 hours each day, and in parts of Tanzania water sources can be at least 5 km from a village. Almost all these journeys undertaken to secure essential daily needs are on foot, and lengthy trips for water can best be eliminated if simple boreholes or wells are sunk closer to housing clusters (Table 9.1).

9.2.2 Rural transport deprivation

Measurement of the difficulties faced by rural societies in gaining access to essential services involves the identification of what are described as 'transport-disadvantaged' or 'mobility-deprived' groups and individuals. In rural areas the basic problems are associated with the dispersed nature of the population and the difficulties experienced in securing acceptable levels of access to services which are only available in certain settlements. In advanced societies the progressive withdrawal of shops, schools, social and medical facilities from smaller villages and their relocation in larger rural centres or small towns creates serious problems for all households without permanent or temporary access to private transport. The distinction between transport need and transport demand is particularly difficult to identify in the rural context, both in developing and advanced countries, but the success of any plan to combat mobility problems depends upon an understanding of this distinction. Rural transport deprivation is but one of several components of a wider set of social and economic problems which face rural populations and which are interlinked in terms of their origins and their possible solutions.

In advanced economies these solutions are almost always based upon the deployment of various forms of motorised transport, but in developing countries there will continue to be a substantial reliance upon hand and animals carts and motor vehicles usually play only a small part in providing access to service centres.

9.3 Rural transport in developing countries

9.3.1 Rural transport infrastructure in developing countries

General issues related to transport and economic and social development have already been discussed in Chapter 4 and in this section the particular problems of rural settlements are examined. In most developing countries the road is the principal mode available, and rail and inland waterway transport generally play a less important role within rural areas. However, throughout the developing world, it is the poor physical condition of the rural road network which is one of the principal constraints upon plans for economic expansion and the upgrading of social facilities. In the mid-1960s the road networks of most African states, apart from South Africa, had less than 5 per cent of their total length with sealed surfaces, and a large proportion of the minor feeder roads were, and still are, tracks beaten out by walkers and animal carts (Fig. 9.1).

Comparative analyses of rural roads in developing states are difficult as there are no universally accepted definitions of route status and the three general categories of seasonal road, track and footpath frequently merge. The definition of a road as 'a route having a bed above the level

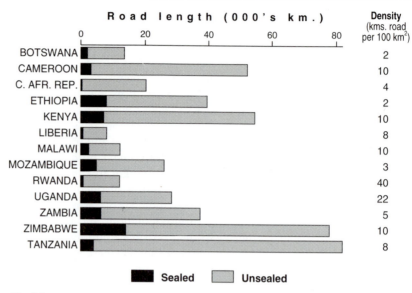

Fig. 9.1
Road densities and percentages of total road lengths that are sealed in selected developing countries in Africa. *Source:* International Road Federation: *World road statistics* (1992).

Plate 9.1
Bridge carrying gravel road over seasonal river in the Mhangoro communal farming area south of Harare. These concrete bridges are designed to cope with flood waters in the wet season between October and March and are an essential part of the plans for improved communications in rural Zimbabwe. Foreground shows part of more recent replacement structure.

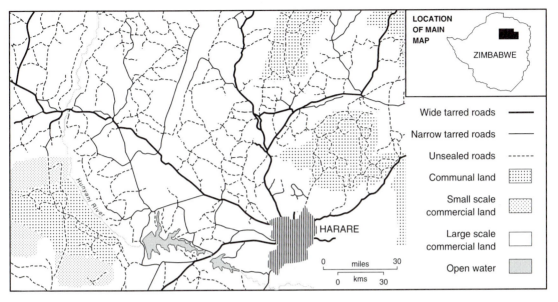

Fig. 9.2
Road categories in Mashonaland, west of Harare, Zimbabwe.

of the surrounding terrain and surfaced with gravel rather than soil' is a useful working description, and such roads carry the bulk of vehicular traffic although they may be flooded during the rainy season. A basic distinction can therefore be drawn between all-weather and dry-weather routes, and also between tracks capable of carrying wheeled vehicles and those open only to pack animals and pedestrians. A recently published Automobile Association road map of Zimbabwe categorises the poorer quality routes as being 'unsuited to use by tourists', a realistic definition in a state where international tourism is of growing importance in the economy (Fig. 9.2).

9.3.2 *Rural transport modes in developing countries*

Roads of varying quality carry the greatest share of all rural traffic, but there are wide variations in the methods of transportation adopted. Headloading is still dominant in many areas both for household and social needs and for the marketing of agricultural produce. An upper limit of about 40 kg per person and a day's maximum journey of 50 km place definite constraints upon the radius of action of head porterage, but this mode is frequently the only practicable means of carriage in areas where no routes other than rudimentary tracks are available. The continued dependence upon this primitive means of carriage makes substantial demands upon the labour resources of peasant farming communities and

Table 9.2
Trips made by farmers in
northern Nigeria

(a) Farmers resident in villages without all-weather roads

Travel time (mins)	Average numbers of annual person-trips/household				
	Trips for acquiring inputs	Trips to farms	Trips for domestic needs	Trips to markets	Recreational trips
0–5	15	31	22	17	—
5–10	9	418	116	41	1
10–15	8	500	41	17	—
15–20	10	365	13	39	2
20–25	15	—	30	20	1
25–30	5	608	77	28	3
30–60	8	365	18	44	3
60–120	—	—	—	—	3
120–180	—	—	—	—	2
Over 180	—	—	—	—	5
Total	70	2787	317	206	20
Average travel time (mins)	18	18	17	21	80

(b) Farmers resident in villages with all-weather roads

Travel time (mins)	Trips for acquiring inputs	Trips to farms	Trips for domestic needs	Trips to markets	Recreational trips
0–5	1	898	72	—	—
5–10	2	939	55	20	1
10–15	2	565	172	26	—
15–20	1	423	189	1	—
20–25	1	—	13	2	—
25–30	2	—	1	30	2
30–60	5	—	—	4	7
Over 60	—	—	—	—	—
Total	14	2825	502	83	10
Average travel time (mins)	25	8	13	19	41

Source: Barwell et al (1985).

women are often assigned to headloading, which is seen as one of their traditional responsibilities within the family structure. Time and effort which could be devoted to tending crops have of necessity to be diverted to headloading and where it is possible to replace it by hand- or animal-carts then the labour resources of the community are substantially increased (Table 9.2). It is in these rural societies of the less developed world that the transition from walking to wheeled transport such as carts or bicycles represents such a fundamental advance in their social and economic lifestyles.

The extent to which animals may be used either as beasts of burden or as tractive power for carts is determined by household income, the conditions of roads and tracks and by the level of demand for the transport of crops. Pack animals are invaluable in areas where wheeled vehicles cannot penetrate and can be integrated into the domestic economy by providing both food and manure. Animal-drawn carts can cope with loads of up to 2 tonnes but their regular use is dependent upon satisfactory routeways. It is at this stage that the distinction between footpaths, negotiable only by pedestrians and animals, and tracks accepting vehicles becomes critical, since a route open to carts can also with suitable upgrading accept the simpler forms of mechanised transport, such as a light truck. However, the progressive transition from ownership of pack animal to cart and to small motor truck depends upon available income and can present severe barriers to the small-scale farmer in remote areas.

The transformation of a rural agricultural society largely based upon head porterage and mule or oxen transport into one which can derive substantial benefits from the use of motor vehicles therefore depends not only upon improvements to the road network but also on increases in household prosperity in order that cheap motor vehicles can be acquired. In contrast to farms in the industrialised world, the tractor is seldom used in peasant holdings in developing countries because of costs and its unsuitability for small fields. Motor transport usually offers most advantages when adopted for social and marketing journeys between villages and towns.

Many of the motorised freight and public passenger services in less advanced rural areas are operated as very small-scale undertakings and one type of vehicle, such as an open lorry or the Kenyan *matutu*, frequently caters for all demands. Operating costs vary greatly with the size of vehicle, road surface quality and maintenance standards, and the penetration of reliable services into the more remote areas most in need of improved accessibility is often restricted by the reluctance of operators to run their vehicles over indifferent surfaces where increased running costs cannot be recouped from revenue. The availability of trucks in developing countries is much lower than in advanced states, India for example having only one truck per 600 persons compared with one to 10 persons in Canada or Japan.

In Africa many of the rural bus services are confined to the better all-weather roads connecting villages with towns, whereas the pressing need, as revealed by surveys in Zimbabwe, is often for the extension of bus routes into remote districts to supply village-to-village links. Extensions of this kind would not always be economic, however, and a further widespread difficulty is that bus design is often standardised and the large 100-passenger vehicles in common use in southern Africa are suited only to running on the better quality roads.

Rural settlements in states with extensive coastlines, such as Indonesia or Malaysia, or with large river systems, such as Zaïre or Brazil, can make use of water transport in conjunction, where appropriate, with roads, but such opportunities are limited. Railways are usually of most value for longer-distance freight and passenger transport and generally play little part in the rural economies.

9.3.3 Rural transport planning and policy in less developed states

Investment in transport facilities is usually just one element of a complex development programme designed to improve standards of living in these rural areas, the main objectives being to increase the quality and areal extent of commercial farming and to ensure a better distribution of educational, health and other social services. The statement in the First Five-Year National Development Plan for Zimbabwe, published in 1986, that 'Adequate roads and transport services in rural areas are a prerequisite to continued economic and social development' typifies the approach to these issues taken by newly independent states in the less developed world.

However, the improvement of roads for economic motives is often a first priority, although the same road can of course serve the needs of farmers, children walking to school and women travelling to a clinic for medical attention. These road-building and upgrading projects are essential if access to local markets, schools and clinics is to be increased and, given the high costs of road-building schemes, most plans are preceded by feasibility studies which investigate the potential advantages and benefits to be gained from road upgrading and, where appropriate, the introduction of mechanised transport.

Rural communities, however, are just one of many sectors which make an urgent call upon funds allocated at national level for transport improvements, and even when an appropriate allocation has been made there is the question of how best to distribute the investment within individual rural districts to secure the most effective return in social and economic terms. The specific transport needs of each area must be identified, followed by the selection of the most effective technology in terms of road surface standard and vehicle type. Finally the extent to which the opportunities offered by new or improved roads for the advancement of living standards are actually realised must be assessed wherever possible. Specific surveys are usually limited to those rural areas with proven or potential agricultural value which are judged most likely to benefit from investment and, given the expense of road building, the issue of cost-effectiveness is often a major criterion (Box 9.2).

A survey of the southern Indian state of Karnataka estimated that if every village was to be provided with an all-weather road the cost would be 60 per cent of the total annual state budget for all economic services. Moreover Karnataka is a state where over 90 per cent of villages already have good road links, but for India as a whole it is estimated that two-thirds of the total 600 000 villages have no adequate all-weather road access. Thus the cost of providing road connections to all isolated

> **Box 9.2 Road investment in Sierra Leone**
>
> The building of access or feeder roads is often an integral part of rural agricultural improvement programmes which also include investment in irrigation schemes, schools and clinics, telephones and power supplies. Sierra Leone began a feeder road construction project in 1975, focusing upon villages with agricultural potential for coffee and cocoa and where farmers had demonstrated a positive attitude towards improvement plans. A subsequent study of travel in these areas found that where feeder roads had been provided 50 per cent of all marketing trips were made by some form of wheeled vehicle, whereas on unimproved trails, headloading, or a combination of headloading and vehicles, accounted for 75 per cent of all journeys to market. It was also found that the laterite feeder roads had been poorly maintained and were unlikely to last for their planned lifespan. The advantages of feeder roads are only fully realised if rates for the carriage of the cocoa and coffee crops are acceptable to the producers, and in Sierra Leone those farmers nearest to feeder roads did not always derive the anticipated benefits because of high transport costs. In fact coffee- and cocoa-growing areas remote from improved roads were in some cases able to market their crops as easily as villages with feeder routes by making use of cheap family labour for headloading over distances up to 10 km.
>
> The Sierra Leone experience therefore shows that areas in which new feeder roads have been provided do not necessarily show a significant increase in commercial crop output in the short term. However, these improved routes can also serve an important social function by attracting teachers, doctors and other welfare workers into areas which previously would have been regarded as inaccessible.

settlements in the nation would be astronomic and some form of selection exercise must be made.

However, the Indian government was responsible for introducing one of the more ambitious and enterprising schemes designed to combat rural isolation by utilising modern communications technology. The satellite instructional television experiment provided 2400 remote villages with receivers linked with a transmitting station in Delhi, which sent out information on education opportunities, farming techniques, health care and weather conditions. Each of the communities selected was already provided with electricity and road access, but after the year-long experiment there was a marked increase in the awareness by inhabitants of services available, as indicated by increased schools attendance and an increase in the number of farmers seeking advice on farming. This Indian initiative does demonstrate that improvements in rural accessibility can depend as much upon the spreading of information to a community about available services as upon the upgrading of roads allowing villagers to reach them.

Other feasibility studies have concentrated upon identifying links between existing road facilities and levels of economic activity, upon measuring vehicle operating costs on different types of road and upon predicting future traffic flows in order to identify requirements for additional roads. All have yielded valuable data upon which road investment decisions can be based, but all are restricted in scope, and for vast areas of the developing world rural road planning still rests upon a very insecure basis with very little background data (Fig. 9.3).

Early studies of the effectiveness of road improvement programmes in Africa and South-east Asia were primarily concerned with improvements in vehicle operating costs and changes in agricultural output and prosperity. For Sabah a relationship was identified between the areas under export crops, standard of available roads and distance between farms and marketing centres, and a model was derived providing an estimate of the area of cash crops likely to be produced per unit length of road. The value of this pioneer study was its success in showing that the main parameters involved in rural road development could be analysed quantitatively and the results used to forecast both the likely effects of road building and the transport requirements of other areas for which development was proposed. In the central region of Ghana measurements of road density, population density and agricultural output were subjected to a regression analysis, which indicated that 84 per cent of the variation in road density was explicable by that of population and by crop output variations. In this region, characterised by extensive agriculture with low yields, much of the farmland is within 6 km of a sealed road and commercial crops such as cocoa can be easily headloaded to roadside truck collection points.

A recent pilot study to establish priorities for road upgrading in the tea-growing area of Kenya faced a situation where the tea leaf must be processed at the factory within 24 hours of harvesting. This requirement involves transport costs on the predominantly earth and gravel road network which are high but unavoidable if tea losses are to be minimised. Priorities for road improvement were then identified after taking into account the costs of both road upgrading and subsequent maintenance, the reductions which could be made in transport costs for tea and other local commercial crops and the changes in other sectors of the economy resulting from better roads.

In areas with low crop yields, which typify much of the less developed world, it is likely that additional investment in road upgrading and new feeder road construction will continue to be highly selective and confined to areas producing the more profitable and higher-yielding crops. In the Ashanti region of Ghana 98 per cent of the population are within 2 km of all-weather roads and motorable tracks, and a recent study concluded that further road improvements would have little effect upon either farm-gate or village market prices since non-transport factors such as marketing practices are of more importance in the local agricultural economy. Schemes to upgrade footpaths to accept vehicles in this region

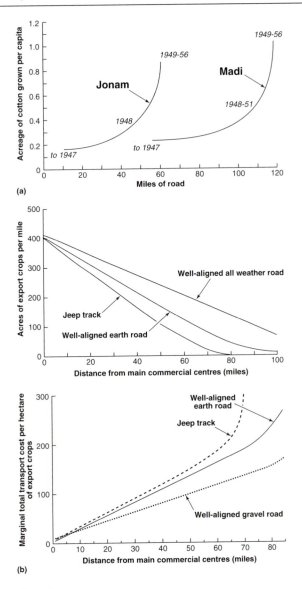

Fig. 9.3
Examples of relationships established between road building and improvement projects, changes in agricultural productivity and transport costs of export crops: (a) Uganda; (b) Sabah, Malaysia. *Source:* based on Figs 1, 3 and 4 in Millard (1967).

would, however, be effective since improvements in the field–village linkages could lead to more efficient farm practices.

It can be seen from these examples that much of the research into rural road improvement has concentrated upon measuring its effects upon local agricultural economies. In Kenya, however, the issue of rural transport needs has been investigated by examining existing relationships between land use and traffic generation to predict future trends and hence additional road requirements. In the rural districts around Mount Kenya the persons-to-vehicle ratio varies from 230:1 to 10 140:1, with one-half of all vehicles in use being private cars and 45 per cent light- and medium-capacity trucks. Only seven out of the seventeen rural zones investigated has any sealed roads and the highest proportion of road length sealed in any single zone was only 28 per cent. The number of vehicle trips generated by different land-use zones varied considerably and the restricted scope of the survey made it difficult to reach any reliable conclusions, but this analysis does reinforce the need for more detailed studies of travel patterns at the local level.

Meeting the transport needs of developing rural societies requires the appropriate recognition of the deficiencies in mobility at various scales. In many agricultural districts it appears that further rises in productivity are most likely to be achieved by improving the condition of paths and tracks, thus enabling the farmer to devote more time to actual cultivation and less to the necessary but time-consuming trips between dispersed cultivated areas. The benefits of building more schools and clinics in rural service centres will only be widespread if access from the more remote villages is improved.

Progressive upgrading of existing all-weather inter-village and village-to-town roads can only yield results if the rural population is able to afford the purchase or operation of the motor vehicles appropriate to these improved routes. It can often be most beneficial to local communities to provide entirely new tracks suited to wheeled vehicles, thus increasing accessibility to these existing roads.

9.3.4 *The intermediate technology approach*

Many transport planners now accept that what has been termed 'intermediate technology' can be the most effective approach to satisfying immediate needs for better accessibility, matching the limited financial means of isolated farming communities with the demands for more efficient but affordable means of transport. Low-cost vehicles such as improved handcarts or animal carts designed to be used on simple tracks and rudimentary roads can provide adequate facilities for the foreseeable demands of many communities where motorised transport is at present not available. Such vehicles are simple to design and build, are easy to maintain and can be constructed at low cost in local workshops, thus saving valuable foreign currency for other more advanced transport requirements. The bicycle and its adapted versions for goods carriage can be of particular value in meeting this need for unsophisticated but reliable transport. Sample studies in northern Nigeria discovered that 60 per cent of households owned bicycles and in parts of central Kenya the bicycle is the second most important mode of transport after walking.

Traffic densities in many parts of rural Africa and Asia are often insufficient to justify the extension of all-weather road networks, and the ultimate goal of securing a more prosperous rural society with higher levels of accessibility may be achieved more rapidly and in a more cost-effective manner by means of less sophisticated transport improvements. Moreover individual local authorities, rather than regional or national agencies, would be capable of initiating and operating these 'intermediate technology' improvements which could then be geared more closely to the transport needs of specific settlements. If commercial agriculture then prospered to reach a level where motorised transport for marketing was justified, a more ambitious programme of road improvement could subsequently be introduced.

In many rural areas, therefore, current investment in the short term is best directed at the lower end of the road hierarchy, converting simple tracks into serviceable roads for the benefit of local communities. Once information on the existing transport requirement has been obtained a decision can be made upon the level of transport technology most appropriate to the situation. It is inevitable that much of the required funding will need to come from external sources but local communities should be encouraged to contribute as much self-help as possible. In developing countries where agricultural output from rural areas is still a very significant component of the national economy it is important to realise that the small-scale transport systems which serve such areas are in turn an essential part of the national system of communications and require as much attention from transport planners as does inter-urban transport.

9.4 *Rural transport in developed countries*

9.4.1 *Rural transport modes in developed countries*

Mobility in the rural areas of advanced countries, in direct contrast to those of less developed regions, is largely based upon privately owned motorised transport, making use of road networks which are generally well maintained and accessible to almost all districts. During the late nineteenth century the spread of railways into rural areas of Europe encouraged the expansion of many villages, but this new transport medium was restricted in its extent and many areas remote from stations continued to rely upon horse-drawn carts for access to market towns.

It was not until the introduction of motor buses in the 1920s that a more widespread public passenger service was first provided, but walking and cycling continued as important means of personal movement. With the growth in car ownership in the 1960s the rural population enjoyed a higher level of mobility than had ever before been possible. New opportunities were presented for employment, social and leisure activities and there were substantial increases in both the number and the length of trips made in rural areas. Inevitably this expansion of car ownership and usage resulted in a decline in the patronage of rail and bus services and the remainder of this chapter is concerned with the problems of rural accessibility which affect those without private transport and the policy and planning responses which have been made to combat these difficulties.

These difficulties and the solutions which have been devised must be seen within a rural context which has been changing in many European countries as a result of counter-urbanisation. This process involves the creation of 'urban villages' where most of those employed commute to towns and cities, the establishment of many new industrial enterprises in rural locations, and an increasing movement of retired urban dwellers into the countryside.

9.4.2 *Components of the rural transport problem*

In this section an attempt is made to identify the transport difficulties which rural populations face in regions as diverse as the western USA, interior Australia and the UK, where different densities and settlement patterns produce different travel needs, which in turn create a variety of mobility and accessibility problems. In regions of the sparsest population, such as the interior of Australia, the widely scattered agricultural stations are situated well beyond the range of any urban centre or any public transport service and almost all travel must be organised on a personal basis. Public services such as mail delivery and collection, education and medical aid are provided at infrequent intervals and often rely upon air transport, such as the Flying Doctor facility (Box 9.3 and Fig. 9.4). Within these remote interior regions of Australia the required levels of daily communication and personal mobility are therefore maintained by a combination of motor transport, light aircraft, the telephone and radio transmissions.

Much of the USA's legislation relating to rural transport has been focused upon the 'transportation disadvantaged' such as the young, the handicapped, the elderly and infirm and the poor. At the federal level legislation has often been in the enabling category and the areas where transport difficulties have been particularly severe have not always received the level of assistance necessary to maintain adequate mobility standards. Social programmes have included what are termed 'human-service' transport systems for the elderly, but most have been concentrated in rural areas of eastern states and the problems of the more widely dispersed population further to the west have been neglected, partly because of the higher costs of operating transport services.

Box 9.3 Airborne medical services in Australia

The extensive rural areas of Western Australia and Queensland are served by a pattern of health services, ranging from large hospitals with all the necessary ancillary services situated in towns with over 10 000 persons to doctors with well-equipped surgeries and nurses in the smaller settlements. These medical specialists will have only a limited radius of action and the Flying Doctor Service was established to provide emergency cover for the more isolated farming and mining communities where land-based access can often be cut by flooding or where the high speed of air transport is essential to transfer the patients to hospital. The service depends upon radio links between the medical staff and patients, and its staff and aircraft operate from twelve bases, the Northern Territory for example being covered by bases at Alice Springs, Wyndham in Western Australia and Mount Isa in Queensland. Each base serves an area of several thousand square kilometres and operational expenses costs are high; wherever possible the service is linked with surface ambulances to minimise costs.

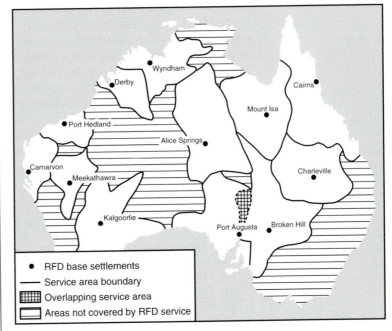

Fig. 9.4
Airborne health services in rural Australia: the Royal Flying Doctor network. *Source:* Fig. 15.5 in Brownlea and McDonald (1981).

Within the UK the contemporary problems can be traced to the increase in car ownership in the 1960s, the consequent decline of bus and rail services and the exposure of several groups within rural communities to transport deprivation. What Dobbs (1979) has described as 'the economic stranglehold' on public passenger transport has two components: the first arises from economic and political forces producing changes in the rural UK which are beyond the immediate control of bus or rail management; the second relates to how the surviving bus and rail services can offer acceptable levels of mobility to a population which is usually unevenly dispersed and whose travel needs vary widely within the day.

Until the mid-1970s, when the process of bus route closures was well under way, only a limited amount of information on their effects upon rural lifestyles had been gathered, but it was clear that many specific categories of difficulty had been created, both for the consumers and for the diminishing number of bus operators that had survived. Bus service withdrawals will affect former users according to their socio-economic status, and the frequency and purposes of the journeys that had previously been made by bus. The adjustments made in travel behaviour and general daily activity patterns by rural dwellers following withdrawal of bus services need to be assessed and those groups of people and sets of activities most strongly affected need to be identified. The perception by rural dwellers of an acceptable and desirable level of mobility and their concept of unnecessary deprivation vary between different groups and areas (Box 9.4). Their reactions to the introduction of alternative means of transport to alleviate the effects of bus service closures may also differ. In the UK the period of declining bus usage has coincided with a change in the distribution of rural services, with a progressive concentration of retail outlets, surgeries and health clinics into fewer centres, thus lengthening the journeys which have to be made to reach these facilities.

The relationship between changing rates of car usage and reductions in public passenger transport is difficult to measure and will vary with living standards. In Sweden it has been estimated that when the ratio of

---

Box 9.4 The National Travel Survey in the UK

The changes in travel patterns which have resulted from recent social trends can be seen in data from the National Travel Survey. Within the UK car ownership levels are higher in rural than in urban districts and one-third of rural households have access to two or more vehicles compared to only 16 per cent in the towns. Although the number of journeys made by each person per week does not differ greatly between the urban and rural UK, distances travelled in the countryside are 40 per cent longer on average than urban trips. This differential is often greatest for schoolchildren as a result of the much larger school catchments which exist in rural areas. Walking is of less importance in the countryside and 85 per cent of the average distance covered per week is by car, compared with 76 per cent in towns and cities. In most areas the rural motorist also encounters more agreeable driving conditions than his urban counterpart, with less congested roads and less restrictions on parking at his destination.

persons to car falls to below 7 then public transport patronage begins to decline, and in the UK every extra car which is available can produce a loss of 300 annual bus journeys, with a corresponding decline in frequency as revenue is reduced. Between 1962 and 1982 the use of buses in the UK fell by 50 per cent and in rural areas today only 8 per cent of all personal trips are by bus compared with about 40 per cent in the early 1950s.

Many factors have contributed to the declining appeal of public passenger transport nationally, but most of them are associated with rising car ownership, and in rural districts, where bus services have always been less accessible than in the town, their influence is even stronger. Rural bus operators initially responded to falling demand by cutting frequencies, but in many areas where not even a daily bus had been available entire services were withdrawn. Small companies were often more vulnerable and more likely to collapse, and larger companies would counter declining patronage by rationalisation, concentration upon the more lucrative inter-urban routes and withdrawing vehicles from the more remote villages. However, the running costs of these larger undertakings are often higher than those of small independents, and at times when patronage is declining it is often possible for the latter to survive on reduced revenues for longer periods whereas the larger operator would cut back services at an earlier stage. The problem of declining rural bus revenue has also been compounded by falling receipts on urban services, since the profits formerly made on the latter compensated to some extent for the loss-making or marginal country routes (Fig. 9.5).

The use made of rural bus and rail facilities, and the extent to which the private car is used, vary with area and social status, and as a result the withdrawal of bus services adversely affects certain groups within the population more severely than others. In the UK transport for pupils living over a specified distance from rural schools is provided by bus companies under contract to the education authorities, so access for this group is guaranteed. Workers living in rural areas but employed in towns are more likely to have the use of a car for their journey to work but, where only one vehicle is available in a household, this leaves the remainder of the family without personal transport throughout the working day. These people, together with the retired, the unemployed and the young below 18 are the groups most likely to suffer from reductions in bus facilities (see Ch. 11). In the case of the elderly the problem is particularly acute for the many former urban dwellers who have moved to rural areas on retirement, where eventually illness, disability or death of a partner has deprived them of use of the car; in many parts of the more remote areas of the UK one-quarter of the population comes into this category.

Although rural bus services have in the past provided a useful link between village and town the essential daily requirements of life could usually be met by locally based facilities such as retail stores, post offices,

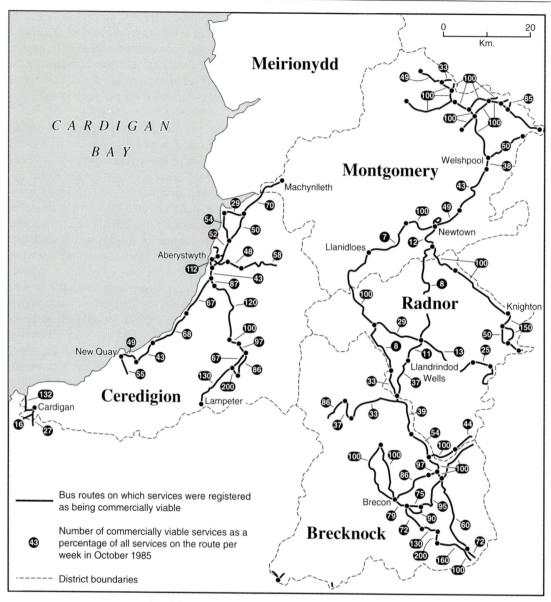

Fig. 9.5
Rural bus services in central Wales, United Kingdom. *Source:* Moyes (1988).

schools and doctors. With declining village populations many of these services have been rationalised and are now available only in the larger rural centres or in towns. It is this process, combined with fewer bus facilities, that has increased the difficulties of securing reasonable access to essential day-to-day needs for all rural dwellers. The existence of a regular bus, putting the nearest town within easy reach, enhanced the range of facilities which existed on a limited scale in the village, but the centralisation of many of these key facilities combined with bus withdrawals has left the smaller communities without any local services or the means of reaching the towns where they are now available. Thus the widespread closures of schools, surgeries, shops and post offices have obliged rural dwellers to undertake longer journeys to reach these vital services.

The problems created in rural communities when public transport services are reduced or withdrawn may be measured by studying trips made in terms of numbers, purpose and length. In some cases the overall total of journeys generated is reduced and in others the same trips continue to be made, but under more difficult conditions. Monitoring this process of adjustment by rural populations to the declining opportunities for travel, other than by car, is an essential exercise if appropriate solutions are to be applied. However, the increasing isolation which is experienced by the transport-disadvantaged elements in rural communities, particularly in respect of access to medical and social services, can also induce personal stress which will affect the overall lifestyle.

Any attempts to redress the balance and provide those without any regular access to personal transport with an acceptable alternative must therefore consider both the needs of village communities as a whole and those of specific groups within them. It is a paradox that during a period when levels of car ownership have risen for the majority of the rural population, accessibility for many groups has continued to decline. To achieve maximum effect, therefore, all policies directed towards the improvement of rural transport in general must take into account this selective pattern of deprivation. The inevitable conclusion of many analyses of rural transport problems is that the car offers the ideal solution but that there will always be those without access to vehicles. Given this situation, the wide range of remedies that have been offered as compromise solutions must now be considered.

9.4.3 *Policy and planning responses*

As in the less developed world, rural transport planning should ideally be seen as part of a wider and more comprehensive programme designed to combat the overall economic and social problems of dispersed communities. In practice, however, these problems are rarely tackled in a co-ordinated way and the efforts which have been made to improve rural accessibility are usually the responsibility of a wide range of different organisations. In the UK the Jack Report on rural buses published in 1961 recognised that their survival would require subsidies to support increasingly uneconomic services if existing levels of mobility were to be maintained. However, as the difficulties experienced by the non-car-owning elements of the rural population became more acute it was realised that effective solutions would need to take into account all the basic needs of village communities and the various means by which they could best be met.

Surveys of six selected rural areas in the UK in 1963 were followed in 1977 by what were termed rural transport experiments (RUTEX) to discover to what extent changes in transport legislation and policy had altered the situation. The traditional approach of ensuring adequate bus links between village and town became just one of several possible solutions involving a redeployment of transport modes, services and possibly people themselves, and rural transport planning needs to address the following broad objectives:

1. planning to improve the mobility of the individual;
2. increasing the mobility of essential services;
3. increasing the level of accessibility to services;
4. a possible movement of certain rural dwellers into villages with a greater range of services;
5. temporal solutions.

Each of these approaches is now discussed in more detail with examples.

Planning to improve the mobility of the individual

This issue has probably attracted more attention than any of the other categories, as it is based upon the adaptation and redeployment of existing means of private and public transport in order to secure a more equitable distribution of opportunities for access to services for all groups within rural communities. Under the 1968 Transport Act county councils in the UK were empowered but not obliged to subsidise those loss-making bus services where at least 50 per cent of costs were still being recouped from fares and where retention of the service was seen to be in the interests of the local community. However, the spiral of decline continued into the following decades with the inevitable fare increases producing some extra revenue but at the expense of further falls in bus passenger traffic.

Over 50 per cent of all passengers on scheduled services are carried in the two daily peak periods and the problem of minimal loadings at other times could only be solved by changes in operational practices. Undertakings within the former National Bus Company rationalised many rural routes after a market analysis exercise but the passenger market continued its decline. From 1972 county councils were required to produce annual transport policy and programme documents in which their plans for rural public transport support were detailed and under the 1978 Transport Act these councils began preparation of public transport plans with the aims of recasting bus services to serve public demand more effectively.

Deregulation of most bus services in the UK in 1986, and privatisation and dismantling of the National Bus Company, was intended to stimulate competition and lower fares but it also involved the progressive phasing out of state subsidies. Between 1987 and 1990 the grant to rural bus undertakings was cut from 6p to 1p per vehicle-mile and it is unlikely that many of the present rural services will survive as financial support is finally eliminated. However, many local authorities still see the maintenance of selected conventional bus services as desirable and in central Wales, for example, where much of the population is concentrated in small towns, it is thought that a daily connecting service run by either a private company or the local council would be viable and form part of a basic network of public passenger transport.

In the USA bus services in the remoter rural areas have only survived where they form a part of inter-urban routes, but by the mid-1970s only about 40 per cent of all rural centres with populations of between 2500 and 10 000 still had bus links. Most of these small communities are in the east and public transport facilities in the rural western states are negligible. Both the 1974 Mass Transportation Assistance Act and the 1978 Surface Transportation Assistance Act authorised cash payments to support rural buses and over 100 federal support programmes had been established by 1976. However, during the 1980s the non-metropolitan areas received only 3 per cent of this federal aid although they contained one-quarter of the national population.

The impact of these programmes in the USA has been proportionally much less than in the UK because of the distances involved and the high operating costs, and even where subsidised routes are still within reach of rural communities the take-up rate has been very low. As in the UK, members of American rural households travel greater distances each year than urban dwellers, both for journeys to work and social and leisure purposes, and the rural population spends a higher share of its household budget on transport than that in metropolitan America. The greatest difficulties in terms of lack of adequate public transport are experienced in the southern and east-central states, and West Virginia has been singled out as a region with exceptional problems because of its above average proportion of retired persons, its low level of car ownership and the fact that public transport is used by less than 2 per cent of the population.

The inability of conventional scheduled buses to meet the needs of rural societies has prompted the introduction of a range of multi-purpose vehicles, which are often run as demand-actuated services. In Alpine Europe the postbus, carrying both passengers and mail, has been an essential and popular feature for many years but its use in the UK dates only from the early 1970s. Although it appears logical to make use of spare capacity in a mail vehicle for passengers there are several difficulties involved. Mail delivery and collection schedules do not often coincide with timings of journeys to work or shopping trips, and in particular mail distribution patterns are the reverse of those associated with village-to-town trips. Routes are circuitous and average speeds low, and only those passengers for whom journey time is not critical are attracted to the postbus. The spread of such services in the UK by the Post Office was encouraged by the receipt of vehicle grants, fuel rebates

Plate 9.2
Postbuses offer a partial solution to the rural transport problem in isolated districts of north-west Scotland. At the station of Achnasheen, on the railway linking Inverness with the Kyle of Lochalsh, mail is tranferred to the postbus serving coastal settlements in Wester Ross, and tourists, especially campers, also make use of the service in summer.

and some local authority support but these were later cut back or withdrawn. Most routes are between 10 and 15 km in length but in the remote districts of the Scottish highlands some are over 50 km long. By the late 1970s one-half of these postbus services provided for two daily return trips between town and village, with the elderly forming a large part of the total market. Fares are similar to those on conventional buses and about 60 per cent of all routes were supported initially by local authorities. After the mid-1970s the rate of expansion slowed down and growth became concentrated in Scotland, where it was estimated that one-third of all rural postal routes had potential for conversion to postbus services. Conversions only took place, however, where the service could be finally guaranteed and where no competition was offered to existing bus operators, a constraint removed by the 1985 Transport Act.

In Australia the remote areas of Queensland and New South Wales contain many small towns separated by distances of between 100 and 300 km and the weekly mail van carries passengers and goods to outlying farms in the absence of any other means of public transport. Despite the drawbacks of these multi-purpose vehicles they do offer valuable services in areas where alternative means of transport have been withdrawn or have never existed in the first place.

In the UK the 1968 Transport Act also allowed school buses to carry

Table 9.3
A classification of unconventional modes of rural transport in Britain, indicating the principal unconventional features

	Mode	Restricted eligibility	Flexible routeing and timing (or D/R)	Multi-purpose transport (vehicles or operations)	Community management and/or operation	Alternative finance	New technology
	D/R diversion (Flexibus)	—	✓	—	—	—	—
	Multiple service bus	—	◆	✓	—	—	—
	Contract bus	—	—	—	◆	✓	—
	Subscription bus	—	—	—	✓	✓	—
	Free shoppers' bus	—	—	—	—	✓	—
	School bus	—	—	✓	—	—	—
	Postbus	—	—	✓	—	—	—
CT	Community bus	—	✓	✓	✓	◆	—
	PSV dial-a-ride	—	✓	—	—	—	◆
CT	'Welfare' dial-a-bus	✓	✓	—	◆	◆	◆
CT	Social car	◆	✓	—	✓	◆	—
CT	Hospital car	✓	✓	—	◆	◆	—
CT	Lift-giving scheme	—	✓	—	✓	◆	—
CT	Car pooling	—	✓	—	✓	◆	—
	Shared taxi/hire car	—	✓	◆	—	—	—
	Passenger/freight service (courier)	—	◆	✓	—	—	—
	Demountable vehicle	—	—	✓	—	—	◆
	Railbus	—	—	—	—	—	✓

D/R = Demand response service.
CT = Community transport mode.
PSV = Public service vehicle.
✓ = Primary unconventional feature.
◆ = Secondary unconventional feature.
Source: Nutley (1988a).

other passengers where practicable, but here again the timings and directions taken by these vehicles do not always suit the needs of other travellers. Employees in the rural UK are also carried to work in contract buses, and surveys have shown that these vehicles and school buses can together account for as many trips as conventional scheduled buses. Some supermarkets also provide free minibus transport for shoppers on a regular basis, but only a few of these initiatives are located in rural areas because of the distances involved (Table 9.3).

In the UK during the 1980s attempts were made to improve individual mobility by enabling car owners in rural areas to make their vehicles more easily available to those otherwise dependent upon public transport. Legal restrictions on payment for car lifts were relaxed and a series of locally co-ordinated services were set up for non-car owners to take advantage of private transport. This community transport took several forms, including 'hospital cars' run by volunteer drivers for conveying outpatients to clinics, and 'social cars' for journeys for which no other transport was available at the time of travel. Cars could also be shared on a privately arranged basis, either for occasional journeys or for regular scheduled lifts such as trips to work or urban shopping centres. By the mid-1980s the use of social cars had spread throughout most of England and Wales but had found little application in Scotland. In some counties, such as Dyfed and Hampshire, almost the entire area was covered by community schemes but elsewhere social car provision was much more limited in extent, focusing upon specific centres (Fig. 9.6).

Locally organised community transport also includes the use of small-capacity buses to meet specific demands of groups of villages. Whereas before deregulation the scheduled bus was restricted in its operation by its licence, the community buses are able to offer a flexible combination of timetabled services, excursions and demand-response journeys, the exact nature of the pattern of trips being determined by the requirements of the community responsible for the vehicle. One vehicle could be organised to fulfil different needs on each day of the week, such as market services, shopping trips and surgery visits, covering an area encompassing a large group of villages and two or three market towns. Some vehicles are designed to serve specific groups within rural communities, such as the elderly and the disabled, carrying them to essential destinations such as shops, post offices, clinics and social clubs.

The use of these community buses in the UK is not as widespread as the social car network, but the numbers of persons served is probably greater and by 1990 at least 200 vehicles were in service. Although these vehicles are operated on a voluntary and non-profit-making basis county councils have met some of the costs of establishing services, at the same time ensuring that they do not compete with other forms of public transport. Limited cash support is also available from the Rural Transport Development Fund and some county councils have appointed transport

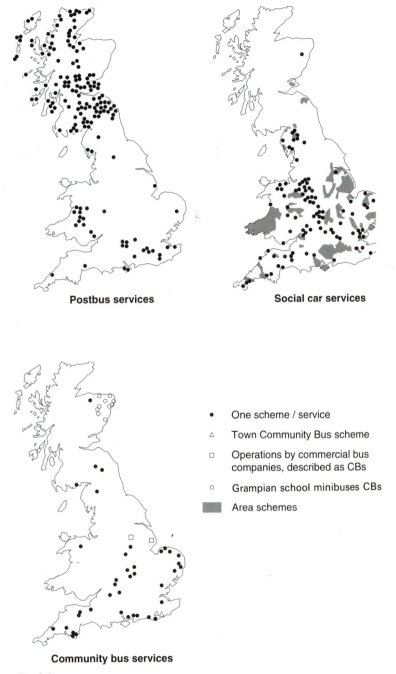

Postbus services

Social car services

Community bus services

● One scheme / service

△ Town Community Bus scheme

□ Operations by commercial bus companies, described as CBs

○ Grampian school minibuses CBs

▨ Area schemes

Fig. 9.6
Postbus services, social car facilities and community bus services in the United Kingdom.
Source: based on Figs 1.5, 1.6 and 1.7 in Nutley (1988b).

Box 9.5 Rural settlement and transport in Strathclyde, Scotland

Although the Scottish region of Strathclyde is dominated by the Glasgow conurbation it also contains extensive rural areas where small settlements experience severe difficulties related to inaccessibility. In an attempt to approach the problem the Greater Glasgow Passenger Transport Executive measured the balance between transport 'supply' and 'demand' in each village, using two indices. The first used current bus timetables to determine the length of time that village dwellers could spend in local towns and the second estimated demand levels by examining village population structures, car ownership and the extent to which local needs could be met within the village community. Each settlement was then assigned a supply/   ▶

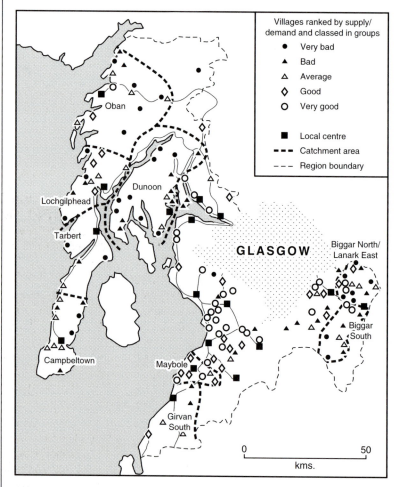

Fig. 9.7
Rural accessibility and transport catchment areas in the Strathclyde region, Scotland.
Source: based on Fig. 16.6 in Pacione (1983).

demand score which indicated the extent to which public transport satisfied existing needs for access to service towns. The resultant map (Fig. 9.7) indicates the distribution of individual settlements ranked by this score and shows that access presented the greatest problems in Lanark and Argyll to the north-west of Glasgow.

The Strathclyde region was then divided into catchment areas, each with a transport co-ordinator whose responsibility was to develop local bus networks. The subsidies for these services were calculated on the basis of the total demand score for each catchment expressed as a percentage of the demand for the entire region. This approach does recognise the variations in accessibility within the region and allows subsidies towards public transport costs to be allocated according to need.

brokers to co-ordinate and develop independent rural transport schemes (Box 9.5 and Fig. 9.7). Extension of these practices, however, depends upon the level of community initiative in villages, particularly with regard to finding volunteer drivers and organisers who must identify the type and size of the demand.

Although the 'social car' and the community bus have helped to combat some of the transport problems caused by the withdrawal of scheduled buses, the extent of their influence is limited and neither type of service can rival the convenience of the family car as a satisfactory means of securing adequate mobility.

Increasing the mobility of essential services
Within several European countries both the private sector, as represented by mobile shops, and the public sector, for example the mobile library, have for many years provided 'services-on-wheels' for rural communities. Mobile banks and dental surgeries are more recent innovations, and other specialist services have been provided in some areas, with one vehicle capable of providing a large group of key villages with a regular weekly visit, the exact number depending upon the distances involved and the level of investment available. Existing services could in future be co-ordinated to ensure that each community in turn became the focus of several of these services, so that the hinterland population need make only one journey into the centre to take advantage of a range of facilities. If necessary a small-capacity bus could operate in conjunction with this mobile or 'periodic market', collecting up passengers from the more isolated settlements and carrying them to and from the key village. The attraction of this scheme is that each village is guaranteed the regular availability of a group of services and that the need for public transport is minimised. However, the success of this approach depends upon the viability of the rural market, and there may be a need for some form of subsidy to encourage private enterprises to participate in the scheme. It is possible that the subsidies which have been applied to ensure the survival

of conventional forms of rural transport such as the bus could be more effective if redirected towards mobile services.

Increasing the level of accessibility to services

Much progress has been made in improving individual mobility and in encouraging the mobility of individual services in rural areas. The related objectives of increasing levels of access to essential facilities and of reorganising the entire pattern of settlement to focus upon 'key villages' involve more fundamental changes in the accepted lifestyle of rural communities, and realising these aims may be difficult and controversial.

Where the size and distribution of the rural population favour the continuation of certain shops, social and medical services in larger settlements then the viability of these facilities can be strengthened by improving levels of accessibility from smaller villages to these key centres, thus reducing the need for lengthy journeys into towns for daily or weekly essentials. The success of this approach depends upon a large measure of co-operation between many disparate interests ranging from the village storekeeper-postmaster to the doctor, and the local health and education authorities, who all have to consider the local and broader implications of maintaining their presence in villages where the market is often marginal. Such key village policies also involve the concept of a new form of subsidy payable to the shopkeeper to keep a business going rather than the more commonly applied subsidy to a bus operator.

Redistribution of the rural population

The chances of survival of essential services in remote areas can be strengthened if the threshold of each facility is increased by concentrating the population within easy access of key settlements. This implies a redeployment of the existing population, and although local planning authorities can achieve some measure of control over the location of new housing schemes it would be difficult to encourage this movement of people into the hinterland of larger centres despite the reduced costs of service provision which would result. Of all the policy alternatives which have been considered this would be the most controversial and the most difficult to implement.

Temporal solutions

Several of the proposals to improve accessibility within rural areas which have been discussed involve ideas which collectively have been termed 'non-transport solutions'. These are based upon placing essential services within easier reach of the rural-based population without increasing travelling activities and preferably by reducing them. Problems often arise because the timing of bus services, working periods and shop-opening periods rarely coincide; if a rural dweller working in the town had more opportunity to shop and make use of medical or social services after completing work and before leaving for home then many of the difficulties in gaining access to such facilities would be solved. These temporal solutions depend upon late opening of shops, council offices and similar services, combined with flexitime working, but they would ensure that rurally based urban workers had access to facilities which would otherwise be available to them only on a Saturday or possibly during their daytime work breaks. The principle could also be applied to services located in larger key villages. The wider issues of this 'time-window' approach have been explored by Nutley (1990), but although the idea has its attractions it is not thought to be cost-effective when compared with other solutions to the rural transport problem. In the wider sense all approaches which involve bringing services back into rural locations, either on a permanent or temporary (e.g. mobile shops) basis can be seen as 'non-transport' solutions since they would reduce effort spent on travelling.

9.5 Concluding summary

Problems of mobility in rural societies involve two related issues. The first is based upon the difficulties associated with personal travel together with the feature of rural deprivation in the wider sense. Secondly, these travel difficulties must be evaluated with greater precision in terms of the measurement of need, which in turn raises the question of establishing minimum levels of mobility for the various groups within rural society. In the developed world attempts to solve the problems raised by steadily declining public transport and the consequent emergence of a set of 'transport-poor' groups have taken many forms, and the difficulties caused by low levels of accessibility may not always necessarily be due to a lack of adequate transport.

A transport policy is just one part of the wider programme necessary to ensure the survival of remote rural communities, and action taken to improve transport facilities must be co-ordinated with the objectives of this overall programme. This is particularly relevant when proposing schemes designed to improve accessibility to services, and the decision on whether to take people to the services or vice versa is critical here. In Europe, North America and Japan there are many households in high-income groups who have been attracted to life in rural areas and who have the resources, such as multiple car ownership, to overcome all the disadvantages of remoteness. In particular the introduction of information technology such as the fax machine enables them to carry out much of their business commitments from their home base.

Rural transport problems in developing countries still need to be appraised and tackled at a much more basic level than in advanced economies. The creation of a network of adequate roads is often a first priority and the subsequent decision as to which means of transport is the most appropriate is determined to a large extent by the resources of the local community and by household incomes. Non-mechanised travel will continue to play a large part in rural economic and social life within Africa and South-east Asia, and the spread of the motor vehicle into these regions is dependent upon the success of efforts made to improve living standards and family income levels.

Further reading

Banister D 1980 *Transport mobility and deprivation in inter-urban areas* Saxon House, Farnborough

Banister D 1989 *The reality of the rural transport problem* Rees Jeffreys Discussion Paper 1, Transport Studies Unit, Oxford

Barwell I, Edmonds G A, Hoare J D, de Veen J 1985 *Rural transport in developing countries* Intermediate Technology Publications

Cloke P (ed) 1985 *Rural accessibility and mobility* Institute of British Geographers

Lonsdale R E, Holmes J H (eds) 1982 *Settlement systems in sparsely-populated regions: the United States and Australia* Pergamon

Moseley M 1979 *Accessibility – the rural challenge* Methuen

Moseley M, Packman J 1983 *Mobile services in rural areas* University of East Anglia

Nutley S D 1990 *Unconventional and community transport in the United Kingdom* Gordon and Breach

Owen W 1987 *Transportation and world development* Johns Hopkins

Yerrell J S (ed) 1981 *Transport research for social and economic progress* Gower

Part Four

Implications, impacts and policies

$I0$ The environmental effects of transport

Energy, pollution and sustainability . . .

The trouble with most forms of transport . . . is basically one of them not being worth all the bother. On Earth – when there had been an Earth, before it was demolished to make way for a new hyperspace by-pass – the problem had been with cars. The disadvantages involved in pulling lots of black sticky slime from out of the ground where it had been safely hidden out of harm's way, turning it into tar to cover the land with, smoke to fill the air with and pouring the rest into the sea, all seemed to outweigh the advantages of being able to get more quickly from one place to another – particularly when the place you arrived at had probably become, as a result of this, very similar to the place you had left; i.e. covered with tar, full of smoke and short of fish. (Douglas Adams, 1980)

This chapter will throw light on the trends lampooned by Adams. We begin by examining the atmospheric pollution from transport and discuss its relationship with impending global crises such as the 'greenhouse effect'. The associated problems of energy consumption are debated and ways of reducing impacts are outlined. Other side-effects of transport on land-use and ecology, noise and damage to the visual environment are examined in turn and potential methods of mitigation discussed. The chapter concludes with an examination of ways in which environmental impacts can be measured, assessed and evaluated alongside the other costs and benefits of transport development.

10.1 Introduction

The phrase 'energy crisis' dates from the early 1970s, when rapid price rises for oil raised concerns of security of supply and the finite nature of reserves. The long-term spectre of fossil energy exhaustion has emphasised the strategic importance of oil reserves, many of which are located in politically unstable areas of the world. Indeed in 1990 the conflict in the Persian Gulf was begun in part due to territorial disputes over oil resources and was pursued by the Allied powers with a determination that reflected their dependence on oil from the Middle East.

However, a new energy crisis is developing in the 1990s related to the environmental cost of energy supply and use. The burning of fossil fuels is now recognised to cause air pollution problems at all scales, from local 'smogs', through continent-wide acid rain deposition, to global problems such as the 'greenhouse effect' The volume of greenhouse gases, which retain heat in the atmosphere, is now growing at an unprecedented rate, leading to predictions of climate change, coastal flooding, ecosystem effects and impacts on the security of food supplies and thus on human habitation.

The new energy crisis is different from the old one because the solutions are much more difficult. It is no longer a case simply of using less oil, but instead of using less of any fuel which contains large amounts of carbon. For example, the battery car charged by electricity from coal-fired power stations may have been a solution to the earlier crisis. Now, though, it is not, because the power stations will produce larger amounts of greenhouse gases than using petrol in cars. The energy crisis has shifted from whether there will be enough fuel in the world to feed into cars to whether the global environment can cope with what comes out of the back-ends of cars.

We now outline this transport–energy–pollution system, before examining and evaluating the other major side-effects of transport – noise, vibration, land use, ecological impacts and visual intrusion. As energy consumption lies at the heart of the system, it provides an appropriate starting-point.

10.2 *Inputs: energy consumption*

Transport requires energy. In the UK in 1988, £16 billion was spent on transport energy, £11 billion on domestic uses and £6 billion on industrial, so that 42 per cent of the UK's energy bill was paid by the transport sector. These sums were spent overwhelmingly on petroleum products – over 500 000 Gwh, compared to only 3000 on electricity, principally for rail use. Petroleum thus accounts for 99 per cent of all the energy used by the transport sector.

However, these figures refer only to vehicle operation, what we might call direct energy consumption. For every vehicle produced, energy is consumed in making steel, glass, rubber and other components, whilst for

Plate 10.1
The movement of fuel for transport is itself an energy-, resource- and land-intensive process. Final delivery is usually by road, but long-distance movement is more often by pipeline or ocean-going tankers which have their own – often catastrophic – environmental impacts. Interim storage sterilises large areas of land as seen here in the Maasvlakte area of Rotterdam. This port's concentration on oil import, storage and redistribution has led it to sprawl for more than 50 km alongside the New Maas waterway, a major use of land given the Netherlands' dense population and need for space for living and recreation.

every kilometre driven, more energy is required for infrastructure in the form of bitumen, stone, concrete and so on. Moreover, producing the energy – whether it be petrol for cars or electricity to run the assembly lines and blast furnaces – itself is an energy-intensive process. So important are these 'indirect' forms of energy consumption that they make up a further 50 per cent of road transport energy use over and above that for direct vehicle operation. This relationship is illustrated in Fig. 10.1. More transport produces greater energy consumption, both directly in the operation of the vehicles themselves (66 per cent) and indirectly in provision for them (34 per cent) and each of these sectors produces pollution. Increases in demand for transport will inevitably result in rising energy consumption and more pollution emission throughout the system.

An examination of trends in energy consumption demonstrates the growing importance of transport. Though total energy consumption, world-wide, has been relatively stable since 1973, consumption in the transport sector increased by 23 per cent between 1973 and 1987, including a 27 per cent increase in road transport alone. These trends are mirrored in the UK, where land transport is the fastest growing sector of energy consumption. It may be helpful to examine the components of these shifts.

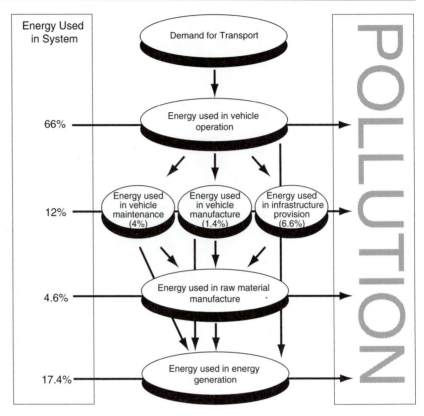

Fig. 10.1
Pollution and the transport energy system. *Source:* Transnet (1990).

Firstly, it is important to be aware of the very significant differences in fuel efficiencies between the various modes. Table 10.1 shows the relative efficiencies of transport modes in London, illustrating that the car is more than ten times as energy-hungry as the most efficient mode, the underground train.

Secondly, there has been a shift from higher efficiency to lower efficiency modes, for in the past 40 years in the UK, car passenger-kilometres have increased nearly tenfold, while rail use has remained static (Table 10.2). Fuel-efficient buses have lost half of their market. A similar tale can be told about freight transport, where rail's tonne-kilometres have halved in the same period that road haulage has quadrupled. Thus the energy needed to move each tonne-kilometre has risen by two-thirds in the 10 years from 1978.

Thirdly, there has been a significant improvement in the fuel efficiency of individual cars, by about 13 per cent in the UK from 1978 to 1988. However, this has been counteracted by the trend towards larger-engined models and towards greater speeds, which increase fuel consumption. It

Table 10.1
Relative efficiencies of
transport modes in London

Passengers (Mega-joules per passenger-km)		Freight (Mega-joules per tonne-km)	
Taxi	5.00	Light goods (petrol)	8.35
Car (central London)	3.73	Light goods (diesel)	4.64
Car (suburbs)	3.15	Medium goods vehicles	1.7–2.3
Motor cycle	2.50	Heavy goods vehicles	1.2–2.8
Bus	0.90	Rail (bulk loads)	0.4–1.2
BR train	0.84	Rail (general merchandise)	0.5–1.7
Underground train	0.32		

Source: Transnet (1990).

Table 10.2
Intermodal shift in the UK,
1952–88

	1952	1964	1976	1988
Passenger-kilometres (billion)				
Cars and taxis	54	204	339	517
Motor cycles	n/a	8	7	6
Buses and coaches	81	62	53	41
Cycles	23	8	5	5
Total road	158	282	404	569
Rail	39	37	33	41
Total land	197	319	437	610
Freight tonne-kilometres (billion)				
Road	31	66	96	125
Rail	37	26	21	18
Total land	68	92	117	143

Source: Transport statistics Great Britain, 1976–88.

is also undermined by the growth in car ownership, which means that the average occupancy per car falls (from 1.91 in 1965 to 1.64 in 1990 in the UK) and thus fuel used per passenger-kilometre rises.

To summarise, it is clear that changing patterns of transport in the UK have resulted in large increases in fuel directly consumed: transport energy consumption rose by 20 per cent from 1984 to 1988, despite a fall in that used by rail. In turn, the logic of Fig. 10.1 implies a further rise in indirect energy consumption, used to expand and maintain the transport system. The energy saving from the use of more fuel-efficient cars has been obliterated by increases in their numbers and use. If the traffic increases of up 142 per cent predicted for the year 2025 materialise, energy consumption will continue to increase substantially, even if unheard-of improvements in fuel economy are achieved.

10.3 Outputs: air pollution

10.3.1 Introduction

For years air pollution from domestic and industrial sources has been falling in economically advanced societies, often under the impact of public health programmes such as the Clean Air Act in the UK. However, the growth of air pollution from motor vehicles has grown insidiously, unregulated and for the most part unmonitored. The impression that vehicle emissions were not a matter for concern was fostered by the lack of surveys: obviously nothing can be found if nothing is looked for. That complacency was shattered in Scotland in 1991 when it was reported that thirteen of fourteen sites monitored in Edinburgh had nitrogen dioxide levels that exceeded EU guidelines.

It is now becoming more widely understood that transport is a major source of air pollution and in many developed countries is the fastest growing source. The number of pollutants involved is very large indeed, with some 400 compounds emitted by petrol and diesel vehicles and from petrol vapour. The main pollutants are outlined below, together with an indication of their health and environmental impacts and the trends of emission in one industrialised country, France.

1. *Carbon dioxide* (CO_2), a colourless, odourless naturally occurring gas, is not strictly a pollutant, but concern arises because of its major contribution to global warming, principally through the burning of fossil fuels.
2. *Carbon monoxide* (CO) can have detrimental health effects particularly in confined spaces and urban areas, but its major impact is its oxidisation to CO_2. Transport produces 71 per cent of the emissions in France and is the only source that is still increasing.
3. *Nitrogen oxides* (NO_x) are nitrogen-based pollutants which have harmful effects on health and the global environment, especially when combined with other pollutants. In combination with sulphur dioxide, emitted from fossil-fuel burning power stations, NO_x is the major contributor to acid rain. Transport is the largest and fastest growing source in France, increasing to 76 per cent of the total by 1991.
4. *Hydrocarbons* (HC) including volatile organic compounds (VOCs) are compounds that result from the incomplete combustion of carbon-based fuels. They play an important role in the formation of photo-chemical oxidants, such as ozone, which irritate eyes in smogs, damage plants and contribute to acidification and global warming. Moreover some HC are toxic in their own right, such as benzene, a known human carcinogen causing leukaemia. In France transport is responsible for 60 per cent of total emissions and, once again, is the fastest growing source.
5. *Other pollutants:* lead compounds added to gasoline have known effects

on IQ and behaviour, especially in children. Particulates, such as soot from diesel vehicles and asbestos from brake linings, are known to cause respiratory ailments. Chlorofluorocarbons (CFCs), responsible for the depletion of ozone levels in the stratosphere, commonly occur in materials used in vehicles, such as plastic foams.

It is clear that very large amounts of pollutants are being emitted from various forms of transport into the air that we breathe. In most of the world such pollution is unregulated and most city governments know little if anything about the impacts on the environment and on their citizens. The health aspects of this will be discussed in more detail in Chapter 11. Suffice it to say here that it is only now dawning on us that we are releasing increasing volumes of toxic materials into living areas without even understanding the short-term effects on human health, let alone the effects of cumulative exposure to a cocktail of interrelating chemicals or the legacy we are leaving for future generations.

10.3.2 Global warming

The global warming issue will be familiar to readers. The Intergovernmental Panel on Climate Change (IPCC) is the most authoritative source on the issue. Its principal conclusions are that:

- emissions resulting from human activities are significantly exacerbating the natural greenhouse effect;
- under present conditions a rate of increase of global mean temperature of about $0.3\,°C$ per decade will occur, resulting in a likely increase in global mean temperature of about $1\,°C$ above the present level by 2025 and $3\,°C$ before the end of the next century.

The consequences will include rising sea levels which over the next 100 years are likely to displace populations, destroy some coastal cities and inundate arable lands. Increasing temperatures will produce greater storminess, latitudinal shifts of crop boundaries and powerful feedback effects which will accelerate global warming. Though opinions vary on exact impacts, many scientists agree that ecological catastrophe is a likely outcome. In any case, the consequences of taking the predictions seriously when they turn out to be too pessimistic are very much less than the consequences of ignoring them when they turn out to be correct. To stabilise the concentrations of greenhouse gases at 1991 levels would require reductions in emissions of the long-lived gases – CO_2, NO_2 and CFCs – of over 60 per cent according to the IPCC. Without question these reductions will demand fundamental changes in lifestyle rather than technical changes and will pose major economic, social, philosophical and political dilemmas. The policy shifts that would be required will be dealt with in Chapter 12: here we will attempt to assess the role of transport as a contributor to the problem.

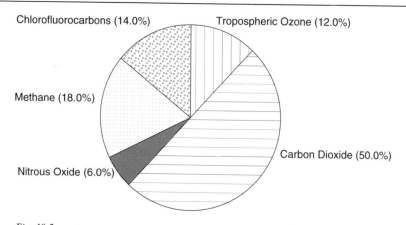

Fig. 10.2
Contributions of different gases to global warming. *Source:* Earth Resources Research (1989).

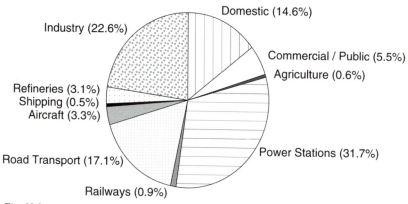

Fig. 10.3
Carbon dioxide emissions by source in the UK, 1988. *Source:* Earth Resources Research (1989).

The relative contribution of different gases to global warming is shown in Fig. 10.2, highlighting the central role of CO_2. For the UK, the sources of CO_2 emissions are shown in Fig. 10.3; though transport is only the third largest CO_2 source it is the fastest growing. By 1988 transport was contributing about 25 per cent to UK CO_2 emissions, an increase of 35 per cent in 10 years. All other sources showed a decline of some 14 per cent over the same period.

The relative contribution of different modes of transport is shown in Fig. 10.4 for the UK, extrapolated on the basis of national road traffic forecasts. The domination of road transport is clear, though the rapid growth of air travel is increasingly significant, particularly because the release of NO_x at high altitudes increases its global warming potential

Fig. 10.4
Transport carbon dioxide
emissions by mode in the
UK, 1988–2020. *Source:*
Earth Resources Research
(1989).

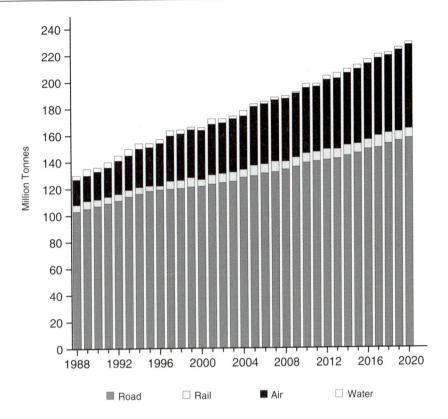

Legend: Road | Rail | Air | Water

Fig. 10.5
Global carbon dioxide
emissions from cars under
different growth rates. *Source:*
Walsh (1990).

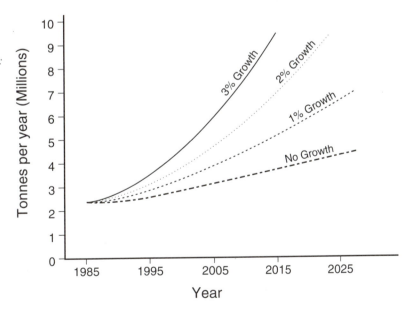

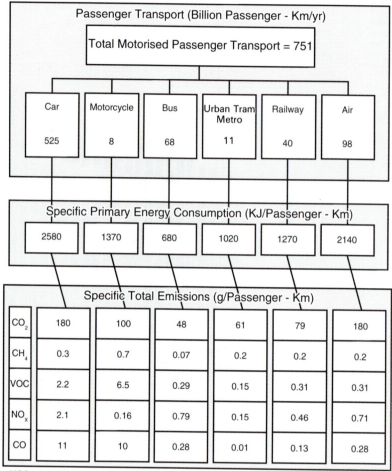

Fig. 10.6
Different modes of passenger transport in terms of energy use and pollution, Germany, 1987.
Source: Whitelegg (1993).

fiftyfold. World-wide, as car ownership grows so will the CO_2 output, rising from less than 3 million tonnes in 1985 up to 10 million by 2015 depending on growth rates (Fig. 10.5).

10.3.3 Summary: transport, energy consumption and pollution

The trends discussed in this section are summarised in Figs 10.6 and 10.7 for passenger and freight transport respectively. If we express the polluting characteristics of each mode in terms of some measure of work done – such as passenger- or tonne-kilometre – it becomes clear that road-based

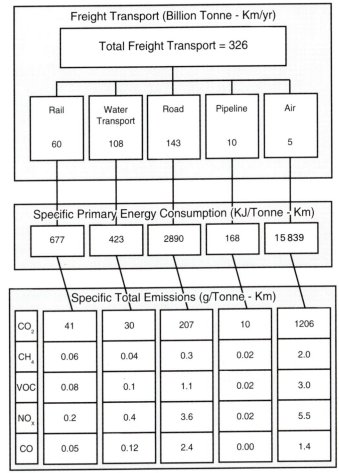

Freight Transport (Billion Tonne - Km/yr)				
Total Freight Transport = 326				
Rail	Water Transport	Road	Pipeline	Air
60	108	143	10	5

Specific Primary Energy Consumption (KJ/Tonne - Km)				
677	423	2890	168	15 839

Specific Total Emissions (g/Tonne - Km)					
CO_2	41	30	207	10	1206
CH_4	0.06	0.04	0.3	0.02	2.0
VOC	0.08	0.1	1.1	0.02	3.0
NO_x	0.2	0.4	3.6	0.02	5.5
CO	0.05	0.12	2.4	0.00	1.4

VOC = Volatile Organic Compounds

Fig. 10.7
Different modes of freight transport in terms of energy use and pollution, Germany, 1987.
Source: Whitelegg (1993).

transport is by far the most damaging land-based mode. For every kilometre travelled by every passenger, cars not only consume double the amount of energy but they also produce more than twice as much CO_2 as rail, four times as much NO_x and seven times as much in the form of VOCs, many of which are carcinogenic and thus unsafe at any level. Similarly, road freight transport is far more polluting per tonne-kilometre than rail or water-based freight transport, whilst air transport is energy-intensive and highly polluting (Box 10.1). In the absence of restraint strategies we can expect energy-hungry road and air transport to increase their modal share in all countries, so that the underlying trend will be for an increase in petroleum consumption and pollution emission. How these trends may be combated is addressed in the next section.

10.4 Reducing atmospheric pollution from land transport

10.4.1 Technical measures

Technological improvements to road vehicles would seem to offer the greatest potential for reducing pollution in the short to medium term if there are no major changes to transport policy or individual behaviour. The three-way catalytic converter can reduce NO_x, CO and HC by 95, 80 and 90 per cent respectively. Not surprisingly therefore, the fitting of these devices to new cars is now mandatory in many industrialised countries.

However, it will be some time before older non-equipped vehicles are phased out. Moreover, the converter is inoperative when the engine is cold, a significant problem in the UK for example, where some 61 per cent of car journeys are less than 8 km. Nor can they be fitted to diesel vehicles, so that most trucks will be unaffected and emissions of NO_x will continue to rise. The most serious drawback, however, is that they do not decrease the emission of CO_2 and actually increase fuel consumption.

For all of these reasons, the catalytic converter cannot be regarded as anything like a full answer to the problem of pollution from the transport energy system. In some ways it actually makes matters worse, for it deludes people into thinking that it is some sort of magical technical fix that deals with all pollution, so that there is no need to change their behaviour in any way. It obscures the fact that the only way to deal with the problem properly, at source, is to burn less of the fuels or to reduce the amount of carbon in them.

Burning less of the fuels may be achieved in a number of ways. Changes to engines and power train, weight reduction and aerodynamic improvements are estimated to be capable of achieving a 28 per cent reduction in energy consumption by cars in the 1989–2010 period in the UK. However, whether this potential becomes reality depends on manufacturers' attitudes and consumer preferences; judging by trends since the first energy shock of 1973, it would seem more likely that whatever technological improvements are made will be used to increase power and performance rather than to save fuel.

A second method of reducing energy consumption per vehicle is to encourage changes in driver behaviour. It is thought that driving smoothly could save up to 20 per cent of car fuel, whilst a similar saving could be made if all those who drove above the UK's 70 miles per hour (110 km per hour) speed limit reduced their speed to the legal limits. However, motorists are frequently less interested in saving money than they are in saving time and many such as van drivers or the users of company cars do not pay the costs of their wasteful driving practices. It would seem that behavioural changes are highly unlikely unless precipitated by legislation or by pricing mechanisms.

One problem with both of these approaches is that making cars more fuel efficient contributes less to overall energy consumption than might have been thought because it does nothing to reduce consumption in indirect uses such as vehicle-making and road building. Indeed, weight-saving materials could actually consume more energy in production and assembly than traditional ones. However, the crucial drawback is that although the potential savings per kilometre seem large, they are overwhelmed by the likely increase in the number of kilometres travelled. In other words, such improvements on their own will not be sufficient to reduce energy consumption in the system as a whole to the level where the output of harmful emissions will be lowered.

10.4.2 *Alternative fuels and power sources*

Rather than burning less fuel, another way of reducing CO_2 emissions is to develop alternative, low-carbon fuels to power road vehicles. Some are already in use, such as methanol, liquefied petroleum gas and electricity (batteries), whereas others such as compressed or liquefied hydrogen have yet to be fully exploited. However, most still face technical problems and performance limitations, especially because of the large fuel storage volume or weight required on the vehicle, which really restricts applications to local delivery services such as the electric milk floats used in the UK. Moreover as far as the longer-term issue of greenhouse gas emissions is concerned it does not appear that any of the alternative fuel sources offer a clear advantage over petrol and diesel when the full fuel cycle is taken into account. For example, the extent to which electric battery vehicles can become a serious option depends on how far the CO_2 emissions associated with electricity production (much of which is in fossil-fuel-burning power stations) can be reduced. Only when relatively clean electricity production can be achieved will electric- and hydrogen-powered vehicles become a lower-pollution option.

10.4.3 *Reducing total emissions*

Increasing appreciation of the seriousness of these trends and of the growing contribution of road transport to global warming is leading to attempts to reduce carbon emissions per vehicle, but given the continuing rise in the number of vehicles and in the number of kilometres travelled by them, this can only be the starting-point. Longer-term goals must thus address the issue of total emissions rather than emissions per vehicle, as in the EU where there is a commitment to overall CO_2 emission reductions, beginning with a stabilisation of emissions by the year 2000. How this is to be achieved is unclear, particularly against the background of forecasts of increasing traffic of up to 140 per cent in the UK and 500 per cent in southern Europe. As Whitelegg (1993: 16) says,

'There is no possibility whatsoever that any targets for stabilising let alone reducing CO_2 emissions can be achieved against these increases in vehicle numbers and activity.' In Chapter 12 we will return to this issue when we consider the way in which national transport policies need to be reorientated in order to incorporate environmental targets such as these.

10.5 Side-effects of transport systems: noise and vibration

10.5.1 Introduction

Noise is any disagreeable sound. Its effect will depend very much on the sensitivity of the individual, on the location, on the time of day and on existing noise levels. It disrupts activities, disturbs sleep, slows the learning process at school, impedes verbal communication and causes annoyance and stress. People at home are particularly susceptible, having to close windows even in summer and not being able to make full use of gardens or balconies. Noisy neighbours are often major reasons for complaint, but transport-related sources of noise constitute by far the major part of the total noise environment, with road traffic the chief offender.

It is estimated that some 135 million people in OECD countries suffer transport noise levels in excess of 65 dB (A), equivalent to the sound of a busy street through open windows (Fig. 10.8). Noise is seen as a high-priority environmental concern by communities, one that can cause property price falls and in the case of new transport proposals is the driving force behind much community opposition. For example, objectors to the building of a new road in Oxford voiced their opposition by using the slogan that it would 'transform Oxford from being the city of dreaming spires to the city of screaming tyres'!

Vibration is part of the noise spectrum, but is felt rather than heard. It is extremely localised close to roads and railways and as a consequence bothers fewer people than noise. The costs in terms of damage to buildings is not known, though it is a problem in heritage zones of old towns.

10.5.2 The sources of noise

The sources of noise from road vehicles are many and varied, including brake squeal, door slam, loose loads, horns, over-amplified music systems, anti-theft alarms and sirens on emergency vehicles. The major sources though are propulsion noises at low speeds – especially for trucks climbing

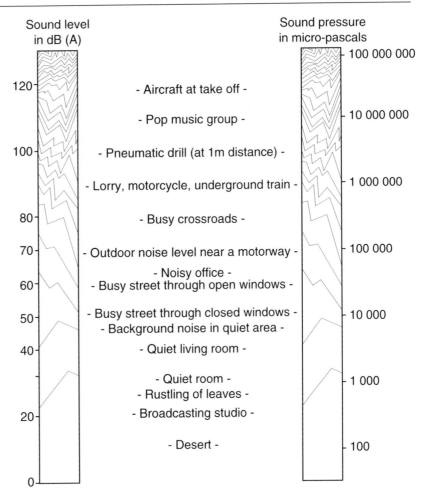

Fig. 10.8
Noise levels from different sources. *Source:* OECD (1988).

gradients – and tyre/road interaction at higher speeds. The nuisance will vary according to a number of factors, including background noise level, time of day and vehicle type. In recent German surveys of urban road noise, the greatest mean levels were heard to emanate from heavy commercial vehicles, followed by motorcycles, smaller trucks and buses (Fig. 10.9). Motorcycles have a very large spread between the noisiest and quietest, whilst cars, though the quietest in the sample, exceed 82 dB (A) at 100 km per hour.

Rail noise depends on the form of propulsion, the nature of the load, the speed of the train and the type of track. In general terms freight trains cause more nuisance than passenger trains and diesels cause more disturbance than electric locomotives. Increasing objections are being

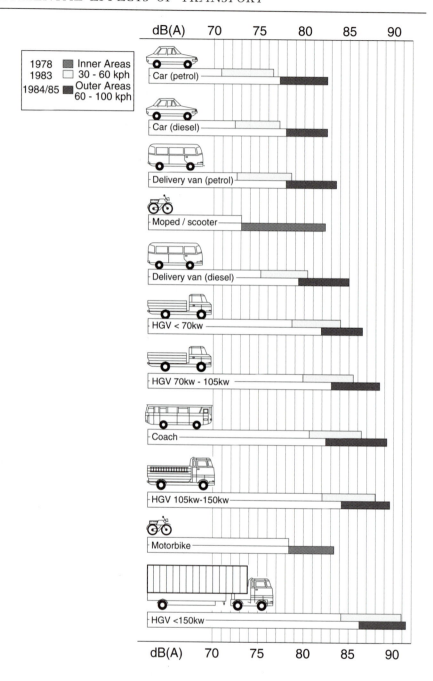

Fig. 10.9
Noise levels recorded for different vehicles under differing speed conditions. *Source:* Steven (nd).

raised to the extension of high-speed rail systems due to the noise and vibration inflicted on line-side communities, well seen in the case of opposition to the proposed route of the new London–Channel Tunnel line in England. Noise from individual aircraft is now less disturbing than in the past since a new generation of quieter aircraft has gone into service. Nevertheless, the growth in air travel has meant more flights from more airports for more hours in the day, resulting in some level of disturbance for more people. These issues are discussed in Box 10.1.

Box 10.1 Air transport and the environment

The noise pollution problems around airports are well known. For example Heathrow, the largest international airport in the world, experiences severe difficulties, with runways orientated east–west so that most of the 200 000 landings per year come in over the heart of the city, disturbing 500 000 people day and night. The airport has expanded from 5 million passengers a year to over 45 million from 1960–93 and the growth continues at around 4 per cent per year. Planning applications have been submitted for a fifth terminal which will bring the capacity up to 80 million passengers a year, of whom 54 million will arrive by road, with all the attendant side-effects.

However, a new menace is now emerging as the realisation grows that pollution from aircraft is much more significant than was previously thought. Though only about 13 per cent of transport fuel is consumed by aircraft, pollutants have different lifetimes, reactions, effects and distribution patterns when injected into the upper atmosphere compared with the lower atmosphere. For example, NO_x destroys high level ozone, allowing so much increased UV radiation to reach the earth that the growth of subsonic aircraft flying higher and faster, together with likely future supersonic transport in the highly sensitive stratosphere seem likely to negate the world-wide phase-out of ozone-destroying CFCs. Again, water vapour emission from engines – not a problem at low levels – is of great concern at altitude because it freezes into clouds which both reflect heat back and forth to earth and react with NO_x to destroy ozone. Additionally, the contribution of aircraft NO_x to global warming may be as much as all the man-made surface emissions of NO_x put together.

These issues are becoming more serious as air transport demand continues to grow, currently doubling every 10–15 years. Annual air passenger-km per person per year range from about 1700 for North America through 480 for Europe to 75 for Asia and 45 for Africa. Given future growth in world population and some equalisation of travel between regions, the potential growth in demand is almost limitless. Fig. 10.10 shows that in the business as usual case (BAU) emissions will rise to more than four times the present level, or more than 10 per cent of global emissions over the next 40 years and that no one mitigation measure is enough on its own to stabilise the position. More than 50 per cent of emissions are due to tourism which cannot continue to grow at current rates without accelerating damage to the global atmosphere and other environments. ▶

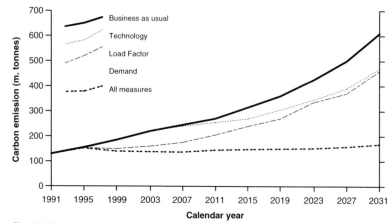

Fig. 10.10
The effect of different control measures on global carbon dioxide emissions from aircraft. The easiest operational improvement to achieve is an increase in the load factor (how full an aircraft is). Other operational improvements may become possible in the future. Demand management and technology can help limit emissions too. No one control measure is enough on its own to stabilise carbon dioxide emissions. *Source:* Barrett (1993).

Management of demand on an international scale must soon move to centre stage as one ingredient of green sustainable tourism. The cheap winter sun holiday in Florida or Majorca, so characteristic of mass-consumption capitalism in the late twentieth century, may one day be seen as a luxury that the earth cannot afford.

10.5.3 Alternative methods of abatement

Noise may be abated by moderating it at its source, ameliorating it along the path of movement, reducing the impact of its existing levels, moving its source elsewhere, or reducing traffic. Each of these will be examined in turn.

Reduction of noise at its source
A number of measures might be adopted under this heading:

1. *Introducing technical measures* such as the redesign of the engine for lower engine speeds; encapsulation of the engine and gearbox in a noise-absorbing tunnel or box; and modification of road surfaces to reduce tyre noise, though this may conflict with the aim of reducing skidding. Associated with this is the need to maintain road surface smoothness in order to minimise vibration.

2. *Influencing driving habits* to produce smoother flow and lower speeds and thus lower noise peaks.
3. *Managing traffic* to keep it flowing smoothly, perhaps by linking traffic lights to produce green waves of traffic and thus reduce stop–start progress.
4. *Controlling motor cycles* perhaps by preventing owner-modification, fitting better exhaust silencers or even controlling top speed and acceleration. As large motor cycles are really items of sports equipment rather than essential means of transport, there is no reason why they should be permitted to emit as much noise as fully laden commercial vehicles.
5. *Lower rail noise* can be achieved by shifting to electric traction, by better track maintenance and by modification of the rail vehicles, particularly the wheels and bogies. Vibration is reduced by much the same package. In the case of the Japanese Shinkansen lines, the use of iron girders has been minimised, vibration-resistant track beds installed and rails polished to reduce corrugations. Even so, noise barriers have had to be erected, homeowners compensated or bought out and much of the line put in tunnel, up to 56 per cent of it on the Okayama–Hakata section.

Reducing noise along the path of movement

Surrounding roads and railways with buffer zones, planting or screening is desirable, though space for this may not be available in urban areas. The most effective solution is to put the road or line in tunnel, but this increases costs dramatically, often to prohibitive levels. In the case of aircraft movements, take-off patterns can be adopted that reduce noise fall-out over residential areas.

Reducing impacts of noise on existing communities

Noise-reduction techniques can be applied to buildings themselves, such as double glazing and noise-absorbing insulation. An alternative or additional tactic may be to ban aircraft movement at night, in the same way that some cities have restricted truck movements to waking hours.

Strategies for removing the source of nuisance

The construction of bypasses around towns is often partly justified in terms of removing the noise nuisance from residential areas. Similarly, noise-reduction strategies influence the location of new airports, which despite their need to be sited close to the city if they are to serve their intended market are frequently banished to more distant locations in order to minimise their environmental impact.

Reducing traffic

One certain way of reducing the nuisance from noise and vibration would be to reduce the amount of traffic in the first place, a strategy which would of course mitigate transport's other environmental impacts too. Discussion of broad goals of this kind is postponed until the whole issue of sustainable transport is examined in Chapter 12.

10.6 Side-effects of transport systems: land use and ecology

10.6.1 Primary land-take

The consumption of land is one of transport's more obvious impacts on the environment, most conspicuous in the case of large projects. Primary land-take for airports, for instance, may involve great loss of high-value agricultural land close to cities, whilst the relocation of marine terminals downstream to estuarial sites has led to large-scale destruction of wetland and intertidal habitats.

However, it is roads that have the greatest impact, taking up about 19 per cent of the surface area in large cities, with rail taking a further 4 per cent. These are significant areas in circumstances where land is in short supply and expensive. In the UK, the total area needed for each car in order to allow it to park at home, at work and at shopping areas has

Plate 10.2
The emergence of the suburban shopping mall in car-dominated societies has accentuated the utility of car ownership and use, thus lengthening trips and increasing the consumption of resources, including land for parking (Buffalo, New York).

been calculated at $372 \, m^2$. This amount is not only three times the size of the average home, but it remains unoccupied for 80 per cent of the time. Space needs also have to be considered in terms of efficiency. Not only does a road network need thirteen times as much space as a suburban rail network to convey the same number of passengers, but some road vehicles will be less efficient than others in their consumption of space for the amount of work done, with private cars performing very much worse than buses, as was seen in Chapter 7 (Fig. 7.4).

In rural areas new roads disproportionately take higher quality agricultural land, since both tend to avoid the steeper slopes. Inevitably they also result in the loss of cherished landscapes and valuable habitats. For example, the proposals contained in the UK's 'Roads to Prosperity' proposals of 1989 will, according to the Nature Conservancy Council, affect *inter alia* four nature reserves and no fewer than 162 sites of special scientific interest (SSSIs).

On the other hand, new habitats have been created along linear routes of rail and road, principally because of their freedom from human interference. Railway verge vegetation is becoming more valuable aesthetically and ecologically in the UK now that it does not need to be intensively maintained in order to minimise the risk of fire caused by steam locomotives. Motorway verges too have developed a rich flora and have been colonised by insects and animals which, because of their high casualty rates, have attracted scavenging bird and mammal species, part of a new linear ecosystem relatively untouched by man. These benefits, however, must be kept in perspective: we should set the habitat gain represented by the $7.3 \, km^2$ of road verge alongside the UK's $3200 \, km$ of motorway against the $135 \, km^2$ loss due to the total land-take for these roads.

10.6.2 Secondary land-take

Secondary land-take is that land which, in addition to primary land-take, is affected by the development of the transport system. It includes land taken for quarrying materials used in transport infrastructure (such as sand, gravel and roadstone), together with land affected by hydrological changes and by runoff, especially of salt and oil. The effect of runoff of de-icing salt is not limited to roadside vegetation: for example, 90 per cent of the salt applied to the snowy streets of Buffalo, New York goes into the city sewerage system and thence into Lake Ontario. About half of the pollution from the mineral oil industry – through tanker accidents and spillages and leaks from oil installations – can be attributed to the production of fuels for motorised traffic. Used oil is extremely toxic and carcinogenic, with a litre having the potential to contaminate a million litres of fresh water. In the USA nearly a billion litres of used oil is spilled, poured down the drain or buried every year: this is about 20 times the amount spilled by the *Exxon Valdez* when it was wrecked on the Alaskan coast. The disposal of scrapped vehicles and used parts is a further source of ground and water pollution, with many of the materials being extremely toxic to humans and nature: asbestos, lead, chrome, nickel, cadmium and grease would be cases in point.

Secondary land-take also includes changes in land use that result from transport developments. For example, completion of new roads increases accessibility of land close to intersections and stimulates speculative development. The completion of London's orbital M25 motorway in 1986 dramatically increased the accessibility of some rural locations and led in that year to proposals for sixteen large shopping complexes on sites adjacent to the motorway. Such developments not only take land themselves but also produce new volumes of traffic, which in turn raise the pressure for further transport infrastructure provision.

10.6.3 *Effects of severance*

The linear nature of roads and railway lines leads to the splitting of natural habitats, thus decreasing habitat size and reducing interaction with other communities, producing a decline in both the number of species and the density of their populations. Inability to cross roads leads to an island effect, threatening species diversity. For animals that are able to cross roads or railway lines, conflict with traffic may lead to significant casualties. In the USA estimated losses are in the region of at least one million wild animals per day, whilst UK casualties are estimated at, for badgers, one member of each family group per year, and for breeding amphibians, some 40 per cent of the population per annum. Bird deaths are estimated at anywhere between 30 and 70 million per annum in the UK, with road accidents the principal cause of death for some species, such as barn owls.

It is plain that very little can be done about land-take for transport once the decision to invest in infrastructure has been taken and the mode of transport selected. There are exceptions: STOL (short take-off and landing) airports may be an appropriate substitute for the full-size version in cramped urban areas as in London's Docklands. Underground facilities such as railway stations may save valuable land. However, for many facilities the consumption of space is actually essential to their efficient operation: container terminals work on the principle of trying to get ships back at sea as fast as possible, so they must have extensive quayside areas to off-load containers and store them awaiting onward movement (see section 5.5).

It must be stressed that the biggest problem is often the need for circulation and access, especially by car. Airports are space-hungry less because of the runway length than because of the huge demand for travellers' car-parking and the same principle governs the extensive nature of out-of-town shopping complexes. There is no mitigation of the consequences here except in not building the facility, but this is part of the feedback effect that we have encountered so frequently in this book. If a car-orientated society is desired, then space-hungry facilities become essential to cope with the car-parking demand. The result is that facilities sprawl so much that car use becomes essential, fuelling the process. Motorised road transport is one of the most space-consuming activities that we know: the 'Los Angelisation' of landscapes is an inevitable consequence of acceptance of the car as the principal form of urban transport.

10.7 Side-effects of transport systems: visual intrusion

There is no doubt that transport has a major impact on the visual landscape. As linear features, the lines of rail, canal or highway can seem discordant with the contours of rural landscapes, whilst viaducts, flyovers, terminals and parking garages frequently dominate the urban scene. How a particular piece of infrastructure is perceived will vary from person to person, according to how much they use it, how far they live from it or how much it intrudes upon their familiar views.

Nor it is just the infrastructure that intrudes. The sheer volume of cars in urban areas may spoil vistas of architectural or historical significance, as when the medieval city wall or the Renaissance square are seen across the roofs of hundreds of parked cars. Such visual intrusion undermines the quality of urban life, for cities should be places that are worth living in, not just travelling through.

Effects of such visual intrusion can be mitigated in a variety of ways. Rural highways can be screened by tree-planting, by earth buffers or by placement in shallow cuttings. Alignments may be chosen that follow the contours of the land and structures such as bridges and service points designed, located and coloured so as to blend in as far as possible with the landscape. Tunnelling may be used too, but only at significant additional cost. It is an unavoidable fact that many transport developments such as airports, container terminals and freeway intersections are too big to hide, so that they must be designed in the first place to be as unobtrusive as possible in their own right.

Transport effects are not all negative, however. In some cases, long-established infrastructure may come to be seen as part of the landscape or heritage of the country, notwithstanding the fact that it may have been condemned as a gross intrusion when it was first built. To date it is mostly railway and canal lines that have preoccupied preservationists, as roads are commonplace and in any case are too new to excite much in the way of nostalgia. It may not seem likely now, but it is entirely possible that a

(a)

(b)

Plate 10.3
Beauty is in the eye of the beholder. The Victorian period of railway building produced structures which were intrusive at the time but which now generate powerful preservationist sentiments. The Ribblehead viaduct in North Yorkshire is a case in point (a) as is Thomas Telford's bridge over the Conwy River in North Wales (b) where at least attempts were made to blend it with existing townscapes. In our own age urban freeway overpasses (as in Albany, New York (c) (facing page)) are perceived as ugly and dangerous spaces but it is possible, that with time, they too make evoke a fondness for a bygone age.

Plate 10.3 continued

future generation may wish to preserve particular motorway bridges, service stations or even multi-storey parking garages!

10.8 The appraisal of environmental impacts

評價

10.8.1 Introduction

From what has been said in this chapter so far it should be evident that environmental factors are assuming greater importance in all societies. However, it is also true that many countries are investing heavily in transport infrastructure in a belief that this will improve their competiveness in world markets. These projects are getting larger: the US Inter-state highway system is the largest civil engineering project ever undertaken and facilities such as the European high-speed train network, the Channel Tunnel and the new Hong Kong airport all illustrate the point. The environmental conflicts become ever more serious and controversial, provoking not only local residents into NIMBY (not in my back yard)-style protests but groups elsewhere concerned about the transboundary nature of the environmental impacts on the atmosphere, the oceans and the earth's resources. The main environmental effects of the various modes are listed in Table 10.3. Because road traffic is the major problem, this section will use as an example the assessment of the environmental impact of roads in the UK.

Table 10.3
The main environmental effects of the various transport modes

Mode	Air	Water resources	Land resources	Solid waste	Noise	Accident risk	Other impacts
Rail			Land taken for rights of way and terminals; dereliction of obsolete facilities	Abandoned lines, equipment and rolling stock	Noise and vibration around terminals and along lines	Derailment or collision of freight carrying hazardous substances	Partition or destruction of neighbourhoods, farmland and wildlife habitats
Road	Local (CO, C_xH_y, NO_x, fuel additives such as lead and particulates), global (CO_2, CFC)	Pollution of surface water and groundwater by surface runoff; modification of water systems by road building	Land taken for infrastructure; extraction of road building materials	Abandoned spoil tips and rubble from road works; road vehicles withdrawn from service; waste oil	Noise and vibration from cars, motorcycles and lorries in cities and a long main roads	Deaths, injuries and property damage from accidents; risk from transport of hazardous substances; risk of structural failure in old or worn road facilities	Partition or destruction of neighbourhoods, farmland and wildlife habitats; congestion
Air	Air pollution	Modification of water tables, river courses and field drainage in airport construction	Land taken for infrastructures; dereliction of obsolete sites	Scrapped aircraft	Noise around airports		Congestion on access routes to airports
Marine and inland water		Modification of water system during port construction and canal cutting and dredging	Land taken for infrastructure; dereliction of obsolete port facilities and canals	Vessels and craft withdrawn from service		Bulk transport of fuels and hazardous substances	

Source: OECD (1988).

Environmental assessments (EAs) began in the USA around 1970 and may be defined as an orderly process for gathering and evaluating information and opinions about the likely environmental consequences of proposed projects, to assist in decision-making. The process aims to:

- identify the relevant environmental factors;
- measure (as far as possible) and predict the environmental effects on the different groups of people affected;
- determine how these effects should be valued and compared;
- specify a format for the presentation of the data;
- integrate the results with those from the economic and operational evaluation to assist decision-makers to determine which of the alternatives should be chosen.

10.8.2 Road building in the UK

The planning process for inter-urban roads in the UK, shown in Fig. 10.11, is initiated by estimating the 'need' for a new road. Existing traffic on the network is analysed and forecasts of the growth of that traffic are made based on the national road traffic forecast, giving a picture of the volume and type of traffic which might be found in the future on the existing network if nothing is done. Then an attempt is made to quantify the economic costs and benefits of different scheme options and compare them with the 'do nothing' option, using a technique known as cost–benefit analysis. In the particular variant used in the UK – called COBA – the variables incorporated are as follows:

1. *Travelling time:* calculated journey times are costed for all travellers using a value for working time based on wage rates and one for non-working time. The summed results can be used to compare different routes.
2. *Valuation of accidents:* the likely incidence of road casualties is calculated for each of the options and the results given a monetary value, calculated from variables such as loss of lifetime earnings, damage to property, cost of police time and a notional payment for pain, grief and suffering.
3. *Vehicle operating costs:* the cost of fuel, oil, tyres, vehicle maintenance, etc. are calculated for the various schemes and are a function of the distance and speed of travel.
4. *Capital expenditure and cost of maintenance:* the cost of land acquisition and construction is assessed and future expenditure on the various schemes – in repairs, lighting, etc. – is estimated. The annual cost of these items is calculated over a 30-year span and discounted to the present. Comparison with the 'do nothing' option gives a net present value (NPV) which may be positive or negative.

At first the COBA procedure did not incorporate environmental impacts,

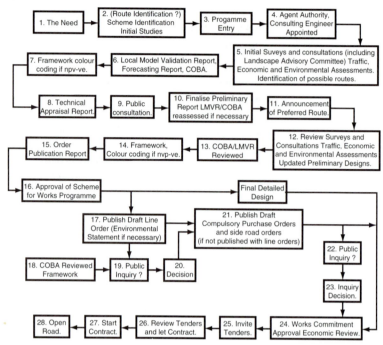

Fig. 10.11
Stages in the evolution of trunk road schemes. *Source:* Department of Transport (1992).

but subsequent changes have meant that a large number of specific environmental effects are formally evaluated even though they are not deemed to be capable of expression in money terms. The *Manual of Environmental Appraisal* gives advice on the assessment of specific environmental effects and these are listed in Table 10.4. They are brought together with the COBA variables in a special tabulation known as a framework, which is now the standard medium for the presentation of all the environmental and economic factors that have to be taken into account.

On average it takes 13 years between the first proposal for a new road and its opening. The environmental assessment is applied at various stages of the project (e.g. stages 5 and 12 on Fig. 10.11) and must, by EU Directive, be exposed to public inquiry in the case of all major schemes (stage 17). This directive requires information on the direct and indirect effects of the project on human beings, flora and fauna; soil, water, air, climate and the landscape; the interaction of the above two groups of factors; and material assets and the cultural heritage.

10.8.3 *An assessment of the evaluation process*

This procedure has been subject to great criticism over the years, relating to almost every single detail of its operation. Travel times have been said

Table 10.4
The Manual of
Environmental Appraisal:
effects, measurement units
and groups affected

Environmental and social effect	Measurement units	Groups affected
Demolition	No. of premises	Occupiers; conservation policies
Land-take	Area (ha)	Occupiers (open space); development policies
Noise change	No. of premises in distance bands	Occupiers; users of facilities; all policies
Visual obstruction	No. of premises in bands of visual obstruction (mS40)	Occupiers (property); conservation policies
Visual intrusion	General description	Occupiers; users of facilities; conservation policies
Community severance	No./type of premises and degree of delay or diversion	Cyclists, pedestrians; occupiers; users of facilities
Impact on agriculture	No. of farms affected and area of land-take by quality	Occupiers (agriculture); policies for development/ protection of resources
Air pollution	Not normally considered	Occupiers; users of facilities
Disruption during construction	No. of premises within 100 m of road	Occupiers
Heritage and conservation	Description of impact	Occupiers; policies for conservation
Ecological impacts	Description of impact	Occupiers; policies for conservation
Pedestrians and cyclists	1. Changes in journey time and numbers affected 2. Changes in amenity	Pedestrians and cyclists; users of facilities
View from the road	Brief description	Travellers
Driver stress	Three-point scale based on type of road and volume of traffic	Drivers

Source: TRRL (1992).

to so dominate the analysis that virtually any proposal that saved large numbers of people just a few seconds each would show, in aggregate, a huge time saving that would be so 'valuable' that the new road would appear to be highly beneficial in economic terms. Again, the valuation of accidents has been said to underestimate the real cost of accidents as it does not include such factors as trauma suffered by witnesses, loss of earnings for carers of victims, trips foregone because of fear of accidents or the cost of parental chauffering of the young because of traffic danger.

Without doubt the rigidity of the original COBA process has been tempered as a wider range of criteria has been included, as environmental appraisal has become institutionalised into the process and as a standard framework has been adopted allowing like to be compared with like. The overall procedure is now much more sophisticated than it was. However, operational problems remain, particularly in relation to forecasting and strategy. As far as forecasts are concerned, any improvements to the assessment process in the future are of little use if traffic forecasts continue to be as unreliable as they have been in the past. An examination of 41 recent road projects in the UK revealed that about half had traffic forecasts within 20 per cent of the flows that materialised, but that the rest had forecasts ranging from 50 per cent above to 105 per cent below. The fixed trip matrix that is used fails to present a reliable picture of the future effects of a proposed scheme, with new roads frequently generating new traffic because the new networks make driving more attractive and bring destinations formerly perceived as inaccessible within reach.

Related to this is the lack of a strategic view. The assessment of individual projects is not the same as scrutiny of the environmental effects of the overall route of which the particular development was one part. For example, a bypass might be justified on the grounds that the peace and quiet gained in the town can be assessed in monetary terms and balanced against the cost of the bypass and the loss of landscape quality along the route. But faster travel around the bypass will shorten the overall time taken along the transport corridor as a whole, thus attracting new traffic. Other corridor towns which could cope with previous levels of traffic now in turn become congested, so that the case for them being bypassed in turn becomes difficult to resist. Each town along the route will have its choices hi-jacked by earlier decisions made on other places, a problem well illustrated by the M3 case (Box 10.2) where the construction of the two ends of the road meant that there were few options available when it came to consideration of the middle section.

Moreover, the process as applied in the UK does not connect decisions on individual schemes with much larger environmental concerns such as global warming. Each new bypass shortens journey times and encourages more use of the road and thus adds to emissions, global warming and health impacts. Regressive outcomes are likely, such as the switching of travel along the corridor from rail to road and the resultant closure of the railway line. The overall result is to push the whole of society another notch up the ratchet of higher levels of car ownership and use.

10.8.4 Summary: valuing the environment

Clearly the appraisal process, as represented by the UK's COBA for roads, is not perfect. The British government in its review of COBA has taken the view that although the cost–benefit analysis 'vehicle' is not operating smoothly, it is fundamentally sound and simply requires a little modification and regular fine tuning for it to continue to give valuable service. A diametrically opposed view, held by the geographers Adams and Whitelegg,

Box 10.2 The battle for Twyford Down

From the *Daily Telegraph*, 1 July 1976:

There were wild and chaotic scenes at the resumed enquiry into the proposed M3 at Winchester yesterday when 100 protestors were led away by strong-arm stewards and policemen ... The protestors were not young with long hair and sawn-off jeans, but teachers, doctors, clergymen and their wives ...

In the afternoon the hearing became a shambles as the audience turned their backs on the inspector and sang from hymn sheets 'Rule Britannia' and 'Land of Hope and Glory' ... More than a dozen burly policemen removed the protestors who refused to sit down and be quiet ...

There was a roar of approval as the Headmaster of Winchester College, Mr John Thorn, marched into the hearing but as he was escorted out he shouted: 'I'll be back.'

The subject of this furore was the proposed extension of the M3 motorway from London to Southampton. At Twyford Down, near Winchester – an Area of Outstanding Natural Beauty (AONB) – the route crossed two Sites of Special Scientific Interest (SSSIs), the chalk downland itself and water meadows at the foot of the hill (Fig. 10.12). Though some objectors argued that the impact of the project on these features could have been reduced by tunnelling, others challenged the need for the road in the first place. But if the need for a road was to be judged then need had to be demonstrated – and this had been done largely through the presentation of traffic forecasts which had become more and more discredited. At this and many other public enquiries objectors repeatedly argued that forecasting traffic growth and providing roads for it is a classic self-fulfilling prophecy, for newly generated traffic will appear to fill the space, justifying the need for yet another road.　　　　　　　　　　　　　　　　　　　　　　　　▶

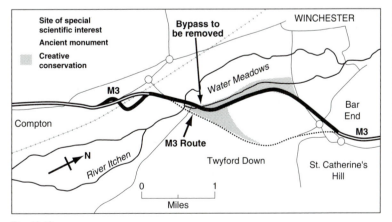

Fig. 10.12
The route of the M3 through Twyford Down.

At Twyford the arguments have raged, on and off, for the best part of twenty years. As protestors lost successive rounds and the arrival of the bulldozers become imminent so a group of young people moved in to live on the site, naming themselves the 'Dongas' after the Iron Age tribe that had lived at the encampment on the Down, the site of which was to be destroyed by the new road. In October 1991 the European Environment Commissioner, Carlo Ripa di Meana, demanded that work be stopped pending the result of legal action by the Commission against the UK Government over the alleged failure to carry out Environmental Assessments (EAs) as required by EU Directive, but in July 1992 the new Commissioner Karel van Miert dropped the case against the UK. This was widely interpreted as a political decision, linked to a post-Maastricht policy of reduced interference in the affairs of member states. By 1993 the Dongas had been forcibly evicted by a security firm employed by the Department of Transport and the battle for Twyford Down was finally over. The road opened to traffic in 1994, shortening the journey time from London to Southampton by about ten minutes.

is that because the vehicle is structurally unsound and impossible to repair the only solution is to scrap it completely. We can examine these views in turn.

The Department of Transport view is that the size and complexity of the trunk road programme make it necessary to have a standardised evaluation procedure to ensure consistency. In reviewing this stance the government concluded that a formal method of appraisal *is* required to furnish decision-makers with all the relevant information in a form which will allow them to give due weight to all the disparate effects of competing proposals. Such a method was felt to be a tool in decision-making and not a substitute for it. Cost–benefit analysis and the use of money values have advantages over other methods, it was argued, but there are limitations and weaknesses which must be acknowledged. These include the valuation of some environmental effects, some of which cannot sensibly be valued. That should not, however, prevent the valuation of those that can be.

A critical view of the cost–benefit approach would be that this search for a comprehensive, systematic, consistent, quantified and centrally controllable decision-making method assumes that large and complex social issues can be broken up into their constituent parts, that all the parts can be weighed in the banker's scales and then reassembled to provide a decision. Such consistency and control are purchased at the cost of arbitrariness, it is said, because COBA embraces only a tiny fraction of the costs and benefits of a major road scheme and omits consideration of effects that are distant in space and time. In spatial terms, for example, should the cost of the bypass include the impact of the mining and smelting for the metals that go into the vehicles that will use it? Should it include the environmental damage of the sand, gravel and roadstone extraction required for the new road? What about the effects on the environment of oil exploration, drilling, production, transport and refining? And in time terms, what about the effects generated by the disposal of the vehicles whose purchase was encouraged by the construction of the new road? Should the secondary land-take issues of tyre and battery disposal be considered here along with the fact that every vehicle during its lifetime produces hundreds of litres of waste oil, brake fluids and antifreeze, many of which end up polluting ground-water? COBA's critics argue that a discussion about the benefit of a new road should incorporate information about the spatially removed impacts as well as the life cycles of the vehicles and the road, but they are sceptical about the likelihood of this being done properly. Whitelegg (1993) has argued that the degree to which these views are considered excessive or ridiculous is inversely related to the importance we attach to the environmental impact of transport.

The argument is not only that COBA excludes many variables that should be included, but also that it puts values on things like views, pollution and life at levels which do not seriously threaten the likelihood of new roads being built. The thrust of these views is that if all of these issues could be captured by the environmental appraisal process, then road construction would not show the positive results that often now emerge from narrower balance sheets. The Automobile Association, with its pro-motoring stance, claims that £15 billion per year is raised in taxation on motoring in the UK set against the cost of the road system of only £6 billion, but Earth Resources Research (1993) has calculated that if accidents and pollution are included the cost figure is £34 billion, which equates to £1000 shortfall per car per year and represents a larger subsidy to road transport than to any other form of movement. What is more, this figure excludes health costs, ecological damage, the costs of the judicial system and virtually all of the spatially and temporally removed impacts discussed above. Similar conclusions can be reached for Germany and are shown in Fig. 10.13.

Cost–benefit analysis and environmental assessment are lengthy and complex processes. It should be obvious from the conflicting opinions expressed above that they are also highly politicised issues, for the whole concept rests on decisions as to what to measure and how to measure it. It should be borne in mind that though evaluations are carried out by 'experts' – transport planners, mathematicians, economists, engineers, etc. – and though the public has its say, the final decision is one made by a politician who weighs up the relevant factors and strikes a balance between competing priorities. The application of environmental assessment becomes, in the final analysis, not so much a technical matter as one of political judgement. As such it needs to be considered in the context of national transport policies, an issue to which we will return in Chapter 12.

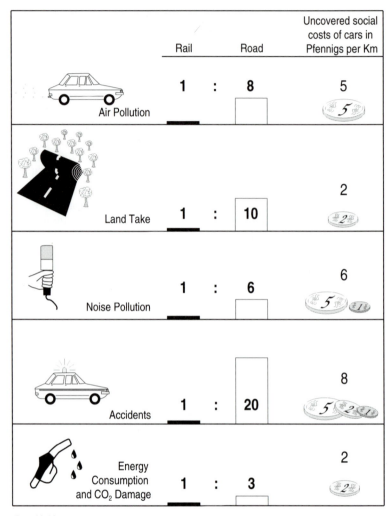

Fig. 10.13
Environmental damage and costs of different modes of transport. *Source:* Seifried (1990).

10.9 Concluding summary

It should be clear from this chapter that transport is a major source of pollutants and that in most cases it is the fastest growing source. There is a similar picture in energy use, where transport consumption is growing rapidly at a time when use by other sectors is declining or stable. Because transport pollution stems for the most part from the burning of petroleum products it is strongly related to energy consumption and the explanations for the increases are similar. Technological changes have led to falling pollution and energy use per kilometre but these have been overwhelmed by a huge rise in the number of kilometres and a massive switch to forms of transport which consume most energy and produce most pollution per kilometre. As impacts escalate at all scales there is growing concern amongst scientists about the ability of the planet's ecosystems to cope with this growing burden and international agreements are being struck to address the problem. But present trends are deeply entrenched in international systems of motorisation, which conflict fundamentally with the necessity to reduce emissions. More developed countries are experiencing great difficulty in stabilising emissions, let alone reducing them to compensate for rising CO_2 output from the newly industrialising nations.

To avoid these cumulative impacts of transport, environmental policy needs to be broadened and intensified. Further transport development is inescapable, but it must be done in a way that meets the needs of the present without compromising the ability of future generations to meet their own needs. This is a working definition of 'sustainable development', which demands that every generation should leave behind good environmental quality and not a negative environmental legacy. It is evident that current patterns and levels of transport are not sustainable, with rapid growth in traffic justified on the grounds that improvements in mobility are essential for human progress and should not be obstructed merely because they have some rather unpleasant environmental consequences. As concern for the global environment becomes a pressing need for all of humankind, so must the assumptions on which such trends are based be challenged, a theme which will be taken up again in Chapter 12.

Further reading

Adams J 1981 *Transport planning – vision and practice* Routledge and Kegan Paul

Banister D, Button K 1993 *Transport, the environment and sustainable development* Spon

Barde J-P, Button K 1990 *Transport policy and the environment: six case studies* Earthscan

Earth Resources Research 1993 *Costing the benefits: the value of cycling* Cyclists Touring Club

Holman C, Fergusson M, Mitchell C 1991 *Road transport and air pollution: future prospects* Rees Jeffreys Road Fund Discussion Paper 25, Transport Studies Unit, Oxford

Kroon M, Smit R, van Ham J 1991 *Freight transport and the environment* Elsevier, Amsterdam

OECD 1988 *Transport and the environment* OECD, Paris

Pearce D, Markandya A, Barbier E 1989 *Blue print for a green economy* Earthscan

Roberts J et al 1992 *Travel sickness: the need for a sustainable transport policy for Britain* Lawrence and Wishart

TEST 1991 *Wrong side of the tracks: impacts of road and rail transport on the environment* TEST

Transnet 1990 *Energy, transport and the environment* Transnet

Whitelegg J 1993 *Transport for a sustainable future: the case for Europe* Belhaven

II *The social impacts of transport*

> **Travel – broadening minds or shortening lives?**
>
> Cars, cars, fast, fast! One is seized, filled with enthusiasm, with joy ... the joy of power. The simple and naive pleasure of being in the midst of power, of strength. One participates in it. One takes part in this society that is just dawning. One has confidence in this new society: it will find a magnificent expression of its power. One believes in it. (Le Corbusier, 1929)
>
> Los Angeles ... Tomorrow's City visible today, the city of multilane freeways, the city of the automobile ... where there are districts where a human biped denuded of any conveyance and primitively ambulating along a pavement is so disconcerting a sight as to call for instant interrogation by the police. (Elaine Morgan, 1976)
>
> Earlier chapters in this book have shown how mass mobility has changed the world irrevocably, rubbing out boundaries, changing lifestyles, opening up opportunities for economic advancement and shrinking the world for its people. But mobility is unequally distributed and has damaging side-effects on the quality of our lives. This chapter examines these issues in general terms and then focuses on specific impacts – those of transport deprivation, accidents, 'livability' of urban areas and the effects of transport on human health.

11.1 Introduction

Everywhere in the world people are travelling more, at lower real cost, than their parents did. The world has shrunk for many people, in time if not in space, as air travel has changed from being the preserve of the very rich to an instrument of mass tourism. Motor cars have made their mark in all but the remotest places and, if present trends are not arrested by environmental or energy limits, ownership looks set to widen still further. Now, in the late twentieth century, as every two new babies are born one new car is made.

However, as we have seen in Chapter 10, people's desires to move around may conflict with economic and environmental objectives and there is no easy way of reconciling them. This is particularly so as we know so little about the value that people place on travel. In what circumstances and to what people is it a friction, a barrier, to be overcome at a cost? And when is it a liberation, an opportunity to get out and about and broaden the mind? In what specific ways are lifestyles affected by transport availability? How, for example, do people change their transport behaviour after changing jobs – or after moving house, getting married, having children, losing a local railway service, or at the onset of old age?

These social influences on transport have only recently come to occupy a prominent position on transport geography's agenda, which was dominated by supply-side issues until the 1970s. Then, as reaction to the quantitative revolution of the 1960s began to set in, so did work begin to take on a more behavioural approach. A series of monographs from London's Policy Studies Institute was instrumental in charting the way forward, initiated by a pioneering study in 1976 by Mayer Hillman and his colleagues, who tried to establish the ground rules for a people-centred approach. *Transport realities and planning policy* was an attempt to provide a comprehensive view of the ways in which mobility and travel behaviour vary between different people living in areas with contrasting physical characteristics. The overall picture which emerged was one in which personal mobility problems are more complex and more localised than had previously been presupposed, demanding a policy response based on access needs rather than on the supply of transport.

Other studies too began to take place at this disaggregated level. Instead of studies – commonplace earlier – that reported on the total amount of car travel (without saying who was doing it) or on household car ownership (without investigating who held the keys), the focus began to fall on the individual, recognising the differing nature and characteristics of people concealed in the generalised picture. In particular it emerged from these studies that cars are not the universal form of transport that most transport planners had assumed them to be and that it is not car ownership which matters but individual access to its availability.

If we were to try to develop a framework for this people-centred approach to transport, it would still be difficult to better that set out by Appleyard in 1971. His set of social goals for transport development for the metropolitan environment of the future were:

1. increasing the availability of transport services to deprived segments of the population;
2. better choice and quality of travel for all of society;

3. lowering the undesirable impacts of transport on the human and natural environments;
4. enhancing environmetal quality through improved transport planning.

Of these four, the enhancement of environmental quality by transport planning and the effects of transport on the natural environment have been the focus of Chapter 10. Human environmental impacts are dealt with in this chapter through the medium of three examples – road traffic accidents, the reduction in the quality of street life as a result of the growth of traffic, and transport and health. This chapter begins, though, by addressing the subject matter of Appleyard's first two goals. No attempt has been made to deal with them separately; rather, the position is taken that availability, choice and quality are complementary aspects of transport provision and that meeting the needs of the transport-deprived will in fact mean that the needs of all members of society are met too.

11.2 The transport-deprived

As Chapter 9 has shown in the context of rural areas, mobility may be impeded by isolation, poor transport services or any number of external factors in the local environment. However, mobility is also relative to internal, individual characteristics – not so much a case of *where* you are as *who* or *what* you are.

11.2.1 Mobility gaps

Kerry Hamilton and Linda Jenkins (1992: 61) have observed that 'Transport disadvantage is not equally or randomly distributed throughout society, but follows the well-established lines of structural social inequality.' What we might call 'mobility gaps' are tending to become bigger and to affect larger numbers of people. At the global scale most of the world's people live in circumstances where low levels of access to mobility are the norm. In industrialised societies on the other hand there are great variations between people in terms of mobility, with the obvious contrast between the jet-setting executive on the one hand and spatially restricted groups such as the poor, the disabled, the old, the young and women on the other. It is helpful to examine the principal characteristics of each of these groups, though of course there is considerable overlap between them.

11.2.2 The poor

It is the less affluent whose communities are most likely to be severed by new roads, who suffer the highest levels of road accidents, who are least able to escape air pollution and to reach shops, services and workplaces that have been decentralised. A shortage of money results in trips that are shorter and made less often by car. However, the primary impact of poverty is to reduce the number of trips made, so that poor people really only travel when it is absolutely necessary. There is some evidence from the USA that poverty and transport deprivation have a mutual cause and effect relationship, in the sense that people are unable to access job locations because of transport deficiencies which they cannot overcome because of their lack of income. For example, in the ghettos of American cities blacks have car ownership rates only one-third to one-half those of whites and are thus heavily dependent on public transport, which is almost always in its worst condition on routes serving the ghetto. The suburbanisation of jobs coupled with the difficulty blacks experience in moving to suburban homes due to racial discrimination in the housing market means that many ghetto dwellers have lengthy, expensive and time-consuming trips on public transport to suburban workplaces. Finding a job and keeping it must be very difficult in these circumstances.

Poverty in turn is associated with ill health especially in a market-led health-care system such as that in the USA. But not only must one be able to afford the treatment, but one must also be able to pay for transport to get to it or to live close enough to be able to walk. There is a connection between ease of access to medical facilities and use of them and thus with mortality and morbidity. The two-way relationships between poverty and ill health are thus exacerbated by geographical inequalities in the distribution of health facilities as Fig. 2.6 showed. A further example is given in Table 11.1 which shows the inaccessibility to health-care facilities for the carless in the ghettos of Los Angeles. That this is not necessarily an inner-city phenomenon is shown by the case of Glasgow – one of Europe's most unhealthy cities – where large areas are beyond the crucial buggy-pushing distance of 0.75 km from medical facilities, especially in the peripheral housing estates where poverty and unemployment are rife (Fig. 11.1).

11.2.3 The disabled

A recent survey in the UK reveals just how prevalent disability is in modern societies. An estimated 6.2 million people – some 12 per cent of the population – are disabled, but they are not a homogeneous group because their needs vary according to the nature and degree of disability, financial status and household circumstances. Simple aids like walking sticks or hearing aids enable some to lead near-normal lives, whilst others

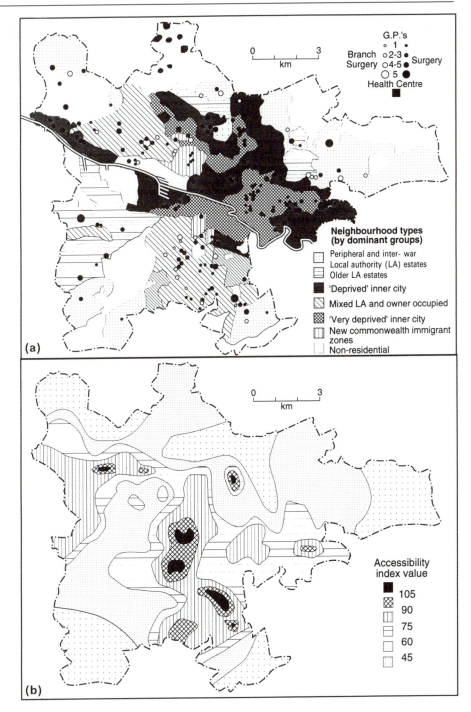

Fig. 11.1
Access to health care in Glasgow: (a) health-care facilities and housing type; (b) index of relative accessibility. *Source:* Knox (1982).

Table 11.1
Accessibility to health-care
facilities in two ghettos in
Los Angeles

	South Central		Bell Gardens	
	By car	By public transport	By car	By public transport
No. of health-care opportunities within 15 mins travel time				
General practitioners	335	11	285	18
Hospitals and clinics	40	2	41	0
Total	375	13	326	18
No. within 30 mins travel time				
General practitioners	1534	112	1529	36
Hospitals and clinics	143	14	149	1
Total	1677	126	1678	37

Source: Wachs and Kumagai (1972).

may be almost totally immobile. Gant (1992) has provided a simple
classification (Box 11.1 and Fig. 11.2) which identifies the policy implications of different disability circumstances.

However, provision of mobility is only one way of ensuring that all
people can live full lives, for in many cases the maintenance of services in
locations where people can reach them unassisted can make just as
significant a contribution. Attention to the design and maintenance
of pedestrian environments is thus essential for safe, local movement and
must include provision of adequate seating, tactile surfaces, shelter,
personal security and the removal of steps, broken paving and other
potential hazards to the frail and poorly sighted. This is vital to the
well-being of all of us, not just the mobility deprived. A dropped kerb is
as valuable for the parent with a baby buggy as it is to the wheelchair user.

11.2.4 The elderly

The ageing process affects mobility in a number of ways. Firstly, health
and personal capabilities decline with age. The most common complaints
are arthritis, rheumatism and cardiac conditions which will in their
various ways make boarding vehicles more difficult, driving a car more
dangerous and walking up hills and steps more testing. Secondly,
retirement from work brings with it a reduction in income so that trips
get shorter and thirdly, widowhood means the loss of a travelling partner,
a particularly severe reduction in mobility when the car driver dies leaving
a partner who is unable to drive.

These general problems become everyday ones when aged people have
to use transport systems which have been designed without consideration
for their special needs (Table 11.2). It helps to think of each trip as a chain

Box 11.1 Categories of transport service related to the needs of different groups amongst the disabled and handicapped

Figure 11.2 shows the classification of transport service for the disabled as derived by Gant. For each of the categories there is a specific set of policy objectives as detailed subsequently. ▶

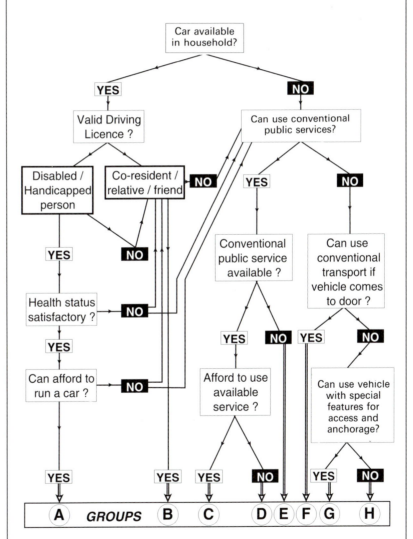

Fig. 11.2
Transport service categories arising from the needs of disabled people.

Policy objectives
Group A: the independent disabled
To enable the independent disabled to use private transport, e.g. through advice services and parking provision

Group B: the dependent disabled
To enable the dependent disabled to use private transport, e.g. through financial help towards car purchase or provision of escort services

Group C: the disabled on public transport
To facilitate journeys by conventional public transport, e.g. making buses wheelchair-accessible to providing large print system maps

Group D: the disabled poor on public transport
To reduce the financial barriers to the use of conventional public transport, e.g. through subsidised travel passes

Group E: the disabled without public transport available
To stimulate innovation in specialised transport services, e.g. through community buses with volunteer drivers or paying neighbours to give lifts in their own cars

Group F: the disabled who cannot access public transport
To provide conventional door-to-door transport services for the transport deprived, e.g. subsidised taxis

Group G: the disabled who can use adapted vehicles
To provide specialised door-to-door services for the severely disabled, e.g. by operating 'dial a ride' adapted vehicles

Group H: the 'chairfast' or 'bedfast'
To provide escorted and specialised services for hospital attendance, e.g. ambulances or specially adapted cars
Source: Gant (1992).

of experiences which begins on the front doorstep and ends when inside the building at our destination. All it needs is for just one of these barriers to intervene and the chain is broken.

There are also changes in society which make the journeys of old people more difficult. To begin with, elderly people seem more likely to be affected by the changing accessibility of services in society, particularly the progressive increase in size – and therefore decrease in number and accessibility – of shops. One of the types of shop that has disappeared most rapidly in the UK is the pharmacist, which older people are more likely to need to go to more frequently. The situation is exacerbated by the break-up of family units, so that sons and daughters increasingly do not live near to parents and cannot act as transporters for them. Furthermore fear of crime leads many elderly people to withdraw behind locked doors; for them travelling on the metro or walking down a poorly

Table 11.2
Travel barriers for the
elderly and handicapped

Physical barriers
Vehicles
    High step required to enter
    Difficult to get into or out of seats
    Seats not available
    Hard to reach handholds
    Cannot see out for landmarks
    No place to put packages
    Cannot see or hear location information
    Non-visible signs
Terminals
    Long stairs
    Long walks
    Poor fare collection facilities
    Poor posting of information
    Poor crowd flow design
    Insufficient seating
    Little interface with other modes
Transit stops
    Insufficient shelter
    Platform incompatible with vehicle
    Inadequate information

Operational barriers
Vehicles
    Frequency of service
    Driver assistance/attitude
    Acceleration/deceleration
    Information presentation
    Schedules maintenance
    Inadequate or inappropriate routes
    Too many transfers
Terminals
    Employee assistance/attitude
    Information clarity and dissemination
    Length of stops too short
    Crowd flow non-directed
    Little or no interface with other modes
Transit stops
    Poor location: for safety or convenience
    Not enough stops
    Information displayed insufficient or confusing

Source: US Dept of Transportation (1973).

lit subway can take on nightmarish proportions. We know little about the impact of fear on mobility, but for many elderly people it may be the major factor influencing their withdrawal from transport use.

In determining policy for the elderly we need to ask, 'What priority should be given to elderly and other disadvantaged groups? Should there be improvement of transport opportunities or changes in land use to reduce the need to travel?' We need more information on the effects and importance of mobility barriers to the elderly (and other mobility-deprived groups for that matter). As people live longer, so the significance of these issues will grow and as formerly mobile people age they will have different expectations about their travel opportunities. Perhaps we should ask 'What sort of access to facilities would *we* like when *we* are old in thirty, forty or fifty years from now?'

11.2.5 The young

Have you come across the riddle about the person who when going out always takes the lift to the ground floor from her fifteenth-floor apartment, but when returning home always takes the elevator only up to the eighth floor and then walks up seven flights of stairs? The fact that few guess the answer to this apparently bizarre behaviour says something about the way we think about and design transport systems for 'average' people: the person in the riddle is a child whose height limits her to reaching the button on the elevator wall marked '8'. Only in a few years will she have grown enough to reach the '15' button!

Of course, being young is only a temporary phenomenon, yet in a way the travel problems that afflict children are a mirror for the difficulties that we all face from time to time in our lives. The pregnant woman, the athlete with the broken leg, the perfectly able man carrying a large suitcase – all find difficulties in getting around, just as children have problems with escalators, doors and steps. That these are temporary disadvantages does not make them any less real for those who are facing them. If our transport systems do not meet the needs of children, then they will not meet our needs as adults either, with our huge variation in size, medical condition and mental capabilities.

There is a distinctive geography of transport for children in that they have trip lengths, purposes, timing, modes and routes that differ from adults in many ways. As many an anxious parent knows, a child is highly likely to deviate from 'logical', straight routes home, attracted by parks, alleyways, ice-cream vendors and friends' houses. However, in industrialised societies this geography is changing as children's safety is increasingly threatened by the car. The authors of a study of children's independent mobility entitled *One false move* have this to say (Hillman et al 1990: 110–11): 'Transport policies in all motorized countries have been transforming the world for the benefit of motorists, but at the cost of children's freedom and independence to get about safely on their own – on foot and by the bicycle that most of them own. This change has gone largely unnoticed, unremarked and unresisted.'

They note that whereas in 1971 80 per cent of English 7- and 8-year-old children were allowed to go to school on their own, by 1990 this figure had dropped to 9 per cent (Fig. 11.3). Though parents have increasing concern

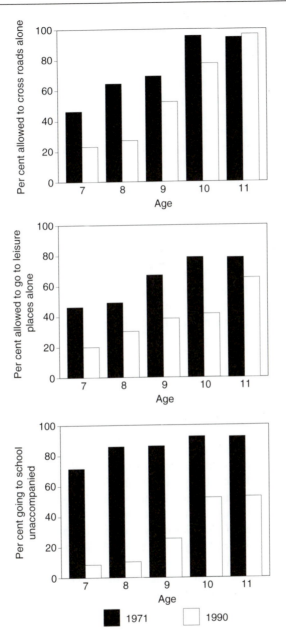

Fig. 11.3
Independent mobility of English junior schoolchildren according to age, 1971 and 1990.
Source: Hillman et al (1990).

now over the possibility of their children being molested or abducted, it is principally the increase in motorised travel that has been responsible for the decrease in children's independence.

The escorting that this implies has very substantial impacts. Not only does it entail considerable cost in resources and contribute to traffic congestion, but it also seriously constrains adult opportunities. The apparent benefit of wider car ownership is being offset by constraints on parents' freedom resulting from their increased need to escort children in conditions of greater traffic danger. The costs of escorting are now reckoned at more then £20 billion per year in the UK, with over one-quarter of all women's journeys being as chauffeurs, which has prompted one commentator to observe that women have escaped being tied to the sink only to become chained to the steering wheel. For the children, it removes a routine means for them to help maintain their physical fitness. It also limits opportunities for the development of their independence, because they are increasingly unable to explore their neighbourhood and develop their sense of space and territory. Children in Western countries used to say 'I'm going out to play'; now they are more likely to say 'I'm going to my room'. They are being conditioned to accept that the car is the only method of mobility, so that their lives are increasingly taking place on isolated islands, accessible only by car. We should not be surprised if there are negative impacts on a child's development if every activity in which it participates begins with a car trip.

11.2.6 *Women*

Women's trips are similar in number to men's, but they tend to be more local or short distance, more off-peak, less often in a car and more likely to be for escort purposes, particularly involving children. Women have to manage complex chains of trips, such as taking small children to the nursery, older children to school, going to the shops, going to part-time work, collecting the children in the afternoon, taking them to sports or social engagements and so on. The different experience of women is a function not only of their roles but also of their differential access to transport. In Germany only about three of every ten women have a car available at any given time, whilst about 80 per cent of men always have a car available. As Box 11.2 shows, women's multiple roles result in a fragmentation of time and their lack of access to cars leads to restricted mobility. Together, they result in women being locked into worlds of severely limited physical space which contrast markedly with the extensive and uninterrupted physical worlds of car-driving men.

Another problem affecting women is lack of personal security whilst travelling or waiting to travel. Lonely and poorly lit streets, subways, unattended stations and nearly empty train carriages have the highest rate of incidents. A survey in London has found that 30 per cent of women would not go out alone after dark, rising to 60 per cent for women over 60 years old. This not only means increased social isolation and dependence but it also creates a vicious circle, with emptier streets an even

Box 11.2 But I'll never make it in time ...

The Swedish geographer Hägerstrand has outlined a framework for the study of time geography which is based on the environment of resources and opportunities which surrounds each individual. If we ▶

(a)

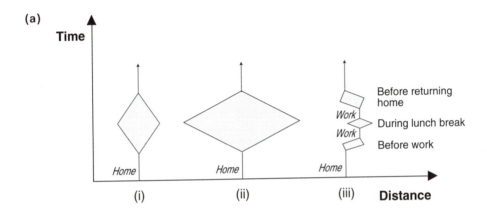

(b)

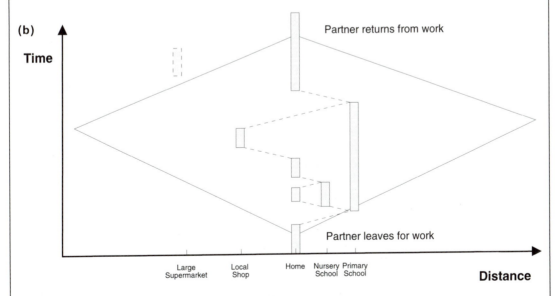

Fig. 11.4
Time–space prisms. (a) Distances which can be travelled: (i) by a pedestrian; (ii) by car in an allotted time; (iii) by a working individual. (b) The daily time–space prism of a woman with a partner and young children.

let the horizontal axis of a graph represent distance and the vertical axis time, the effective range of a person during the day can be drawn as a 'prism'. As Fig. 11.4(a) shows, the extent of the prism depends on the modes of transport available. Every stop – at shop or library for example – results in the prism shrinking in proportion to the length of stay.

The time budgets of women are very different from men because of their more complicated activities, particularly in relation to escorting children as Pred and Palm (1978) have shown. Moreover, their reliance on slower modes of transport such as feet or bicycles, or on less flexible ones such as buses, restrict the range of destinations that can be reached in the time available between essential activities.

Figure 11.4(b) shows the prism of a woman with a partner in full-time employment and two children, one in school and the other at nursery. The family car is taken by the partner with him to work, leaving the woman to deliver the oldest child to school on foot, dropping off the toddler at the nursery on the way back home. After a short time at home completing domestic chores she has to return to the nursery to collect her youngest and bring her home for lunch and her afternoon sleep. Some essential shopping is then done, but only at the local store, because there is insufficient time to get to the bigger (and cheaper) supermarket and still get back in time to collect the elder child from school.

It is obvious from this imaginary case that the woman's time–space prism is highly constrained. If she had a car she would clearly be able to engage in a wider range of activities, but her gender-role constraint would still be very powerful. For example, imagine the effect on her day if she had to wait in to receive a delivery or let in someone to repair the television. More significantly, if she was raising the children without a partner, she would need to fit paid employment into her day, with the likelihood that the choice of workplaces – and thus the amount of job satisfaction and pay – would be severely constrained by proximity to home, school and nursery.

greater threat to the remaining people who are not locked in their own cars.

The situation is self-perpetuating: women's lesser mobility and lack of transport options reduce their access to resources which in turn reduces their chances of changing the existing balance of power between the sexes. The sprawling of suburbs undermines the viability of public transport and increases the separation of home and workplace, placing even greater limitations on women's employment options. Road construction has had adverse effects on the pedestrian environment and thus on women (and the elderly) creating, for example, dangerous no-go areas at night such as multi-storey car parks or subways. The rationalisation of services – shops, hospitals, leisure complexes and schools – has created larger but fewer units which are thus on average further away, impacting particularly on women who make more service-orientated trips than men. This is because they are more likely to be responsible for escorting a child or relative to school or hospital and are more likely than men to work in such places. However, they are less likely to have access to the cars that are needed to get there. These trends all assist car users at the expense of those without and thus in general aid men at the expense of women. The mobility gaps identified earlier in class and age terms thus can also be seen to exist along gender divisions: in all cases the motor age is widening them.

In developing countries too, there are a series of gender assumptions that underpin the provision of transport. As in the advanced nations, there

Plate 11.1
A woman walks home from the fields, carrying sugar cane across the Birchenough Bridge in SE Zimbabwe. African women frequently have greater need for transport then men but much inferior access to it.

is often a mistaken assumption by planners that the man and the woman have equal power and control over resources. It is also assumed that the household consists of a male breadwinner and a woman homemaker, when in reality one-third of the world's households are headed by women, with the figure reaching 50 per cent in some Latin American and African cities. The productive work that women do as income earners and in community management, such as in organising the provision of basic services and education, is often also underestimated. For all of these reasons, women's requirements are frequently ignored by transport planners. Indeed, where data are collected, women's current travel patterns are confused with their travel needs. Yet it has to be understood that women travel less than men in part because current transport provision makes it difficult for them to travel more.

There may also be cultural impediments to greater mobility for women.

In Africa women bear the brunt of haulage, whether of food, fuel, water or babies, yet they are often excluded from utilising any technological assistance with this task, such as carts or bicycles. In many places it seems to be 'improper' for women to ride bicycles, but does this evolve from traditional values, Victorian Western etiquette, or is it 'cultural'? True, in some Muslim societies, even if accessible and safe transport enabled women to fulfil their roles in the company of a male relative, it does not seem likely that they would take advantage of these services without changes in attitudes of both women and men. Yet two African cities where cycling by women is more common – Ouagadougou (Burkina Faso) and Maroua (Cameroon) – are predominantly Muslim.

The provision of transport may have its own part to play in changing attitudes. However, it remains true that much can be achieved by way of a 'gender-aware' planning process, which is necessary to remove the biases which currently provide inadequate transport for women, limiting their access to work and constraining their ability to use services necessary for their reproductive and community roles. The losers in the end are not only women themselves, but the children and others who depend on them.

11.3 Accidents

We now turn away from considerations of transport availability, quality and choice and towards Appleyard's third social goal, that of lowering the undesirable impacts on human environments. Our first example is that of transport accidents.

11.3.1 The statistics: who, what, how, where

World-wide more than a quarter of a million people are killed in transport accidents every year and more than 13 million injured. Your level of safety depends very much on what mode you are using. For example, rail transport is so safe that fatality rates of one per billion passenger-kilometres are not unusual; in most countries deaths from trespassing on the track exceed those in train accidents. Accidents on water tend to be frequent and small scale, so that numbers are uncertain. Crew members are most at risk because of the high personal accident rate on board ship, but there are occasional large ferry disasters involving much loss of life amongst the passengers. Aircraft crashes tend to be fatal but rare: with some 2600 deaths per annum, the risk of death to an individual air traveller within a year is about 1 in 200 000.

In all of these modes occasional disasters may inflate the figures and distort public perceptions of safety, as their spectacular nature and large loss of life mean that they receive more media attention. But the realities are that more than 90 per cent of casualties are due to road traffic accidents (RTAs) and that road travel is by far the most dangerous means of movement. World-wide, the chance of you dying on the road is about 1 in 20 000 every year, some 10 times greater than the risk you take when flying. Obviously any form of movement carries an element of danger, but as road travel is so much more dangerous than any other commonly used form of transport, this chapter will concentrate on the causes and consequences of RTAs and on ways of ameliorating them.

It is inconceivable that the car would have been adopted had it been known in 1885 what we now know – that it would kill some 15–20 million people in its first 100 years. It is hard to grasp the fact that world-wide there are 50 million RTAs per annum, of which 6 million result in injury. By the time you have read this paragraph there will have been seven more accidents and another person injured or killed. In the past it has been a tragedy that has most afflicted the more developed countries, but the trauma is now being exported to developing countries. Of those who die in the latter between the ages of 5 and 44, 10 per cent are killed in RTAs, a higher proportion than from any disease listed in the World Health Organisation's statistics. For many countries the cost of these accidents amounts to more than 2 per cent of GNP.

In most industrialised societies (though not the USA) annual fatalities are now decreasing: over the past 20 years Germany has reduced its toll from 20 000 to 8000 whilst the UK's has fallen from 8000 to 4000. In general it seems that fatality rates are very high at early stages of motorisation, decreasing as motorisation advances. In the 1990s they range from less than 2 fatalities per 100 million vehicle-kilometres in the USA, through 9 in the CIS, to 20 for several African states and 30 for India. This reflects the growth of safety cultures and the changing composition of the vehicle fleet, as the ratio of vulnerable two-wheelers falls. Everywhere, though, the proportion of pedestrian casualties increases.

Relatively few RTAs happen on high-speed roads, with the great majority occurring in urban and suburban areas. A minority of the victims are car occupants and a very high proportion are pedestrians, cyclists and motor cyclists. Per distance travelled, motor and pedal cyclists are 20 and 10 times respectively more vulnerable than occupants of private cars. The safest vehicle is a heavy truck, provided that you are in it. If you are not, it is the most dangerous, as Table 11.3 shows with respect to the UK.

Necessary though these statistics are in setting the scene, it has to be stressed that there are serious doubts about their reliability. The British Medical Association reckons that 30 per cent of RTA casualties seen in hospital are not reported to the police and that at least 70 per cent of cyclist casualties go unreported. This is bound to have the effect of underestimating the significance of these injuries when it comes to policy

Table 11.3
Fatality rates by vehicle user
in the UK

	Bicycle	Heavy truck
Fatalities per 100 000 km	4.9	0.2
In fatal accidents percentage of fatalities not in or on this vehicle	4.0	93.0

Source: Department of Transport (1991).

for casualty reduction. The statistics are also flawed in the sense that they are an unreliable measure of safety. We have to ask, what do we mean by 'safety'? Something is dangerous if it has the potential to harm us, like an airport runway or a freeway. If these are recognised as dangerous and avoided by people on foot, they do not become 'safe', even though the figures may show that no pedestrians have come to harm there. This problem surfaces repeatedly in campaigns by communities to slow traffic or get a crossing built on a busy road in a residential area. Parents, knowing that the road is dangerous, refuse to let their children out to play and escort them whenever they have to cross the road. Because of this the road has relatively few pedestrian accidents and thus appears from the statistics to be so 'safe' that there is no need for any special treatment, when in reality it is deadly dangerous.

11.3.2 Reducing the toll

There are two broad approaches to accident amelioration: prevention and protection. The prevention (pre-crash) phase attempts to reduce the conflict which leads to collisions. This focuses on vehicle design (e.g. on components such as brakes, lights and steering that can fail disastrously) and on effective maintenance; on road design and layout to reduce collision risks (e.g. non-skid surfaces, crash barriers, better alignment); and on traffic management and control to minimise conflict by separating different types and streams of users, such as through-road signs, traffic lights, cycle paths and so on. Modifying road-user behaviour also comes into this category, with many countries adopting programmes of public education, such as those on drinking and driving.

The second approach to accident amelioration is to introduce protective measures in order to reduce the severity of injuries caused by accidents, the post-crash phase. The main emphasis here is on vehicle features such as side-impact bars and seat belts for cars, and leg guards and helmets for motor cyclists. Measures may also help other road users, such as smoothing of vehicle shape to minimise injuries to pedestrians and the fitting of under-run guards on large trucks.

Plate 11.2
The separation of transport modes from people can achieve traffic safety, but often at the cost
of social safety. The large amount of space required for the various transport modes in such
schemes forces up population densities on the remaining non-transport land, producing an
unsatisfactory social environment, with personal safety endangered by mugging in the parks
between the flatblocks, on the walkways and in the elevators. The traffic-segregated design of
the Bijlmermeer, Amsterdam has produced just such a set of social problems.

11.3.3 Risk compensation

The problem with these approaches is that they assume that drivers will
not change their behaviour in the light of the new, 'safer', lower-risk
conditions. However, it can be argued that there will be at any given time
a level of risk that an individual is prepared to tolerate; if the level of risk
is reduced by a safety measure, its effect will be nullified by changes in
behaviour that re-establish the level of risk with which people were
originally content. It is like the rock climber who would attempt much
more dangerous (risky) climbs when provided with ropes and a helmet
than he or she ever would dream of climbing barefoot in a tee-shirt! An
example related to road accidents could be that of a vehicle which can
take a particular bend at 60 km per hour with a known risk (say 10 per
cent) of skidding out of control. If better tyres with twice the grip are
developed and fitted does the driver continue to take the bend at 60 and
reduce his risk to 5 per cent? Or does he maintain the level of risk with
which he was previously comfortable and take the bend faster, say at 100?
In other words, would he accept the innovation as a safety benefit or
would he trade it in as a performance benefit? And what now happens
to the cyclist hidden round the bend that the car could have avoided at
60 km per hour but cannot at 100?

Table 11.4
Casualty rates per billion
person-kilometres by mode
of travel, 1978 and 1988

	1978	1988	Percentage change 1978–88
Walking			
Killed	97	70	−28
Seriously injured	757	645	−15
Cycling			
Killed	62	46	−26
Seriously injured	867	938	+8
Car users			
Killed	7	4	−41
Seriously injured	92	53	−42
Bus or coach passengers			
Killed	1	0.3	−70
Seriously injured	21	20	0.8

Source: Calculated from volumes of Department of Transport, *Road accident statistics Great Britain*, HMSO.

This theory is known as 'risk compensation' and is supported by abundant evidence, though rarely can cause and effect be definitely proved. The geographer John Adams (1985) has cited the example of the introduction of compulsory seat-belt wearing for car drivers in the UK, which he argues has led to drivers feeling safer and compensating by driving more recklessly, thus causing more injuries among the rest of the road users not so protected, i.e. back-seat passengers, pedestrians and cyclists. The accident rates for these groups have indeed risen at the same time as those for drivers have been reduced, evidence of the migration of accident victimisation from less vulnerable to more vulnerable groups. Table 11.4 shows that although casualty rates have been falling in recent years in the UK, the relative riskiness of the 'soft', unprotected modes is growing. For example, in 1978 the fatal and serious injury rate per kilometre for cyclists was nine times higher than that for cars whereas by 1988 it was seventeen times higher.

The proponents of risk compensation theory argue that its effects will tend to undermine the traditional methods of tackling accidents, the 'three Es' of education, enforcement and engineering. In education, for example, British motorcyclists who have not undergone training actually have less accidents than those who have been trained, because it seems that the latter group now take more risks in the belief that they are better equipped to deal with them. As engineering makes road surfaces safer, so drivers go faster and, as we have seen, enforcement of seat-belt laws has shifted the burden of risk to vulnerable road users outside.

Where does this leave us? It is true that RTA fatalities are declining sharply in many developed countries, but it appears that this is more a consequence of improved emergency medical procedures, increased separation of traffic types (freeways, etc.) and continuing withdrawal from conflict situations by those who are most at risk. In other words the amount of walking and cycling is reducing, especially amongst children. They are increasingly confined to home, garden and playground; they are escorted *en route* between these places; and increasingly this takes place in cars. The reduction in fatalities in industrialised societies is thus not a technical achievement but a shift in spatial behaviour. We are yielding territory to the car. On that territory, car drivers encased in steel and protected by their crumple zones, impact bars and air bags can career into each other almost with impunity, walking away from horrendous accidents almost without a scratch. But woe betide the soft human pedestrian who gets mixed up in this real life game of bumper cars. The sensible thing is clearly to get out of the way and let them get on with it. And so we do – if we can. As roads get busier so are they avoided the more. If children played in the street in the way that they did only a generation ago, there would be slaughter on a massive scale. The reduction in fatalities is welcome, but what is the cost to our growth and development, to our humanity, to our knowledge of the places around us?

11.3.4 Improving road safety: where does geography fit in?

These traditional, 'three Es' approaches examine faults in the system and attempt to correct them – but do not examine the operation of the system itself. They do not address the fact that there is historically a close relationship between the number of accidents and the level of mobility, so that attempts to reduce one without reducing the other are unlikely to be successful. Yet mobility is promoted as a desirable future for us all; new roads, it is said, save lives because they are safer than old ones, per kilometre travelled. This is true, but new roads also encourage more travel, thereby increasing the population at risk of accident and thus they do not reduce total casualties. Opening a new freeway that halves the accident risk per kilometre does not make safer the person who now buys a house twice as far from work as before in order to gain from the speed advantages of the new road.

Without wishing in any way to imply the superiority of geographical approaches over others, it is evident that the 'three Es' have, to date, collectively failed to remove the scourge of RTAs from society. We are left with the prevailing view that 'accidents are accidents', an inevitable, inescapable consequence of human frailty in a complex world, a regrettable but acceptable tragedy (as long as it does not happen to *you*). But accidents are 'space–time events', collisions between vehicles and people going to various destinations at various times in order to meet some need that could not be satisfied at the places where the journeys started. In other words, accidents are a result of movement which is spatially and temporally explicable. These factors are ignored in the traditional view –

Plate 11.3
Temporal measures may be used to achieve safety from traffic danger as well as improve
the environment for pedestrians. This is illustrated by the market days which characterise
many European towns – as here in France – in which motorised traffic is excluded from the
market area

yet they can actually be influenced much more by policy than can human
error or behaviour.

Little can be done to modify the time element of accident causation,
though the daytime vehicle bans in some city centres that make shopping
on foot safer and the practice of operating temporary speed limits outside
schools at the times that children are entering and leaving are useful
strategies. In northern countries there is also evidence that RTAs could
be reduced by changing the clocks so that evenings are lighter throughout
the year and the most vulnerable groups are able to get home before
darkness falls. However, generally speaking, the opportunities for accident
reduction are greater through spatial actions than through temporal
measures. Modifying the urban fabric so that new spatial structures
generate patterns of movement that are inherently safer tackles the
problem at its root rather than treating its consequences.

	Scale	Policy response
Table 11.5 Relationship between spatial scale and policy responses for road traffic accidents	Local/particular Neighbourhood Sector of city City-wide	black spot eradication; small scale engineering residential design; traffic calming traffic management/routeing public transport system; land-use planning

Source: Whitelegg (1987).

The geographer John Whitelegg (1987) has championed this approach, suggesting that there is an appropriate type of spatial policy for different scales of consideration (Table 11.5). If the problem is seen to be very local then the remedy can often be found in the manuals of engineers, who may eradicate a black spot through better road surfacing, improved signs, gentler curves and so forth. At the neighbourhood level, such as in a residential area around a school, traffic-calming measures may be instituted in order to slow traffic and change the balance of priority in favour of softer modes of travel and vulnerable residents, as Chapter 8 showed. If an attempt is being made to reduce accidents over a whole sector of a city, then the policy must aim to reduce traffic speeds and volume over a wide area. Small-scale measures would not be appropriate at this level and instead area-wide traffic calming and other traffic restraint measures could be used, as outlined in the discussion on environmental traffic management in section 8.7.6.

At the large-scale, city-wide level of analysis there are two interconnected approaches. Firstly, there is a very clear relationship between modal split and the level of RTAs and this is clearly amenable to policy manipulation. The casualty rate per passenger-kilometre for buses is about a quarter of that for cars in the UK and that for trains is lower still. The case for public transport can thus be made in terms of its accident-reducing potential just as strongly as it can be made on grounds of reduction of congestion or environmental damage. Secondly, land-use planning can bring origins and destinations closer together which reduces distance travelled and therefore the opportunities for conflict. Moreover, keeping journeys short increases the chance of them being made on foot or by bicycle, neither of which cause danger to other road users.

11.3.5 An overview: policy for amelioration of RTAs

Given that, on average, human factors contribute to over 95 per cent of all accidents, it is obvious that total road safety is not achievable. It is also clear that there is no simple, single approach to events that vary so much in their scale, participants, location and causation. In developing countries, for example, the situation is worsening and due to the lack of money improvements in road maintenance and police enforcement will be difficult. Road user behaviour, knowledge and attitudes in these countries are different so there will be uncertainty over transferability of counter-measures from advanced nations. Nevertheless, it seems that improved education, training and enforcement provide greater potential for improving road safety than they do in MDCs. Needed too are better data collection, improved vehicle maintenance and testing, traffic training for children, firmer road safety legislation and improved emergency medical services.

Mayer Hillman – whose work has already been encountered in this chapter – has identified a range of measures that are needed to reduce danger on the road in the industrialised countries. Writing with Stephen Plowden (1984), he lists the principal ones as:

- limitations on the speed and power of motorcycles;
- lower general speed limits and the use of modern technology to help enforcement;
- recognition of the bicycle as a major mode of transport and investment in the provision of safe cycling facilities;
- fairer treatment of public transport, both in management and in investment, recognising among other things its low accident rates;
- land-use and locational policy to keep journeys short and encourage the use of the safer modes of transport;
- management of residential neighbourhoods and redesign of streets to keep motor traffic subordinate to pedestrians.

It is interesting to note that there is a clear recognition in these recommendations that geographical approaches should be taken on board in any broad attack on accidents. For the past 100 years, though, we have had a non-spatial approach – an apparent concern to reduce accidents without reducing mobility – which has had scant success. Such 'traditional' ways of tackling the problem may make each kilometre safer, but do nothing to stop the increasing number of kilometres being driven and therefore the total casualties. The adoption of spatial approaches holds great promise for reducing automobility and increasing the safety levels of the smaller number of kilometres travelled. Geography and geographers have a critical role to play in the future, for we are now dealing with a global trauma, what Adams (1981: 123) has described as 'the biggest game of chance in the world'.

11.4 Loss of the street to traffic

Our second examination of the impact of transport on the human environment takes the form of a brief examination of the loss of the street to traffic. The street has always been a meeting place; indeed democracy was born in the Agora, the market-place, in Athens. It has, too, been a rallying point, a community place; it was the street that toppled the Bastille, that resisted the tanks in Budapest and Prague and Tien An Men Square. It has always been the place for children's games and sport, buying and selling, celebrations and processions, music and parades, friendship and love, community spirit and neighbourliness, gangs and warfare, where you learned about life and became 'street-wise'. But modern high-speed travel has, in a very short space of time, changed all that. One of the consequences of the increasing flows of vehicles is that, in order to protect themselves, people have stopped trying to use the street in the old ways and, except where cars are banned or controlled, have moved their meeting places inside to cafes and halls, or abandoned meeting altogether. Pushkarev and Zupan (1975: Frontispiece) put it simply:

> When everything was finished, when our beloved planet assumed a fairly habitable look, motorists appeared on the scene . . . gentle and intelligent pedestrians began to get squashed. Streets, created by pedestrians, were usurped by motorists . . . roadways were widened . . . sidewalks were narrowed . . . pedestrians began to cower in fear against the walls of buildings.

The classic work by Appleyard and Lintell (1969) in San Francisco (Box 11.3) showed the destructive influence of increasingly heavy traffic on the lives of people and their use of the street. Because the residential street environment is a matter of continual concern to residents but of only fleeting interest to the passing driver, there is a strong argument that it should be designed primarily for living, not travel. In other words if 'livability' is to be promoted in residential areas the city street has to be recaptured from the car.

Box 11.3 What is your street like as a place to live?
In the late 1960s the American researchers Donald Appleyard and Mark Lintell questioned residents on three San Francisco streets that were differentiated mainly by their traffic flows, and which they christened 'Heavy', 'Moderate' and 'Light' Streets. They were interested in finding out what effect traffic had on people's sense of territory and privacy. They found that people living on Light Street considered all or part of the street to be 'their' territory. This was not so on Heavy Street where, for some, territory was even confined to their own apartment and no further.

When they asked about social interaction the patterns in Fig. 11.5 appeared. On Light Street, people had three times as many friends and twice as many acquaintances on the street itself as those on Heavy Street. Contact across the street was much less frequent on Heavy Street and this particularly affected children and the elderly who had far fewer social contacts on Heavy Street than on Light Street.

The authors commented that:

> On the one hand Light Street was a lively close-knit community whose residents made full use of their street. The street had been divided into different zones by the residents. Front steps were used for sitting and chatting, sidewalks by children for playing, and by adults for standing ▶

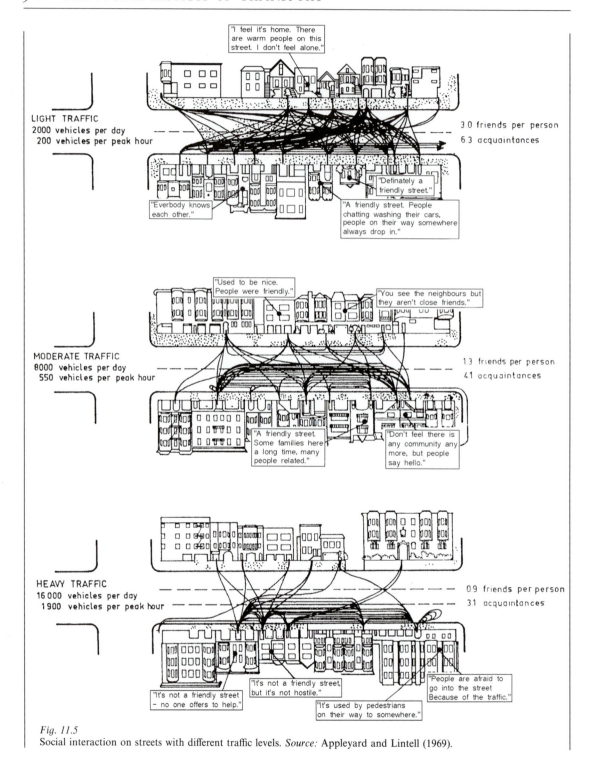

Fig. 11.5
Social interaction on streets with different traffic levels. *Source:* Appleyard and Lintell (1969).

and passing the time of day, the roadway by children and teenagers for playing more active games like football. However, the street was seen as a whole and no part was out of bounds.

Heavy Street, on the other hand, had little or no sidewalk activity and was used solely as a corridor between the sanctuary of individual homes and the outside world. Residents kept very much to themselves so that there was no feeling of community at all . . .

and they concluded:

Heavy traffic is associated with much less social interaction and street activity. Conversely, a street with little traffic, and many families, promotes a rich social climate and a strong sense of community. Heavy traffic is associated with withdrawal from the physical environment. Conversely, residents of the street with low traffic show an acute, critical and appreciative awareness of and care for the physical environment.

11.5 Transport and health

11.5.1 Basic relationships

Any motorised means of transport will cause an element of danger and nuisance. The noise and vibration from trains and lorries are a major irritant to many people whilst the interruption to sleep and fear of accident that comes with living under the flight path of a major airport is well known. Much of the grime on our civic buildings comes from traffic and most people instinctively turn their noses up when lorries or buses belch out diesel smoke. All of these things are unpleasant, but until recently there was no widely understood connection between them and human health. Unpleasant, yes – but dangerous? Unfortunately the evidence is beginning to accumulate and it is not good news for any of us exposed to traffic.

The broad influences of transport on health have been summarised by the Transport and Health Study Group in England and are shown in Table 11.6. The health-promoting influences concern access to places or facilities, but in many urban areas communities are entombed in localities with inadequate access to jobs, recreational facilities and health care, shown in the examples of Los Angeles and Glasgow in section 11.2.2. On the health-damaging side, heavy traffic results in withdrawal from the street and social isolation, a factor shown by research to raise mortality rates, so that even before we consider the specific impacts of air pollution from exhaust emissions it is evident that heavy traffic is bad for your health.

Type of effect	Results
Health promoting	Enables access to: employment, education, shops, recreation, social support networks, health services, countryside; provides recreation and exercise
Health damaging	Accidents; pollution: carbon monoxide, nitrogen oxide, hydrocarbons, ozone, carbon dioxide, lead, benzene; noise and vibration; stress and anxiety; danger; loss of land and planning blight; severance of communities by roads

Table 11.6
Ways in which transport influences health

Source: Transport and Health Study Group (1991).

11.5.2 *Air pollution*

Most of the health-damaging effects of transport result from the fact that vehicles are a major and growing source of air pollution. The threats to the natural environment and thus to human health on a global scale posed by greenhouse gases, ozone depletion and acid deposition have been discussed in Chapter 10 and need not be repeated here. However, evidence is now emerging that air pollution at the local scale is also a significant health hazard, causing a number of basic adverse health effects such as aggravating cardiovascular and respiratory illness, reducing lung capacity and contributing to the development of diseases including bronchitis, emphysema and cancer. These effects are not evenly distributed amongst the population, impacting particularly on children, pregnant women, the elderly, asthmatics and people with existing chest, heart or lung problems, groups which, for example, form some 38 per cent of the UK's population. A connection between vehicle emissions and asthma is strongly suspected if not yet proven, with the number of hospital admissions for childhood asthma increasing fivefold in the 1979–89 period in the UK.

Evidence to support the link between air pollution and ill health on an individual basis has been outlined by John Whitelegg (1993). He cites studies in Colorado indicating an association between traffic density and childhood cancer; in Switzerland implicating particulates and polycyclic hydrocarbons in adult tumours; in Hamburg and Montreal associating exhaust and combustion products with various cancers; and a study which showed that reducing ozone pollution to meet federal health standards would eliminate every year in the Los Angeles area 107 million headaches, 180 million sore throats and 190 million eye irritations. A very cautious and conservative interpretation of the American Lung Association's calculations indicates that vehicle emissions prematurely kill between 10 000 and 24 000 Americans annually.

One specific and well-known pollutant from transport is lead, which has long been used as an additive to vehicle fuels despite its known toxicity; more than 90 per cent of the lead in the urban environment has been derived from this source. It is a major health hazard at any level of concentration and is associated with high blood pressure, kidney and liver damage and retardation in cognitive development in children. In the USA one study has found that three-quarters of children have blood lead concentrations over the level linked with adverse biological effects. These have included inability to concentrate, hyperactivity and lowered IQs. Despite the fact that its use has been reduced in many countries recently, half a century of emission has contaminated soils for probably hundreds of years to come. Plants grown in such soils are toxic; 20 per cent of outer London's land is so heavily contaminated with lead that it is unsuitable for growing food and even rural areas of the UK have lead deposition rates some 800 times those of a car-free environment such as Greenland.

11.5.3 *An overview of transport and health*

We have seen here that the health-damaging potential of transport is extremely large. Urban air quality is deteriorating and the principal source is emissions from vehicles. The primary impacts are on the elderly, the sick, pregnant women and children. It is the effects on the latter that are the greatest cause for concern, for no transport system can be thought of as sustainable if it causes health damage to future generations. But the impacts of transport on health go way beyond air quality as shown by the case studies in this chapter of inadequate access to health-care and health-nurturing environments; of the scourge of accidents and the stress and anxiety caused by fear of them; and of the destruction of social relationships and communities by heavy traffic. Nor should it be forgotten that it is traffic that is the biggest reason why more people do not walk or cycle and thus get the exercise and fitness benefits of these activities. In the UK, even raising cycle use to only 20 per cent of all trips would save £217 million per annum in reduced heart disease and working days lost, augmented by £140 million in pollution reduction and £191 million in accident savings (Table 8.2). It seems that reducing motorised traffic would not only be good for our health but for our welfare too.

11.6 *Concluding summary*

No apology is made for the concentration in this chapter on road traffic as a source of nuisance and blight on people's lives. Of course, around the world, all forms of transport have impacts on the human environment, ranging from the stress caused to people living under aircraft flight paths, to the smell and disease risk associated with animal-powered transport in cities. But almost universally – and certainly in every city – it is road traffic that is the source of noise, intrusion and danger, that damages health, undermines community spirit, alarms parents and intimidates pedestrians.

Clearly universal car ownership would lead to insurmountable environmental and social side-effects. Public transport must therefore have a major role to play, yet in most motorised societies it is deteriorating in quality as its competitiveness is eroded and its service is undermined by congestion. Moreover, many people have personal characteristics and limitations that make it very difficult for them to use public transport in its conventional form. Obvious examples would include disabled people who find that buses and trains are difficult to get on or ride in; the frail and elderly who cannot endure long waits for unreliable services and worry about the absence of necessary facilities such as seating, lighting, timetable information and toilets; small children who cannot manage escalators and stairs; women who need staff to be present to protect them from harassment or assault; the unemployed who cannot afford the high costs of fares; and the rural dweller who cannot get to work because of the infrequency of services or their inappropriate routeing.

All of these groups may find that their mobility is seriously constrained and for many the solution will be seen in terms of getting a car. Other social impacts of transport that we have seen in this chapter also have the effect of encouraging the further growth of car ownership and use. For the parent anxious about the child's journey to school and the cyclist scared by growing volumes and speeds of traffic, the temptation is to join the motorised ranks and use the car as protection. For the resident of the formerly quiet and friendly street whose life is being ruined by the noise and intrusion of traffic, the temptation must be to move out to a quieter, suburban neighbourhood. But to do that a car is essential, for suburban patterns of shopping, friendship, schooling and work involve movements over distances too great for non-motorised means.

However, attempting to solve growing social pressures through buying a car is not only not possible for all groups – such as chairfast people, the blind and children – but in any case simply makes the travel environment more intolerable for those remaining without personal access to a car and increases the pressure on them to find some way of making the same decision. And so it goes, that with each decision to buy a car – a personal solution – everyone else's problem becomes that little bit worse. It is not the first time in this book that we have encountered the paradox of 'the tyranny of small decisions' – that what are perceived individually as rational, sensible decisions lead to collective disaster when taken by everyone. What kinds of policies might assist us to avoid such an outcome will be explored in the next Chapter.

Further reading

Adams J 1985 *Risk and freedom* Transport Publishing Projects

Appleyard D, Lintell M 1969 The environmental quality of city streets: the residents' viewpoint. *Journal of American Planning Association* **35**: 84–101. Reprinted in de Boer E 1986 *Transport sociology: social aspects of transport planning* Pergamon, pp 93–120

Hillman M, Adams J, Whitelegg J 1990 *One false move . . .: a study of children's independent mobility* Policy Studies Institute

Hillman M, Henderson I, Whalley A 1976 *Transport realities and planning policy: studies of friction and freedom* Policy Studies Institute

Plowden S, Hillman M 1984 *Danger on the road: the needless scourge* Policy Studies Institute

Preston B 1990 *The impact of the motor car* Brefi Press

Williams H 1992 *Autogeddon* Jonathan Cape

$I2$ *Transport policy*

> **Transport problems? Planning? . . . now Policy**
>
> Previous chapters have described urban, rural, environmental and social issues in transport and discussed ways in which transport *planning* might provide appropriate solutions. It is time now to take a broader view and examine *policy* responses at national and international scales. Regulated and free-market stances are compared and a series of examples – of the UK, the USA, the Netherlands and the European Union – used to illustrate the diversity of existing policies. We conclude that the dominance of free-market approaches in the late twentieth century is creating problems which the market seems unable to solve, thus calling forward imaginative but untried ideas for managing our transport systems with sustainability in mind.

12.1 The evolution of policy

12.1.1 The four 'ages' of transport policy

Transport policy may be defined as the process of regulating and controlling the provision of transport to facilitate the efficient operation of the economic, social and political life of the country at the lowest social cost. In practice this means assuring adequate transport capacity and efficient operations to meet the needs generated by the nation's geographical array of activities.

How transport is regulated and controlled is currently one of the most controversial areas of public policy, with two contrasting approaches in conflict. One argues that state intervention is a prerequisite to an effective policy and the other that intervention is a barrier to any effective policy;

in simple terms it is the role of the free market versus the public interest vested in the state. The late twentieth century has seen an unprecedented attack on regulation but this cannot properly be appreciated without an understanding of why controls were established in the first place. The evolution of policy can be examined by reference to what Button and Gillingwater (1986) have called the four 'ages' of transport policy. In essence the argument is that after the establishment of regulation in the nineteenth century, there was general agreement for more than 50 years in the UK, USA and in other countries such as Canada about the need for a formal policy to regulate transport, about its objectives and its content. In the 1950s the consensus began to be questioned and by the 1960s and 1970s was the target of open assault.

12.1.2 *The railway 'age'*

For 100 years in the UK from 1830 policy and legislation were directed almost solely at a single industry, that of the railways. It was sparked by increasing concerns over the monopoly power of the railway companies and sowed the seeds of many future aspects of transport policy. Up to 1875 most legislation concerned safety and the need to ensure efficient through services between the different companies, but thereafter control began to be extended to prices and culminated in the management of the railways for the public good during the First World War by a government-appointed Railway Executive Committee. This experience led to the enforced amalgamation of the 120 separate companies into four groups under the Railways Act of 1921 in an attempt to eliminate rivalry, reduce duplication of facilities and gain significant operating economies.

12.1.3 *The 'age' of protection*

The period from 1918 to 1945 was characterised by the growth of new transport modes and the demand for them to be regulated in order to protect the public interest in the way that the railways had been. In the UK that meant national quantity and price controls over buses (1930), public ownership of London's buses and underground railways (1933) and quantity control of road haulage (1933) amongst others. In the USA regulations were introduced between 1925 and 1942 to cover bus operations, road haulage, airlines, coastal shipping, inland waterways and freight forwarders. The public mood on both sides of the Atlantic was fundamentally anti-competition, favouring mergers and rationalisation of companies and, in the interests of safety and the public good, licensing by the state – but not at this stage state ownership of undertakings.

12.1.4 The 'age' of administrative planning

After the rationing of fuel and requisitioning of public transport carriers for military ends during the Second World War, planning of transport administration was seen as a logical extension of state intervention. The goals were to restructure railway operations to increase efficiency; to concentrate on the structure and organisation of public transport in the big cities; to see transport networks and infrastructure in a fresh light, as a means of directing urban growth; and to aid economic recovery after the Second World War. Though the detail varied – from nationalisation in the UK to regulation in the USA and Canada – the outcomes were very similar. The Urban Mass Transportation Act in the USA in 1964, in providing federal grants for public transport services, had explicit social objectives, while the UK's 1968 Transport Act brought about a major reorganisation of public transport mangement in an attempt to halt the disintegration of public transport services in the face of the growth in car ownership. The Act created the Passenger Transport Authorities and Executives in the major cities, charged with the duty of securing or promoting the provision of a properly integrated system of public passenger transport.

A difficulty for both the USA and the UK was that roads issues were not regarded as part of transport policy, but of traffic policy. As a result there were separate financial programmes and legislation and eventually autonomous bureaucracies dedicated to road building. The advent of scientific urban transport planning in the USA legitimised supply-led initiatives to solve what were seen as capacity problems. In the UK, new road-building proposals met considerable opposition in cities as well as in inter-urban locations as the Twyford Down case study in Chapter 10 showed, and served to raise a generation of individuals who no longer believed that the state could be trusted to serve the public interest in transport or, by implication, to protect the environment. The concentration on road investment also helped to undermine public transport services, so that subsidy needs rose in times of increased pressure on public finances brought about by recession.

As a result the age of administrative planning was on the retreat almost from its inception. The balance between planning and permissible competition within a regulatory framework gradually shifted towards competition with, for example, the taking of British road haulage out of public ownership in 1953 and the abolition of the British Transport Commission in 1962 with the loss of the integrating role that it was supposed to play. Under these circumstances the administrative planning approach was in no fit state to withstand the advent of changed political philosophies in the late 1970s. The movement began in the USA and was enthusiastically adopted and promoted by the incoming Thatcher administration in 1979. The 'age' of contestability had begun.

12.1.5 The 'age' of contestability

The organising principles in the age of contestability are deregulation and the goal of efficiency in transport operation. Instead of the state protecting the public interest via planned competition and regulation, the role of the state is redefined as one of creating conditions for efficient transport operations and allowing maximum competition. The ideologically related but distinct processes of achieving this are twofold – privatisation and deregulation.

Privatisation involves the transfer of previously nationalised transport undertakings from the state to the private sector. Since 1982 in the UK this has included the British Transport Docks Board, British Airways, the National Bus Company, the British Airports Authority and the franchising of railway services. Deregulation aims at removing state control over quantity and quality of services in pursuit of genuine competition as a way of serving the public interest. It continues to gather pace around the world, with recent legislation involving road freight in New Zealand and Eire, inter-state bus services in Australia, railway services in Sweden and long-distance coach services in China. The political changes in the former Soviet bloc have also heralded a new round of transport deregulation and privatisation, while the introduction of the Single European Market (see section 12.5) has resulted in the liberalisation and harmonisation of former national transport regulations. Two case studies of deregulation – the US airline industry and British bus services – are given in Box 12.1.

Box 12.1 Deregulation of transport: case studies of planes and buses

Air travel in the USA
Pressure for deregulation of air transport operations began in the USA in the 1970s. The 1978 Airline Deregulation Act phased out the licensing and fare control responsibilities of the Civil Aeronautics Board (CAB). The principal changes that have occurred are as follows:

1. For the long-haul passenger, fares are lower and more discounts are available, but travel is now more expensive over short distances. Safety has not been compromised.
2. For the airlines, costs are lower, productivity is higher, but few have survived unscathed. Of the eleven trunk airlines in existence prior to deregulation, only six had survived by the early 1990s and several of those were bankrupt. The market share of the eight largest airlines rose from 81 per cent in 1978 to 90 per cent by 1990.
3. The major spatial change has been the development of 'hub and spoke' operations, whereby flights from small centres are fed into major interchange airports which are connected to other hubs by frequent, high-capacity services. For example, Chicago and Dallas are major hubs for American Airlines, while Houston, Denver and New York serve ▶

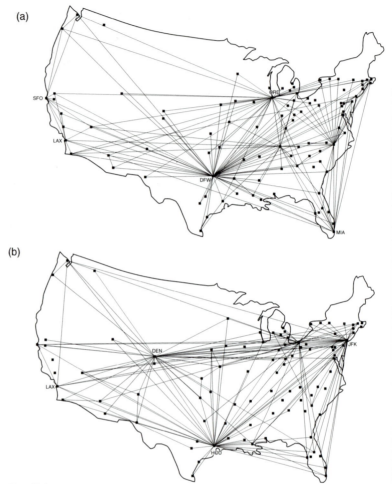

Fig. 12.1
Domestic airline networks in the USA. (a) American Airlines. (b) Continental Airlines. *Source:* based on Figs 2 and 3 in Shaw (1993).

the same purpose for Continental, as the maps in Fig. 12.1 show. This spatial concentration lowers costs for the airlines and fares for the passenger, but does make new entry to the industry very expensive, as new operators struggle to generate enough density of feeder traffic to permit the exploitation of economies of scale on the hub–hub routes. As a result, the industry has become more monopolistic since deregulation and the entry of new competition has been stifled.

The British bus industry
Since 1980 the bus operating industry in the UK has undergone the most dramatic of revolutions. The world's largest bus grouping – the National Bus Company – has been broken up and sold; route competition ▶

Plate 12.1
Buses crowd the streets of central Glasgow in 1988 shortly after deregulation, as the battle for the most profitable routes began. The result was massive congestion and air pollution problems on such routes, paralleled by the reduction – or even loss – of services on routes where profits were less attractive.

has been introduced on commercial services, including inter-city routes; co-ordination schemes in cities have been discontinued, in some cases dismantling 50 years of planning and integration; public subsidy has been slashed; and city streets have seen the decline of the old double-decker and the appearance of mini- and midi-buses, often in eye-catching liveries and operated by previously unheard-of companies.

Some of the consequences are similar to those in US airlines, in that fares are lower and frequencies higher along the major corridors. Elsewhere and outside peak times, fares are higher and buses fewer, or have disappeared entirely. The government has saved a large amount of money in reduced subsidies, but the main effect of deregulation has been a reduction in the number of people using buses, with patronage falling faster after deregulation than before, especially in the large cities. Commuters on main routes with faster, cheaper and more frequent services have benefited, but this is offset by the losses suffered by those who wish to travel off-peak, on less used routes, or who wish to interchange with other urban transport services. Women and the elderly comprise the majority in this category.

There is little doubt that in the late twentieth century contestability theory has dominated national transport policies, especially in the more developed countries but increasingly in developing countries too. That is not to say that it will do so for the foreseeable future, because there are major emerging challenges to transport which may involve global solutions that are not best administered by the private sector. Already there is evidence from many countries that the outcome of privatisation and deregulation has been oligopolistic control of particular transport markets rather than more competition. Some researchers have suggested that a combination of franchising and compulsory competitive tendering would stimulate competition but prevent the misuse of oligopolistic power and thus control service quantity, quality and price. In this way operating costs and public subsidies may be reduced without undermining the integration of public transport services.

12.1.6 *Summary: future transport policy*

Future transport policy is likely to be considerably influenced by the past. In some cases commitments have been made which are not easily abandoned whilst in others specific policies – for particular modes for example – are retained because they are effective and are likely to command considerable political support into the future. There is also the so-called 'supertanker effect', in that policies often lag so far behind significant shifts in public opinion that there are sometimes periods of total discord between the two. The analogy is with the supertanker doing an emergency stop with its engines thrown fully into reverse, yet its momentum forcing it remorselessly in the original direction.

The 1990s are such a period of discord, for just as market forces have replaced state intervention and control as the guiding principles of policy in the past 20 years, so is the emerging environmental crisis ready to precipitate equally dramatic changes from the early years of the twenty-first century. In the future, the key issues of policy for transport are likely to come from government departments responsible for environmental management rather than from those concerned solely with transport. Indeed, the 'Earth Summit' – the United Nations Conference on Environment and Development held in Rio in 1992 – will influence transport policies, just as the changed philosophies of the Thatcherite era did. The confidence in the 1960s and 1970s that technology would come up with the answers has been replaced by a growing realisation of the finiteness of resources and of the fragility of the planet under attack by the products of over-consumption of transport.

Naturally, changes of this magnitude do not proceed smoothly and simultaneously throughout the world. Some places lead and others lag, whilst some make progress on specific fronts. The rest of this chapter illustrates the point by way of four examples: the UK – which has adopted the principle of sustainable development but continues with practice which makes it unachievable – is the starting-point. After a sideways look at the potential for travel demand management and new communication technologies in the USA, we compare 'sustainable development' (in the Netherlands) with 'sustainable mobility' (in the European Union) and conclude with a discussion of the roles of planners and politicians in pursuing sustainable transport policies.

12.2 An absence of policy: the example of the UK

12.2.1 Introduction

It has long been a truism amongst commentators on the British transport scene that there has never been such a thing as a transport policy, merely a series of separate policies for each of the transport modes. John Adams (1993) has typically disagreed with the majority view, arguing that there *is* a policy in place and that it is clear, coherent and powerful. Citing the tenfold increase in car travel since 1952, he observes that 'the old-fashioned forms of travel – bicycle, bus and train – are being phased out. The liberation of the cyclist from his labours is now almost complete. Complete liberation from the bus is also at hand.' He reserves special 'praise' for the Department of Transport's 'most impressive achievement', the reduction of walking, acidly observing that it has gone undocumented only because the Department's statisticians did not consider walking to be a form of transport and therefore did not collect any information about it.

12.2.2 The 'new realism' in transport policy

Adams reserves much of his scorn for official responses to the national road traffic forecasts which were published in 1989, with estimates of road traffic increases of up to 142 per cent by the year 2025. Such traffic could not be accommodated on the existing road network or indeed any conceivable future network. The roads programme announced in response was to cost £23 billion in order to add a mere 5 per cent to road space. As a result, for the first time in the UK, there grew a widespread realisation amongst planners and the public that there is no possibility of increasing road supply at a level which approaches the forecast increases in traffic. This crucial change in public attitude in the UK has been referred to as the 'new realism', which is the title of the Rees Jeffreys Road Fund 'Transport and Society' project, an influential review of policy in the UK published in 1991. It defined the principal problems in this way:

- there is an intolerable imbalance between expected trends in road-based mobility and the capacity of the transport system;

(a)

(b)

Plate 12.2
In a society where car use is growing quickly, new road construction only reduces congestion
for a short period before the attractiveness of the new route generates so many new journeys
that the road becomes overloaded. The M25 London orbital motorway is shown here in
1986 shortly after its opening (a) and in 1990 (b), by which time it was jokingly referred to as
'the largest car park in Britain' and had stimulated plans to widen it, initially in the western
section near Heathrow.

- this is causing problems to industry, to the environment and also to the ability of people to lead comfortable and fulfilling lives;
- the main problem is the growth in reliance on car use, which no longer succeeds in realising its own objectives;
- it is not possible to provide sufficient road capacity to meet unrestrained demands for movement.

This, it argued, has two logical consequences:

- whatever road construction policy is followed the amount of traffic per unit of road will increase, not reduce. In other words all available road construction policies only differ in the speed at which congestion gets worse; and therefore
- demand management must force itself to centre-stage as the essential feature of future transport strategy.

Two main objectives in transport policy were identified:

- to match demand to supply, given the infeasibility of matching supply to demand;
- to encourage the use of environmentally beneficial and economically efficient methods of achieving personal access and freight distribution;

And one constraint:

- it will only be possible to achieve such objectives if the policies offered are capable of making life better not worse. Otherwise there is little chance of political support.

The report argued that policies to accomplish these objectives are technically possible, provided they are properly harmonised. They will include land-use planning; extensive use of traffic management; substantial improvements to the scale, reliability, comfort and cost of public transport; traffic-calming schemes; the encouragement of walking and cycling; and ensuring that all modes pay their full costs, perhaps using methods such as road pricing. Expansion of road infrastructure should not be at the core of transport policy.

The conclusion is a powerful argument for control, for policy direction towards a more sustainable future:

> We are in the middle of a process which is based strongly on entirely legitimate human aspirations, built from individual decisions each of which, on its own, makes perfect sense. But the overall effect is uncontrolled and self-defeating. Our argument is that it is now necessary and possible to choose a different path: necessary for environmental and economic advantage; and possible because of the unprecedented breadth of understanding that it is not possible to provide for unlimited car use (1991: 149).

The UK in the 1990s is a classic case of policy discord due to the supertanker effect, with the momentum of past policies forcing it against a tide which is now running strongly in the opposite direction. Professional and academic opinion is now heavily critical of central government policies and it is local authorities that are taking the high ground when it comes to pursuing sustainability, for example in regard to catering for the national resurgence in cycling. Public attitudes are changing (though behaviour is slower to shift) with ever-increasing realisation of the absolute necessity for restraint of motorised traffic. In the midst of this the government presses ahead with what its own Secretary of State for Transport has claimed is 'the biggest road-building programme since the Romans'.

12.3 Demand management, communications technology and future travel patterns: the example of the USA

12.3.1 Transport demand management

If managing demand is to be part of future transport policy in the UK, innovations in the USA in 'TDM' – 'transport demand management' – may well be of interest. TDM is the art of modifying travel behaviour in order to reduce the number of trips or modify their nature. It may be categorised according to whether it mainly affects trip generation, trip distribution, modal choice or route selection. As Table 12.1 shows, some implementation strategies rely on changes to the transport system, others on land-use policies and still others on alterations to employment conditions and societal values. It is this latter category which most interests us here, for without doubt the USA has done more in this field than any other developed country. The specific strategies of alternative work schedules (flexitime, employment-based ride-sharing and telecommunications) will be examined here.

Persuading a number of large companies to introduce flexible working hours ('flexitime') is a logical way to reduce congestion at peak periods. However, many people are 'locked in' to travelling at certain times by family commitments: children have to be taken to and from school, spouses have to be dropped off or picked up or met and so on. Under these circumstances the room for manoeuvre of one family member in terms of changing working hours may actually be very small indeed. What is more, flexitime does not reduce the overall amount of vehicular traffic and may actually impede some other TDM strategies, such as efforts to introduce car-pooling.

The most innovative TDM measures are the ones that require employers to reduce the number of peak-period car trips made by their workers. In the USA at least 20 suburban communities have enacted such programmes, half of them in California. In greater Los Angeles businesses with more

Table 12.1
Transport demand management strategies in the USA

Transport planning process	Management objective	Main implementation strategy		
		Transport	Land use	Other
Trip generation	Eliminate trip entirely			• Telecommunications • Shortened work weeks
Trip distribution	Shift locations of origins or destinations to modify spatial distribution of trips		• Zoning policy • Reurbanisation • Mixed use development • Transit-friendly design • Growth management	
Mode choice	Shift from low occupancy mode to higher occupancy mode	• Transit, bicycles, walking facilities • High-occupancy vehicle facilities • Gasoline taxes • Licensing policies	• Parking policies	• Employment based ride-sharing
Route selection (spatial)	Shift trip from a more congested route to a less congested route	• In-vehicle navigation systems • Road pricing		
Route selection (temporal)	Shift trip from a more congested time period to a less congested time period	• Congestion pricing		• Alternative work schedules (flexitime)

Source: modified after Ferguson (1990).

than 100 workers are required to implement trip-reduction programmes such as employer-sponsored van-pools, preferential parking for car-poolers or subsidised mass transit tickets, with the aim of increasing average vehicle ridership from 1.5 to 1.75 for commuting trips. Some 8000 employers with 40 per cent of the area's workforce are affected by the regulations and if they fail to at least submit a plan to remedy the problem they face a $25 000 fine and 6 months in prison. However, results to date have not been particularly impressive. There are no penalties for non-attainment and most assessment is in terms of inputs (e.g. providing information to employees on bus timetables) rather than outputs (e.g. getting a quarter of the workforce to ride-share). An evaluation of one Californian programme found that the proportion of workers driving alone had only been reduced from 87.9 to 86.6 per cent.

12.3.2 Telecommuting

Because journeys to work account for over 30 per cent of urban travel, any opportunity to reduce the frequency of work trips must be of interest to those wishing to see reductions in the harmful environmental consequences of transport. One such development is 'telecommuting' or 'teleworking', the substitution of electronic communication for the physical movement of employees. The transmission of data, images and information by electronically based devices has made it technically feasible for many people to carry out their work at home, so that the daily trip to a city-centre office has become, for them, a thing of the past. The more imaginative forecasts of these developments have predicted that former CBD-orientated rush hours would be replaced by a more varied pattern of movement to suburban neighbourhood work-centres. Shopping would also be done from home via the television set, which would provide home education, entertainment and medical services. In this brave new world, there would seem to be little need to get out of the armchair, let alone do battle with a city's congested transport system.

However, the degree to which communications can be substituted for transport is subject to great predictive uncertainty. Actual substitution depends on the cost-effectiveness of technology and external factors such as energy shortages and rising transport costs. In 1987 3 million Americans telecommuted, but only 12 per cent of these worked at home for more than 35 hours per week. It appears that at present telecommuting is not fulfilling the promise of a decade ago, but some feel that a second wave is likely, based on recent changes in business practices and the widespread adoption of technologies suitable for teleworking, including fax machines, laptop computers, voice messaging and digital telephone networks (see Fig. 4.2).

Many questions remain to be answered. Will new communication technologies encourage new functional and spatial relationships between home and workplace? Will they alter the journey to work, the frequency of trips or the choice of travel mode? The fact of the matter is that we do not know, if only because we understand very little about the value of travel: is it seen as a cost to be reduced? Or is it a benefit with its own intrinsic attractions? Nor do we know to what extent this depends on journey time, length, mode or purpose: after all, a closed circuit computer conference is hardly the same as a stroll in the moonlight.

Empirical evidence shows that in California telecommuting has reduced work trips but has not led to an increase in non-work trips: on a telecommuting day travel by telecommuters was only about 20 per cent of the distance they normally travelled on commuting days. With household members using the car less too, it seems as though telecommuting has stimulated a more local lifestyle, contradicting fears that it might encourage people to move from clustered settlements to more rural ones where all of their travel would be by car. How significant this Californian experience is for other societies remains to be seen, particularly as history does not support the notion of technical change stimulating travel reduction. For example, although the telephone is in theory a substitute for movement, the first century since its introduction has been associated with dramatic increases in the volume of physical movement of people.

While it is true that these are difficult and poorly understood relationships, there is at least an argument that electronic information exchange does not reduce movement, but in fact engenders wider associations and activities that raise the level of interaction and thereby generate yet more surface trips. In other words, electronic exchange innovations may actually be counter-productive in terms of trip generation.

12.4 Policies for 'sustainable development': the example of the Netherlands

12.4.1 National Environmental Policy Plan

Our third sample study of national policies brings us back to Europe and one of the most imaginative approaches to transport and environmental problems, that of the Netherlands' National Environmental Policy Plan, or NEPP for short. Despite its green and relatively healthy image, the Netherlands is actually the most heavily polluted country in western Europe. It has greater densities of people and cars and more energy consumption per capita than any other industrialised nation. As environmental awareness has grown in developed countries in the 1980s, so have the Dutch – with their particular vulnerability to sea-level rise from global warming – been in the vanguard of concern. The 1987 Brundtland Commission report *Our Common Future* crystallised environmental anxiety and provided the impetus for the preparation of the NEPP for the Netherlands, published in 1989.

The NEPP recognises that safeguarding environmental quality on behalf of what it calls 'sustainable development' will be a process that will last for several decades. The NEPP is the first step in this process: it contains the medium-term strategy for environmental policy which is directed at the attainment of sustainable development over the longer period.

The objectives of the NEPP are that:

- vehicles must be as clean, quiet, safe and economical as possible;
- the choice of mode for passenger transport must result in the lowest possible energy consumption and the least possible pollution;
- the locations where people live, shop, work and spend their leisure time will be co-ordinated in such a way that the need to travel is minimised.

The approach of the NEPP is shown in Fig. 12.2. As pollution from road traffic is seen as a three-step process, these objectives are to be met through a 'three-track' response, the tracks being those of technical vehicle standards, reducing 'automobility' and instigating urban traffic measures.

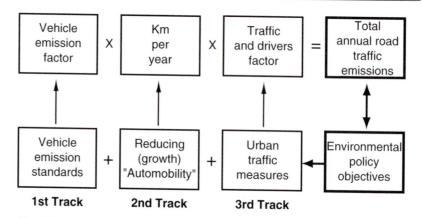

Fig. 12.2
The Dutch three-track approach to the abatement of environmental pollution by motor vehicles. *Source:* Kroon (1990).

12.4.2 The three-track approach towards 'sustainable development'

The first track consists of a series of measures to convert the vehicle fleet into one that is the cleanest possible. This is to be done by initially subsidising the fitting of catalytic converters, establishing targets for the reduction of emissions from lorries and buses and pushing for stricter European standards on emissions.

The second track, of reducing car use, aims to shift people from cars to public transport for the longer journeys and to cycling or walking for the shorter ones. This is to be achieved through provision of more and better facilities for cycling and public transport, more subsidies, better fare and ticket integration and publicity campaigns. However, it is recognised that if the policy is to seek a balance between individual freedom, accessibility and the environment, the only way to achieve this is to control the use of cars. Therefore the strategy is to increase variable motoring costs through fuel taxation and road pricing. Car commuting will be discouraged through a variety of TDM measures including 'kilometre reduction plans', whereby companies and institutions will have to draw up and then implement plans to reduce the distance travelled by employees in the course of work and in commuting to it. Additionally the second track will improve the transport of freight by rail and water and will tighten up physical planning policy, to ensure that businesses which are labour-intensive or amenities which attract numerous visitors are not permitted to locate at places which are not well served by public transport (Box 12.2).

As well as having cleaner vehicles which are used less, the NEPP recognises – the third 'track' – that further measures are necessary to alleviate the problems at a local scale. These include stricter enforcement

Plate 12.3
The Dutch National Environmental Policy Plan aims to increase rail passenger traffic by 15 per cent through improving bicycle parking at stations as seen here at Hoorne.

Box 12.2 Co-ordinating location and accessibility: the Dutch policy of putting the 'right business in the right place'

Dutch government policy is aimed at cutting down the avoidable use of cars and stimulating public transport and the use of bicycles. One way of doing this is by means of a location policy which sites new businesses and services in locations which can easily be reached by public transport and bicycle and not only by car. The process is a three-step one:

Step one: draw up mobility profiles for each business

- Each company has different transport needs. Just as a distribution centre should be properly accessible to freight vehicles, so should a university be reachable in the first instance by public transport and bicycle. For this reason mobility profiles are drawn up for various types of companies by considering their transport needs. In general the more a business is 'people-intensive' (i.e. has a large number of employees and visitors), the greater its need for public transport.

Step two: draw up accessibility profiles for each location

- An accessibility profile indicates the quality of a location's accessibility by public transport and car. So-called 'A' locations are best accessible by public transport; 'C' locations primarily by car, with 'B' locations accessible to both. ▶

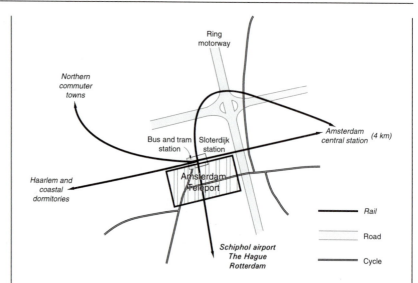

Fig. 12.3
Amsterdam teleport: an 'A' class office location, highly accessible by public transport.

Step three: match mobility profiles to accessibility profiles

- Accessibility profiles of locations and mobility profiles of businesses and services are co-ordinated with one another, so that each business can be sited at a location which is best suited to its own transport requirements. In this way the use of public transport can be stimulated and accessibility can be guaranteed.

This procedure can only work through co-operation between the governments at state, provincial and municipal levels, ensuring that effective physical planning policies match the investment in infrastructure. The guidelines set out in the UK's Planning Policy Guidance on transport (PPG13 in 1993) echo the sentiments of the Dutch policy, but have nothing like the structure and organisation. Guidelines are better than nothing, but are quite inferior to a nationwide policy allied to practical procedures and publicised via a catchy title, all of which seem much more likely to get 'the right business in the right place'.

A good example of the policy in action is the Amsterdam 'teleport' at Sloterdijk, some 4 km from Amsterdam city centre and shown in the map in Fig. 12.3. In an attempt to staunch the exodus of offices from the congested CBD to suburban or rural locations, the city developed this site in the 1980s. Granted it is close to the ring motorway, but its main locational asset is that it is supremely well serviced by public transport. There are nearly 20 bus lines, a cross-town tram line and railway lines from Haarlem and the commuter villages to the north and west, as well as a direct rail connection with Amsterdam central station and the CBD in one direction and Schiphol airport in the other. This is a prime 'A' location and is being developed as a high status 'teleport', with a great concentration of firms in the 'people-intensive' knowledge and informatics sector.

of parking controls, traffic management to influence drivers' choice of routes, circulation schemes to slow traffic and similar measures to improve road safety and increase environmental protection.

The most noticeable feature of the NEPP is the way that its individual measures reinforce each other, to produce an integrated package which links environmental, transport and land-use policy. Yet even this impressive, comprehensive approach comes nowhere near solving the problems. Without the NEPP, car-kilometres had been expected to rise by 72 per cent over the period 1986–2010. With the NEPP this increase is lowered to 48 per cent, a worthwhile reduction but still a very long way from a sustainable level of transport use. The NEPP must be seen only as the first stage in a long-term drive towards sustainability: it serves to illustrate what a difficult task lies ahead of the Dutch (and indeed all motorised countries).

12.5 International transport policy: the example of Europe

12.5.1 Failure to achieve a common market in transport

We continue in Europe for our last example of policy in action, but shift scales to a supranational level with a study of the European Union. Within the EU there are differences between member states in their philosophical position towards transport, with the 'Anglo-Saxon' approach focusing on economic efficiency and contrasting strongly with French–German–Scandinavian attitudes in which efficiency is more usually seen as secondary to the wider role of transport within economic and social planning. Additionally, individual countries have different modal specialisms as Tables 6.1 and 6.2 showed. This conflict between the interests of member states produces an unstable policy environment and one which is far from ideal for the task of producing profound insights or long-term goals. There are further complications as a result of the fact that the EU does not form a contiguous geographical space, with Greece physically separate and routes having to pass through third-party countries such as Switzerland in order to connect two members such as Germany and Italy. Not surprisingly, by the time the Single Economic Market (SEM) came into existence in 1993 a common market in transport still had not been achieved.

The main policy objectives at the European level are now:

- an economic and regulatory framework for transport, including harmonisation of fiscal policies and fair comparison and assessment of different transport projects;

- new research and development initiatives;
- standardisation and technical regulation, e.g. road pricing technology;
- development of trans-European networks;
- information exchange, including better quality transport statistics, which will assist the objective of 'sustainable mobility'.

12.5.2 The goal of 'sustainable mobility'

We may well ask what is meant by 'sustainable mobility' here. By its very nature the SEM seems certain to instigate a large increase in lorry traffic, because it has not only removed physical barriers to freight movement in the form of border controls, but also has done away with technical barriers of standardisation which have excluded one country's products from the market of another and has dismantled the fiscal barriers that formerly inhibited trade between member states. Moreover *cabotage* – the ability of non-resident hauliers to collect and deliver loads within the boundaries of another state – is likely to drive down road freight rates and make quantitative restriction on the number of trucks irrelevant. There will also be more car sales as a consequence of the downward pressure on car prices exerted by harmonisation of technical standards and fiscal practices. It is hard to see how this growth in freight and passenger mobility can be called 'sustainable'.

Further environmental deterioration will come as a result of other EU policies that are not directly concerned with transport. The economic growth generated by the larger internal market will lead to more transport of goods and people, not least because economies of scale will increase plant size and reduce plant numbers, thus increasing length of haul as manufacturers increasingly serve the entire European market from a limited number of locations (see section 2.5). These economic goals are to be pursued without consideration of the transport impacts and can only assist further the domination of road-based movement. In just 10 years, the EU intends to add 12 000 km of new motorways to the 1992 total of 3700 km, 40 per cent of them in Spain, Portugal, Greece and Eire in order to tie these countries more tightly to the economic centre of the EU. The result is likely to be dramatic increases in traffic in these countries and on routes to them.

For all of these reasons – the removal of barriers, the liberalisation of road haulage, economic growth and the greater economies of scale – Europe is likely to see a very large increase in road freight in the near future. How large this will be is not estimated in the official documentation of the EU because to do so would expose the contradictions inherent in EU transport and environmental policies. In 1992 the European Union published a long-awaited White Paper on its transport policy, and subtitled it 'a community framework for sustainable mobility'. It then proceeded to fail to define sustainable mobility, going only as far as to say that the tools for achieving it are 'efficient safe transport under the best possible environmental and social conditions' – a phrase which could (and probably does) mean the best that the economy and society can achieve without disrupting its current practices. The point is that the whole economic and spatial logic of the SEM, together with legislative changes in terms of deregulation and liberalisation, means that sustainability cannot be achieved.

12.5.3 The EU as a negative environmental force

The problem within the EU is simply not being faced. There are parts of the Union where policies seem more enlightened than others – the Netherlands rather than the UK, Denmark rather than Spain – and there are towns whose particular policies would, if followed across Europe, put the continent on to a sustainable trajectory. One could cite the bicycle-dominated city of Groningen, the traffic-restrained Freiburg, the public transport integration of Zurich, the area-wide traffic calming of Buxtehude, near Hamburg and many others. But these are islands in the traffic and unless and until EU policy goes into reverse these local successes cannot counteract the structural trends of increasing motorisation.

Overall the EU is a negative environmental force. Its perception of transport is dominated by views of making the economy bigger and encouraging more development and is characterised by attitudes which continue the commitment to existing patterns of resource consumption whilst installing 'end of pipe' technologies in the hope that serious changes in behaviour can be avoided. Not surprisingly the Common Transport Policy has many critics, among them John Whitelegg, who dismisses it as 'an ineffective policy, pursued in an indifferent manner, based on simplistic economic notions' (1988: 201).

12.6 Policy for sustainable transport: no examples yet!

12.6.1 Sustainable transport

We have already in this book defined a sustainable society as one 'that meets the needs of the present without compromising the ability of future generations to meet their own needs'. Specifically, for transport to be sustainable it must satisfy three basic conditions:

1. its rates of use of renewable resources do not exceed their rates of regeneration;

Table 12.2
Guiding principles for
sustainability in transport

1. Transport is a vital element in economic and social activities but must serve those activities rather than be an end in itself.
2. The consumption of distance by freight and passengers should be minimised as far as possible whilst maximising the potential for locally based social interaction and locally based economic activity.
3. All transport needs should be met by the means that is least damaging to the environment.
4. There should be a presumption in physical land-use planning against those activities which by nature of their size and importance attract car-based users from a large area.
5. All transport investment plans should be subjected to a full health audit notwithstanding the uncertainties surrounding epidemiological proof. Proposals which are potentially health damaging should be rejected.
6. All transport investment plans should have clear objectives designed to cover social, economic and environmental concerns and be evaluated by an independent authority with sufficient expertise to comment on value for money, costs and benefits and the availability of alternative strategies to achieve the same objectives.
7. All transport investments should be monitored over their lifetime to check on the degree to which they meet their objectives and their contribution to environmental damage.
8. All transport policy matters should be dealt with in a transport policy directorate that has no direct responsibilities for the management of individual modes. The responsibilities of the directorate are to deliver sharply focused polices that minimise danger, minimise air and noise pollution, maximise social interaction and urban quality of life and oversee the non-policy-making executives (for road, rail and air) whose role is to implement the directives of the transport policy directorate.

Source: Whitelegg (1993: 157).

2. its rates of use of non-renewable resources do not exceed the rate at which sustainable renewable substitutes are developed;
3. its rates of pollution emission do not exceed the assimilative capacity of the environment.

It is evident from the discussion in Chapter 10 that transport currently does none of these things. If the concept of sustainability is to be applied effectively to transport it requires a set of guiding principles. These have been put forward by Whitelegg and are listed in Table 12.2. These principles would represent a starting-point for a new approach to transport policy and set out an agenda for transport planners and politicians.

12.6.2 The task of the transport planner

Critics of the transport planning process point to the fact that it has resulted in an increase in the number and length of motorised trips. They refer to this trip generation as the 'black hole theory' of road transport

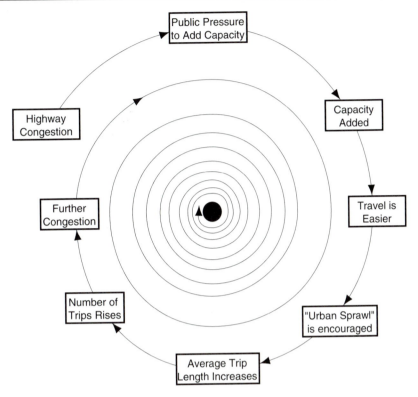

Fig. 12.4
The black hole theory of highway investment. *Source:* Plane (1986).

investment, where the attempt to satisfy demand for travel by increasing the supply of roads forms a positive feedback loop that has been likened to throwing public money down a black hole (Fig. 12.4). From this perspective, the most fundamental task for the planner is to reverse the process, to 'de-generate' trips.

One way to achieve this is through the planning system that controls where activities are located. Recent trends have been towards dispersal of houses to far-flung suburbs, of shops to out-of-town centres (Box 12.3), of employment to green-field business parks and so on. As a result, journeys have become longer as decentralisation has proceeded and energy consumption has necessarily risen. One recent survey of motorisation in cities in developed societies has shown how strong this association is. When the index of centralisation is low – as in the sprawling cities of Australia and North America – energy consumption is high, as Figure 12.6 shows. Conversely, cities with the highest degree of centralisation have transport energy consumption levels that are only about 20 per cent of those of the cities that are most decentralised.

It is clear that planning systems that are able to group school, work, shops, leisure and health facilities around housing have many advantages.

Box 12.3 'Gone shopping: took the car'

Though there are many much earlier examples in North America, 'out of town' development in the UK is a phenomenon of the 1980s. It was only then that all of the favourable conditions came together – new bypasses, growing car ownership especially amongst the young, rapid suburbanisation and ex-urbanisation, and a relaxation of planning controls by a government that had unleashed a 'loadsamoney' culture, in which shopping played an important role. By 1990, as the idealised map of the development of 'Southeastown' shows (Fig. 12.5), many towns had at least one peripheral retail   ▶

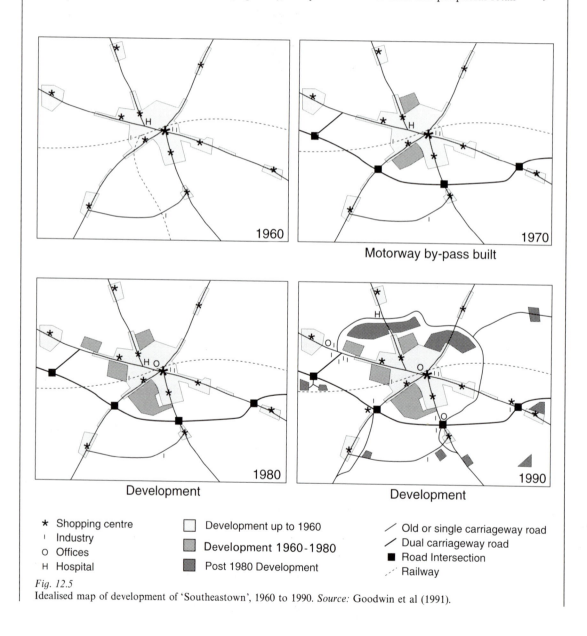

1960

1970
Motorway by-pass built

1980
Development

1990
Development

★ Shopping centre   ☐ Development up to 1960   ╱ Old or single carriageway road

ı Industry   ▨ Development 1960-1980   ╱ Dual carriageway road

o Offices   ▨ Post 1980 Development   ■ Road Intersection

H Hospital   ⋰ Railway

Fig. 12.5
Idealised map of development of 'Southeastown', 1960 to 1990. *Source:* Goodwin et al (1991).

park as well as office and business parks, leisure centres and hospitals, whose common characteristic in transport terms was that they could only be reached by car.

It is evident that the new spatial arrangement involves substantially more travel, but it is rarely appreciated quite what a dramatic change this is. A typical out-of-town superstore requires seven 38-tonne trucks to supply it for one day. Nett, these would deliver 26 tonnes each, giving a total delivery of 182 tonnes. We know that the average customer takes away 23 kg of goods, so this store will have 8000 customers per day, who – given the shopping modal split of at least 70 per cent by car – will arrive in more than 5500 cars. As the average length of a vehicle shopping trip is 8 km each way, the 5500 cars will clock up nearly 90 000 km between them each day in the course of visiting just this one store.

Of course these figures are illustrative, but they do make the point that the sheer volume of travel generated by out-of-town developments is absolutely enormous. It is critical to understand too that these are brand new motorised trips: in 1970 people shopped locally or in the town centre, to which they mostly walked or took the bus. With the decline of the bus service and the closure of town centre stores, now people have little choice but to join in this motorised consumer society.

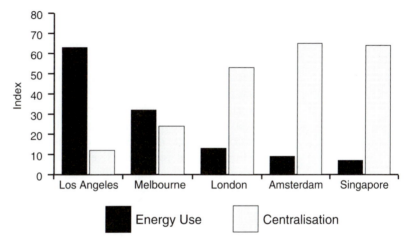

Fig. 12.6
Relationship between energy use and centralisation in selected major cities. *Source:* Transnet (1990).

Public transport is more viable where population densities are high, so that switching to more energy-efficient modes such as buses and LRT may take place. If trips can be made to be really short then people can walk or cycle, thus consuming virtually no energy at all and also reducing the impact of many other adverse side-effects of transport such as noise and accidents. Compact cities are also efficient shapes for the provision of services and infrastructure, especially when new developments can be kept below what might be called a 'socially critical' size, so that 'densification' may result in lower energy costs for activities like food delivery, refuse collection, water distribution and road building.

However, such mobility-reducing planning is likely to be expensive and difficult to implement, particularly because current trends towards increasing decentralisation and increasing trip length are so firmly established. It will be just as important to reduce people's inclination to travel long distances as it will be to reduce the necessity to do so: perhaps the old wartime slogan of 'Is your journey really necessary?' ought to be revived! (Fig. 12.7). Partly for this reason, and partly because people make decisions about where to live and work relatively infrequently, changing land-use patterns must be seen as a long-term contribution to reducing environmental impacts of transport. It is, however, an absolutely fundamental step, without which few other strategies can work.

12.6.3 The task of the politician

Setting out on the road to sustainability is a less difficult task for planners than it is for the politicians, who feel that to get re-elected they must grant people's wishes to become car owners and to have uncongested roads to drive on. Adams (1992) has suggested that there is a way out of this dilemma if politicians stop fostering unrealistic aspirations by pursuing transport policies which push society further along the road to full motorisation. The question must not be: 'Would you like more cars?' (to which the answer is always yes) but: 'Would you like to live in the sort of world that would result if everyone's wish came true, i.e. a dirty, noisy, dangerous, socially polarised, fume-filled greenhouse?' – to which the answer would be no. Better still would be to ask the more positive question: 'Would you like to live in a world where it was safe for children to play in the street and walk or cycle to school on their own, where women and old people were not afraid to venture out at night, where there were local shops, etc., i.e. in a safer, more beautiful and harmonious world?' As Adams comments (1992: 333), 'It ought not to be beyond the wit of politicians to persuade people to trade in their cars for such a prize.'

12.6.4 Making the polluter pay

The point has already been made in Chapter 10 that cars and lorries do not pay their way when environmental and social side-effects are taken into account. This subsidy to road-based movement produces major distortions in the transport sector (Box. 12.4) and represents a misallocation of resources which could be recovered by the imposition of a tax on fuels, in much the same way as a carbon tax has been suggested as a way of reducing fossil fuel consumption. However, such a tax would have to be applied internationally to be effective, requiring considerable inter-country co-operation, as for example within the EU.

Moreover, costs are but one element in modal choice, with many motorists valuing speed – on which the car often scores over public transport – more highly, so that this measure alone seems unlikely to lead

Fig. 12.7
Is your journey really necessary? Wartime exhortation to reduce demand on the British railway system.

to significant modal diversion. In addition, it is unclear how big a reduction in car-kilometres can be achieved, because there are great uncertainties concerning the elasticity of demand and thus the actual effect of price rises. Some have suggested that price rises might have to be very

Box 12.4 Whose travel is subsidised? The company car tax perk in the UK

In the UK major distortions of the transport system are taking place as a result of the company car system. As a result of concessionary tax laws which mean that employees pay less tax on a car than on salaries, somewhere between 50 and 70 per cent of new cars are company cars, provided as a perk more often than as a necessity.

The result is that congestion is worsened:

- company cars are over-represented in the ranks of car commuters. More than one-third of drivers entering London during the peak period in one survey in 1989 had a company car and a further 50 per cent some form of company assistance, though only 39 per cent of public transport users had any subsidy;
- there are more cars on the road than there would otherwise have been.

There are severe environmental side-effects too:

- distances travelled are higher than they would have been without the system, for company car users tend to travel more kilometres per car on personal or family business than owner drivers. Three-quarters of the distance travelled by company cars is actually for private motoring;
- company cars have bigger engines which are less fuel efficient;
- they have accident rates 50 per cent higher than other cars;
- they are responsible for 22 per cent of the CO_2 emissions from cars on British roads.

The result is a tax subsidy to company car users estimated at £2.8 billion in 1991. This is:

- larger than the subsidy to all forms of public transport put together;
- equivalent to a tax payment of £140 by every household;
- a subsidy of over £800 per tonne of CO_2 attributable to them.

Though recent taxation changes have reduced the benefits to company car users, the system continues to provide an incentive for them to travel excessively at someone else's expense. It would seem that the abolition of the company car tax perk should be at the forefront of any transport policy that has pretensions to sustainability.

high indeed to encourage significant fuel saving: one study has forecast that reducing fuel consumption in the USA by 20 per cent in the 1993–2005 period would require price increases in the area of 800 per cent. One could also cite the fact that petrol prices in the UK more than doubled between January 1974 and mid-1975, yet average mileage travelled in 1974 was only 5 per cent down on 1973. It seems that people in developed societies have adopted lifestyles which both exploit and require a high level of mobility, producing great inflexibility in travel patterns. In the long term these may alter, especially as a result of land-use changes, but the short-term elasticities may be very low.

A more effective way of using fiscal and economic methods to confront environmental problems would be to develop a long-term comprehensive strategy which tackles all sectors – such as air pollution, noise, waste disposal and energy use – and which neither increases overall taxation nor causes distortions between different sectors of the economy. One such approach is von Weizsäcker's 'ecological taxation reform' (ETR), which is based on the principle that taxes should be heaviest on those activities that produce the greatest environmental damage (von Weizsäcker and Jesinghaus, 1992). Such taxes would replace taxes on labour and capital and would be phased in over a period of around 30 years. They would steer the economy so that gradually environmentally harmful activities such as road freight haulage would be replaced by environmentally friendly alternatives, such as the train. In time, it is argued, the space economy would evolve into a pattern which would require less energy and produce less waste and pollution. As space would be used much more efficiently, distances would be shorter and public transport, walking and cycling would become more feasible. The application of such an idea would doubtless meet with considerable opposition and may be seen as a potential 'vote loser' threatening the well-being of the party that introduced it. Nevertheless, it has great merit, particularly since it attacks the problem at a fundamental level, ensuring that every private and public expenditure decision is a choice between sharply different options: cheap, clean, energy-efficient and pollution-free on the one hand, or expensive, dirty and damaging on the other.

12.7 Concluding summary

In the concluding chapter there will be an opportunity to take a wider look at global transport development and to discuss the distinction between changing attitudes and changing behaviour. The present chapter has provided substance for that discussion by examining different policies in different locations driven by different policy paradigms. We began by reviewing the conflict between policies based on the free market and those based on public intervention and we observed the growing significance of market approaches involving a minimum of regulation. These trends have been world-wide in the late twentieth century, though naturally they have gone further in some societies than others, if only because some policy paradigms were noticeably closer to market orientation at the outset. One could obviously contrast the continental European tradition of transport as a means of improving society, with American attitudes which regard transport almost as an end in itself, doubtless a reflection of the American love of newness, movement and constant change.

However, throughout the world the growing reliance on the invisible hand of the market is generating new problems which are not easily remedied by conventional market mechanisms, the global pollution from the automobile being the most obvious example. The new challenge for policy-makers is twofold: to produce a fundamental change in the way that we perceive the role of transport and personal mobility and then to translate these changed attitudes into sustainable behaviour. How the geographer may make a contribution to that process is discussed in our final chapter.

Further reading

Button K J, Gillingwater D 1986 *Future transport policy* Croom Helm
Dunn J A 1981 *Miles to go: European and American transportation policies* MIT Press, Cambridge, Mass.
Goodwin P et al 1991 *Transport: the new realism* Transport Studies Unit, University of Oxford
Ministry of Housing, Physical Planning and Environment 1989 *National environmental policy plan – to choose or to lose* The Hague
Roberts J et al 1992 *Travel sickness: the need for a sustainable transport policy in Britain* Lawrence and Wishart
Tolley R S (ed) 1990 *The greening of urban transport: planning for walking and cycling in western cities* Belhaven
Truelove P 1992 *Decision making in transport planning* Longman
Von Weizsäcker E U, Jesinghaus J 1992 *Ecological taxation reform: a policy proposal for sustainable development* Zed
Whitelegg J 1988 *Transport policy in the EEC* Routledge
Whitelegg J 1993 *Transport for a sustainable future: the case for Europe* Belhaven
Wistrich E 1983 *The politics of transport* Longman

13 Conclusion: where do we go from here?

Overcoming the tyranny of distance?

'Glorious, stirring sight!' murmured Toad, never offering to move. 'The poetry of motion, the real way to travel! The only way to travel! Here today – in next week tomorrow! Villages skipped, towns and cities jumped – always somebody else's horizon! O bliss! O poop poop! O my!' (Toad of Toad Hall on seeing a motor car for the first time) (Kenneth Grahame, 1908)

The miles, once the tyrants of the road, the oppressors of the travellers, are now humbly subject to the motor car's triumphant empire. . . . It flattens out the world, enlarges the horizon, loosens a little the hands of time, sets back a little the barriers of space. And man who created and endowed it, who sits and rides upon it as upon a whirlwind, moving a lever here, turning a wheel there, receives in his person the revenues of the vast kingdom it has conquered. . . . (A. B. Filson Young, 1904)

Yet this distance, all those abysses unbridged and then unbridgeable by radio, television, cheap travel and the rest, was not wholly bad. People knew less of each other, perhaps, but they felt more free of each other, and so were more individual. The entire world was not for them only a push or a switch away. Strangers were strange, and sometimes with an exciting, beautiful strangeness. It may be better for humanity that we should communicate more and more. But I am a heretic. I think our ancestors' isolation was like the greater space they enjoyed: it can only be envied. The world is only too literally too much with us now. (John Fowles, 1969)

Our final chapter examines the range of future choices open to us in terms of technology and development. We can anticipate major global impacts from the telecommunications revolution and developments in rail and air travel, but the key role in testing the limits of sustainability is that of motorisation. We examine the views of the philosopher Ivan Illich and conclude with a research agenda which points the way ahead for the next generation of transport geographers.

13.1 Transport trends for the future

13.1.1 Technology and the urban transport problem

The development of transport in the future will be conditioned by economic, social, political and environmental factors as earlier chapters have shown, but advancing technology will define what is possible (Fig. 13.1). The geographer Alan Williams (1992) cautions us against being too euphoric or too pessimistic in predicting the future, which may lie, as he reminds us, anywhere between a global reality 'shaped by too many people, conspicuous consumption, depleting resources and the millstones of poverty and hunger (1992: 269)' on the one hand, and the dream of 'Ecumenopolis', the interconnected global city, on the other. Whether or not future breakthroughs will be translated into transport realities will reflect social, economic and political choices, which are even harder to predict than technological change. As he concludes (1992: 270), 'What its effects will be and what ends it will serve are not inherent in the transport technology, but depend on what people will do with it.'

As geographers we are interested less in the mode or technology itself than in the spatial consequences of its use. The technological break-throughs shown in Fig. 13.1 all have repercussions on where and how we

Plate 13.1
The Docklands Light Railway, opened in 1987, is a fully automated rapid transit line connecting the Isle of Dogs, in London's former docklands with the City. It represents one of the first examples of the large-scale application in a large urban area of the concept of automated trains, first introduced within airports such as Birmingham International and London Gatwick.

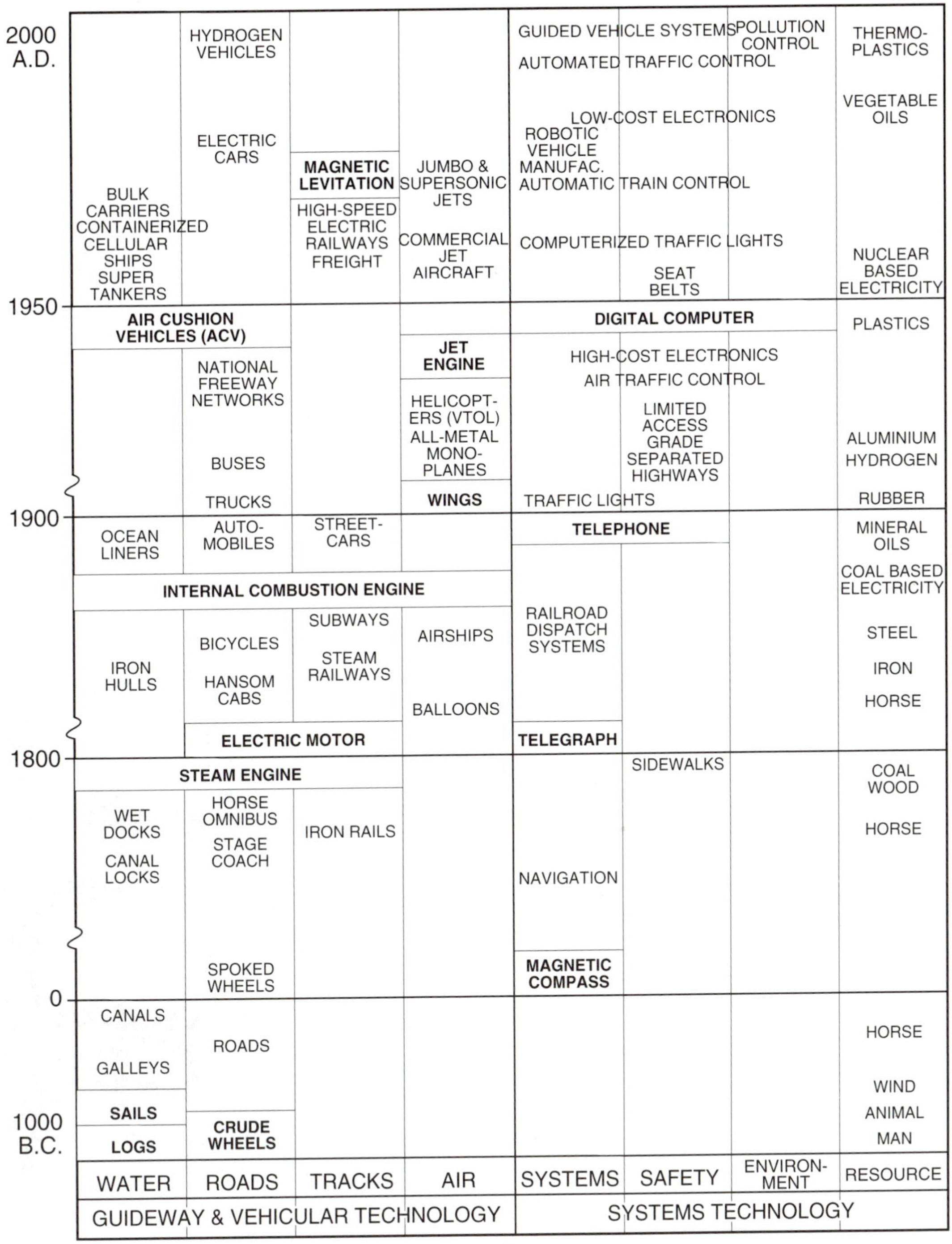

Period	WATER	ROADS	TRACKS	AIR	SYSTEMS	SAFETY	ENVIRON-MENT	RESOURCE
2000 A.D.		HYDROGEN VEHICLES			GUIDED VEHICLE SYSTEMS / AUTOMATED TRAFFIC CONTROL	POLLUTION CONTROL		THERMO-PLASTICS
		ELECTRIC CARS			LOW-COST ELECTRONICS / ROBOTIC VEHICLE MANUFAC.			VEGETABLE OILS
	BULK CARRIERS CONTAINERIZED CELLULAR SHIPS SUPER TANKERS		MAGNETIC LEVITATION / HIGH-SPEED ELECTRIC RAILWAYS FREIGHT	JUMBO & SUPERSONIC JETS / COMMERCIAL JET AIRCRAFT	AUTOMATIC TRAIN CONTROL / COMPUTERIZED TRAFFIC LIGHTS	SEAT BELTS		NUCLEAR BASED ELECTRICITY
1950	AIR CUSHION VEHICLES (ACV)	NATIONAL FREEWAY NETWORKS		JET ENGINE / HELICOPTERS (VTOL) ALL-METAL MONO-PLANES	DIGITAL COMPUTER / HIGH-COST ELECTRONICS / AIR TRAFFIC CONTROL / LIMITED ACCESS GRADE SEPARATED HIGHWAYS			PLASTICS / ALUMINIUM HYDROGEN
		BUSES / TRUCKS		WINGS	TRAFFIC LIGHTS			RUBBER
1900	OCEAN LINERS	AUTOMOBILES	STREETCARS		TELEPHONE			MINERAL OILS / COAL BASED ELECTRICITY
	IRON HULLS	INTERNAL COMBUSTION ENGINE / BICYCLES / HANSOM CABS	SUBWAYS / STEAM RAILWAYS	AIRSHIPS / BALLOONS	RAILROAD DISPATCH SYSTEMS			STEEL / IRON / HORSE
		ELECTRIC MOTOR			TELEGRAPH			
1800		STEAM ENGINE				SIDEWALKS		COAL WOOD
	WET DOCKS / CANAL LOCKS	HORSE OMNIBUS / STAGE COACH	IRON RAILS		NAVIGATION			HORSE
		SPOKED WHEELS			MAGNETIC COMPASS			
0	CANALS / GALLEYS	ROADS						HORSE
	SAILS							WIND / ANIMAL / MAN
1000 B.C.	LOGS	CRUDE WHEELS						
	WATER	**ROADS**	**TRACKS**	**AIR**	**SYSTEMS**	**SAFETY**	**ENVIRON-MENT**	**RESOURCE**
	GUIDEWAY & VEHICULAR TECHNOLOGY				SYSTEMS TECHNOLOGY			

Fig. 13.1
New technology in guideways, vehicles and systems: a historical progression. *Source:* A F Williams (1992).

live, well seen in the case of the relationship between technology and urban form discussed in Box 4.5. However it is interesting to note that city transport systems are basically the same now as they were 75 years ago. If the problems of congestion are so severe and so persistent, why have radically new systems failed to emerge? The principal explanation is that technology has been unable to break the triangular link between high speed, maximum flexibility and low cost, all of which are desirable ingredients of urban transport. Typically, new public transport technologies have met one of these needs, but only at the cost of the other two. For example, automated high-speed trains require long distances between stations and are expensive: lowering the cost or increasing the number of stations to increase flexibility both reduce the speed. The private car *does* fit the bill, however, for as long as new highways are continually provided for it at public expense, it has the advantages of speed and flexibility, together with low costs to the user (although these are very high for society).

It could be argued that the thrust of public transport investment over the past half-century has failed to capture the car driver because it has been misdirected towards providing *mass* public transport. There has been an underlying belief that public transport would have to compete with the car on price. Public transport thus typically bulk-loads passengers into high-capacity vehicles in order to save on labour costs, which even so account for 60–70 per cent of operating expenses. But this does not provide anything like the quality of service given to its driver by the private car. The car provides personal transport in privacy and security, whereas mass public transport does neither of these things. On closer examination therefore, it seems that the main disadvantage of mass public transport is not that it is *public*, but that it is *mass*, prompting the thought that if *personal public* transport could be provided, it could do much to attract the car driver.

This is where new technology may have a role. As rapid transport has become automated, it has dispensed with the need for a driver, the most expensive element of operating costs. Henceforth, vehicles no longer have to be large to maximise the benefit from the driver's time, so there is no need for public transport to be orientated towards mass movement. Public transport vehicles could be as small as existing private cars and thus be personal, private and secure. Such vehicles would need only a lightweight track, which would be less expensive and visually intrusive than the structures required for mass transit and could thus be provided in a dense net across the city. One could imagine users walking the short distance to their nearest stop, boarding a three- or four-seat electric vehicle and requesting the destination via keyboard entry. The vehicle would then be automatically routed through the system while the user's charge card is debited for the fare.

Such a system would offer many of the benefits of the private car, including what one might call 'demand actuation', whereby departure time, destination, number of stops, etc. are under individual control, rather than dictated to the passenger by the needs of the system. On the other hand, it does not have many of the disbenefits, such as causing pollution, accidents or congestion, or requiring parking space at the destination. As public cars, such vehicles could serve hundreds of users during the day, instead of being parked for most of the time awaiting the owner's return. This combination of vehicular, guideway and system technology, to use the phraseology of Fig. 13.1, is of course highly speculative, but something like it will be needed if personal mobility is to be maintained and 'traditional' cities are to survive.

13.1.2 *Technology and spatial change at regional, national and global scales*

The role of technology in altering spatial relationships is influential at all scales, with technological advances in transport frequently having differential spatial effects, conferring advantages on particular places and thereby disadvantaging others. So it is with electronic, satellite and computer-based telecommunications networks, the new trade routes of the twenty-first century, which are attracting information-based industries to the global cities at their nodes, particularly London, New York and Tokyo. Spaces between will be hooked into the networks at speeds that reflect their economic position, with early connections achieving cumulative advantage over later ones. One of the major spatial impacts of these dramatic changes in communication technology is thus likely to be the perpetuation and reinforcement of uneven development.

This is true also of future rail travel, where standardisation of track and vehicles, automated control and internationalisation of systems will enable speeds of 225 km per hour in high-speed trains (HSTs), 300 km per hour in derivatives of the French TGV and possibly 500 km per hour if German or Japanese experiments with magnetic levitation (Maglev) are successful. The time penalties that result from deceleration and acceleration are such that intermediate timetabled stops undermine the advatages of the high top speeds. This encourages uneven development of different places by reducing the accessibility of bypassed points *en route* and enhancing the centrality of terminals. New rail systems are thus effectively land-based airlines and it seems likely that they will be able to replace air travel on the principal international land axes of up to 2700 km, such as the European 'main lines' shown in Fig. 13.2.

The major role for air travel will be for intercontinental movement. In 1976 Masefield predicted the evolution of passenger rockets that would bring New York within 30 minutes of London and Williams (1992) envisages a new generation of supersonic transports (SSTs) for business travel and foresees that every major city around the globe will be linked

Plate 13.2
New rail systems in Europe seem likely to replace air travel on principal national and international land axes of up to 1000 km (TGV Atlantique at Brou, near Le Mans, France).

to every other by 1000 seat airbuses. He also sees a role for local air travel, describing a day in the life of one London resident in the year 2050, who commutes daily to Brussels by train, pops over to Amsterdam for the afternoon by 'VTOL company car-plane' and 'indulges himself' by flying home.

Such predicted levels of land and air mobility rest on assumptions about energy supply and pollution control which are not supported by evidence available at present. Williams forecasts that fossil fuels will be replaced by synthetics because 'Such is the investment in our inherited transport systems that it is far cheaper to alter the fuels than to make major changes in vehicle technology.' However, it is only cheaper if long-run and large-scale issues of global warming, atmospheric pollution and the continuation of unsustainable practices are ignored. Of course, technology will produce some answers, for there will be insistent pressures to improve transport performance in terms of cost, unit size, frequency, safety, speed and stage length. However, it would be extremely unwise, even foolhardy, to assume that the environmental limits as at present understood will be extended by technological advance and it would thus be more prudent to plan for levels of movement that we know are sustainable. This means that we cannot duck the linked issues of motorisation and development, for it is from this relationship that the most profound economic, political, social and environmental tensions are emerging.

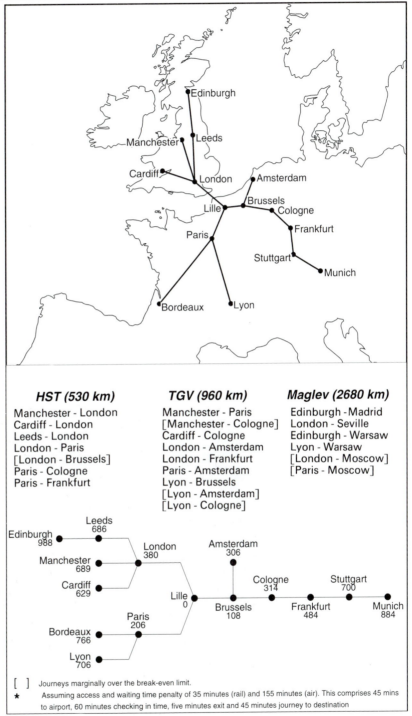

Fig. 13.2
High-speed rail in Europe: journeys faster than by air. *Source:* Hall (1991).

13.2 Motorisation and development

13.2.1 How many cars?

As affluent individuals and societies have achieved 'freedom' in transport by increasing motorisation, so one could argue that this is a benefit that should not be withheld from the less affluent. One government statement in the UK in 1976 put it thus: 'Car ownership should increase, for personal mobility is what people want and those who already have it should not try to pull the ladder up behind them.'

But how many can get on the ladder? Already the former eastern Europe is expected to increase its car-kilometres by 600 per cent by 2010 as suppressed demand is liberated. At present 8 per cent of the world's population owns a car: if only modest levels of ownership were to be achieved by India and China, each would have over 300 million cars, doubling the number in the world. This would make demands on energy, the atmosphere, raw material resources and space which the earth could not sustain – and this for the benefit of only 16 per cent of its population.

The poverty of the South and the consumerism of the North are all part and parcel of the global sustainability issue, nowhere more graphically illustrated than in transport, where hypermobility in the North contrasts with a fundamental lack of accessibility in the South. As long as gross inequalities in transport behaviour exist, so will the have-nots aspire to

Plate 13.3
Despite its widespread availability, the bicycle in developing countries is often regarded in a negative light, as an impediment to motorised travel. Motorisation in China to levels typical of developed countries would increase the world's population of cars by 50 per cent.

trading up to whatever levels of mobility are characteristic of the developed world. If we accept the upward trend in vehicle use in industrialised societies, we cannot logically deny it to developing countries. On the other hand, if we accept that the diffusion of Northern levels of mobility to the South is not sustainable but we fail to reduce levels of consumption in the North, we are expressing a commitment to global inequality and exploitation which is unsupportable.

Sustainability is a global concept, not an issue that can be compartmentalised in Europe or North America without any consideration of the impacts of interregional transfers of pollution, energy and waste. For example, it has been suggested that sustainability in the North could be achieved based on vehicle conversion to biomass fuels such as ethanol. But the raw materials would have to come largely from developing countries by encouraging them to move away from independent foodstuff production in favour of cash crops, whose prices would be dictated by countries in the North and whose production would result in the destruction of fragile environments through intensive agricultural methods, heavy fertiliser applications and other unsustainable practices.

Transport is a key to sustainability because it can either be used to accelerate resource depletion and global atmospheric damage or it can be the focus of living arrangements that can be serviced by environmentally tolerable practices such as walking and cycling. However, reducing dependence on motorised transport and changing the organisation of land uses and activities are obstructed by the persistence among decision-makers of the view that transport investment is an essential prerequisite to economic growth, which is itself essential to development. In times of recession the argument is difficult to resist: it is a Keynesian response to poor trading conditions, putting people back to work on construction projects such as road building and in the vehicle-making industries. But research has shown that spending on road building rather than housing or public transport is a very inefficient way of creating jobs and of cutting costs to industry. It is often implied that the benefits of such schemes are self-evident, but as Chapter 4 has shown, there is no necessary correlation between transport investment and economic growth. British regional policy has long attempted to revive the fortunes of flagging regions through the provision of transport infrastructure such as the Severn Bridge to Wales and the extension of the M6 motorway towards Scotland. However, the impact of these schemes on the economic balance between regions has been found to be at best marginal and at worst negative, with business shifting towards locations favoured on the basis of other criteria besides transport availability. Nevertheless, as long as these views hold sway there will scarcely be a local authority that does not lobby for transport infrastructure as an aid to economic growth, and we are likely to continue to see prestige transport projects initiated by politicians and supported by governments which have powerful vested interests in maintaining the status quo of transport investment.

Other vested interests collude in this process. For example, the ten largest companies in the world all produce either oil or cars and, not surprisingly, have used their enormous power to ensure the continuance of motorisation. One well-documented case concerns the way that General Motors (GM) was able to follow a systematic policy of purchasing streetcar systems in American cities, ripping them up and replacing them with GM-made buses, which were in turn eliminated as alternatives to the private car, of which GM was the world's largest producer. By 1949 this devastating assault had virtually destroyed the USA's formerly extensive city streetcar systems including that of Los Angeles, once the largest in the world.

Post-Rio, the motor industry lobby is selling the idea of the 'clean, green car' as its answer to environmental degradation, but catalytic converters and lead-free fuel do not make a vehicle green even in emission terms, let alone in terms of noise, accidents, space consumption or resource use. The electric car gives an illusion of cleanness but only shifts the emissions problem from one sector to another and does not address the traffic congestion issue. It certainly is possible to produce a vehicle which is almost totally green in the sense that it produces no emissions, consumes no fuel, poses little threat to other road users and takes up minimal amounts of space – but it would be a bicycle, or a rickshaw.

The importance of such 'intermediate' transport technologies should not be underestimated. There are twice as many bicycles in the world as cars: in Asia alone they transport more people than do all the world's cars and in China annual bicycle sales are actually higher than the world's car sales. The developing world thus has many elements of sustainability in place, yet these are under attack from Northern notions of car-based 'modernisation'. These do not depict the bicycle as versatile, inexpensive, resource-efficient and environmentally benign, but deride it as a poor man's transport which obstructs the movement of the car-owning élite, to which most planners belong. That propaganda may sell more cars, but leads in the direction of environmental catastrophe. In both South and North, if genuine sustainability is to be achieved, the bicycle – or something very like it – has to occupy a much more central role than the car in future transport policy.

13.2.2 Ivan Illich on energy and equity

These ideas have been debated by the twentieth-century philosopher Ivan Illich, who has argued that the growth of motorisation actually obstructs traffic by interfering in its flow, by creating isolated destinations and by increasing the time we must spend on travel. These views are worth examining in more detail.

Firstly, Illich (1974) states that the growth of motorisation actually blocks mobility by cluttering up the environment with vehicles and roads.

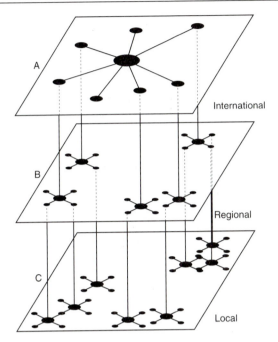

Fig. 13.3
Multi-layered geographies: hierarchies of levels of accessibility. *Source:* Hägerstrand (1967).

In many cities in developing countries the severe traffic congestion is a function of car use by a few who blocking streets for all, whilst in cities in industrialised countries it is no longer possible to walk in anything like a straight line to where you want to go, because of the need to cross roads, avoid parked vehicles or because certain types of road cannot be used. Secondly, he argues that the growth of larger, but more widely distributed shops, schools and hospitals means longer journeys which are therefore less able to be made on foot, so that people without cars become yet more isolated. Transport opportunities become polarised, with a wider choice of destinations available to car users while those without can no longer even get to the places that they used to before lots of cars appeared on the scene. In Illich's view we are seeing the establishment of what might be called 'multi-layered geographies', whereby some people pack unlimited mileage into a lifetime of travel whilst others can no longer reach the simple services they require. Figure 13.3 shows a hierarchy of levels of accessibility, with the 'jet-setter' able to operate across all the scales from local to international in contrast to the low-mobility individual, locked into highly localised geographical areas and wholly dependent on local resources for well-being. Illich (1974: 42–3) puts it thus: 'Beyond a certain speed motorised vehicles create remoteness which they alone can shrink. They create distances for all and shrink them for only a few.'

Illich's third argument is that the car does not help us to save time.

One study has shown how transport expenditure by German workers more than doubled in real terms between 1968 and 1985, yet the time they spent travelling to essential services did not go down because the services had become further away from each other. Buying a car then does not lead to less time travelling, because the time we save by travelling faster has to be spent on more travel, trapping us in a daily loop of longer, unwanted trips, precisely the opposite of what the car beguilingly promises.

Admittedly life may be more colourful as John Fowles says (see beginning of this chapter), but this is at the price of becoming prisoners of transport. At the places we can still reach on foot there is nothing happening. To reach important destinations we have to drive or be driven. Illich describes this as 'compulsory consumption of high doses of energy'. Douglas Adams, in his famous 'Restaurant at the end of the universe' book takes an equally critical – if more amusing – view of the whole system of motorisation as we saw at the beginning of Chapter 10.

Speed also pollutes spaces and places. The faster we go the more space we need, for at low speeds tight curves can be tolerated and the headway between vehicles can be small, but high speeds make greater demands on space, often in areas where it is shortest supply. As Fig. 7.4 showed, a car at 40 km per hour requires 60 m^2 per person, compared to only a third of that at 10 km per hour. Travelling at high speeds blurs our appreciation of places, symbolised by the increasing adoption of the American use of time as a substitute for distance, whereby somewhere is said to be x minutes away rather than y kilometres. The implication is that the journey is a function of the consumption of time and that the space across which you travel is irrelevant. Illich suggests that we have a superficial view of a greater expanse of territory, but we lose sight of the variation in the grain of the land and the subtleties of the landscape – and with it some of our humanity. To be fair, this is not just a problem with the car; it is a philosophical issue that concerns the nature of space and man's relationship to it. A reading of some of the quotations at the beginning of this chapter serves to show the eternal nature of the debate over the conquest of the 'tyranny of distance'.

In the twentieth century, society has put mobility above accessibility and associated higher speeds with progress and lower speeds with antiquated attitudes or poverty. Yet the advantage of the car over non-motorised modes is illusory in so many ways. For example, Illich has shown that if we include the time spent earning money to buy a car, to maintain it, to pay for fuel, taxes, fines and parking fees, the typical American spends 1600 hours per annum to travel 12 000 km, less than 8 km per hour on average and about the same speed achieved in most developing countries by people on foot or bicycle. An alternative approach to more motorisation would be to recognise that the problems caused by excessive consumption of transport are soluble under conditions where transport is minimised. This would mean promoting ways of living that do not encourage increases in travel and which could thus be served by environmentally beneficial transport systems. How people can be persuaded to adopt such fundamental changes in their lifestyles is one of the most important questions facing us as we enter the twenty-first century.

13.3 Final thoughts: a research agenda for transport geography

The twentieth century began with the airship and ended with the space shuttle. The global village has become reality as tourism has grown into the world's biggest industry. But transport development is socially and spatially uneven; many Europeans travel further in a month than some Indians do in a lifetime, and whilst some Americans drive 100 km or more to work in air-conditioned comfort, many African women have to walk for hours carrying heavy burdens of water and firewood on their heads. Our world is part affluent and transport-rich and part hungry, poverty stricken and transport-poor.

But it is a rapidly changing world too, with many interesting questions about transport and spatial change posed by the political and economic developments of the late twentieth century, such as technological innovation in transport and telecommunications; reduced constraints on trade and the related emergence of a world economy; deregulation and privatisation in transport; the move from command economies towards market economies in eastern Europe; the widening mobility gap between rich and poor; growing urban congestion; increasing scarcity of energy resources; and an increase in environmental awareness.

With such stimulation, it is not surprising that recent years have seen an international resurgence of research in transport geography (Table 13.1), though almost inevitably this throws up yet more questions that require further investigation; the mating cry of the transport geographer is 'more research' – or better still 'more money for more research!' As ever, some of the most tantalising questions are theoretical and methodological ones, such as these listed by Knowles (1993):

- Does efficient international and national movement of resources and products, people and information depend on both successful technological innovation and the lifting of regulatory constraints?
- Are new transport routes in less developed areas civilising influences or do they give rise to new tensions and inequalities?
- How can the effectiveness of the more market-orientated policies of transport deregulation, privatisation and subsidy limitation be measured on local, national or international scales?
- How can transport needs be assessed?

(a)

(b)

Plate 13.4
Transport development is socially and spatially uneven, producing a world which is part
affluent and transport-rich ((a) Incline, California) and part poverty-stricken and transport-
poor ((b) Harare, Zimbabwe).

Table 13.1
The research agenda in transport geography: a consensus of opinions of transport geographers from around the world

1. Policy practice and analysis
The impact of political changes on transport systems:

- central and eastern Europe and the CIS
- super-national markets
- democracies in the developing world
- policy typology as a framework for analysis

Deregulation, privatisation and investment:

- effects on networks
- investment issues

2. The impact of infrastructure provision
- the range of effects
- large-scale projects and uneven development
- traffic congestion
- terminals
- redevelopment of infrastructure

3. The declining friction of distance
- location and transport innovation
- intermodal and bulk freight transport
- telecommunications
- logistics

4. The mobility gap and differential accessibility
- inequalities at global and local scales
- less developed countries

5. Demand modelling
- searching for more refined and sensitive forecasting techniques

6. Transport, environment and energy
- impacts
- assessment

7. Travel, recreation and tourism
- implications for transport in the 'leisure age'

8. Challenges in theory and methodology
- developing concepts in transport geography

9. Information for transport planning and operation
- improving the quality and quantity of transport data

Source: adapted from Knowles (1993).

- Do we understand enough about human behaviour to model and predict transport demand?
- How do we place value on environmental gains and losses?

As we search for ways of raising levels of accessibility for all of the world's people it would be silly and pompous to claim that the geographer has a key role, any more than that of the economist, engineer, the educationist or any other needed specialism. But the underlying reality is that transport is a means of overcoming the barriers provided by distance and requires the geographical perspective if it is to be seen in its comprehensive, holistic role. There is a job out there for the geographer to do; we hope that reading and using this book will have done something towards encouraging you to join in.

Further reading

Hall P 1991 Moving information: a tale of four technologies. In Brotchie J et al *Cities of the 21st century* Halsted, pp 1–21

Illich I 1974 *Energy and equity* Calder and Boyars

Knowles R 1993 Research agendas in transport geography for the 1990s. *Journal of Transport Geography* **1**(1): 3–11

Williams A F 1992 Transport and the future. In Hoyle B S, Knowles R D (eds) *Modern transport geography* Belhaven, pp 257–70

Bibliography

Chapter 1

Hoyle B S, Knowles R (eds) 1992 *Modern transport geography* Belhaven
Ullman E L 1956 The role of transportation and the basis for interaction. In
   Thomas W L (ed) *Man's role in changing the face of the earth* University
   of Chicago Press, pp. 862–88
White H P 1977 The geographical approach to transport studies. Discussion
   Papers in Geography No. 1 April, Salford University
Williams A F 1981 Aims and achievements of transport geography. In Whitelegg
   J (ed) *The spirit and purpose of transport geography* Transport Geography
   Study Group, pp 5–31

Chapter 2

Appleton J H 1967 Some geographical aspects of the modernisation of British
   Railways. *Geography* 52:356–76
Barwell I, Edmonds G A, Hoare J D, de Veen J (eds) 1985 *Rural transport in
   developing countries* Intermediate Technology Publications
British Railways Board 1963 *The reshaping of British railways* British Rail
Bunge W 1962 *Theoretical studies in geography* Lund Studies in Geography,
   Series C
Burns L D 1979 *Transportation, temporal and spatial components of accessibility*
   Lexington Books, Lexington, Mass.
Department of Transport various dates *National travel survey* HMSO
Fowkes A S 1987 Forecasting freight mode choice in Great Britain. *Proceedings,
   Seminar G*, PTRC summer meeting, pp 43–54
Fullerton J 1982 *Shopping travel patterns in Tyne and Wear: a before Metro
   profile* Report LR 1045, Transport and Road Research Laboratory

Haggett P, Chorley R 1969 *Network analysis in geography* Edward Arnold

Hay A 1976 *Transport for the space economy* Macmillan

Hopkin J M 1978 *The mobility of old people in Guildford* Report LR 858, Transport and Road Research Laboratory

Hupkes G 1982 The law of constant travel time and trip rates. *Futures* **14**:42

Hurst E E 1974 (ed) *Transportation geography: comments and readings* McGraw-Hill

Independent Commission on Transport 1974 *Changing directions* Coronet Books

Jones P M 1975 Accessibility, mobility and travel need. Paper presented at Transport Geography Study Group conference, Salford University

Jones P M et al 1983 *Understanding travel behaviour* Gower

Jones S R 1981 *Accessibility measures: a literature review* Report LR 967, Transport and Road Research Laboratory

Jones S R 1984 *Accessibility and public transport use* Report SR 832, Transport and Road Research Laboratory

Kaira C 1985 In Barwell I, Edmonds G A, Hoare J D, de Veen J (eds) *Rural transport in developing countries* Ch 10 Intermediate Technology Publications

Lomax D E, Downes J D 1971 *Patterns of travel to school and work in Reading* Report LR 808, Transport and Road Research Laboratory

McCall M 1985 Accessibility and mobility in peasant agriculture. In Cloke P (ed) *Rural accessibility and mobility* Dept of Geography, Lampeter University, pp 42–63

McKinnon A 1983 The causes, costs and benefits of increasing delivery distances. In Turton B J (ed) *Public issues in transport* Transport Geography Study Group, pp 93–111

Moyes A M 1974 Accessibility to general practitioner services on Anglesey. Paper presented at Transport Geography Study Group conference, Salford University

Nice D C 1989 Stability of the Amtrak system. *Transportation Quarterly* **43** (4): 557–70

Page P A 1974 Mobility and the demand for transportation. Paper presented at Transport Geography Study Group conference, Salford University

Patmore J A 1983 *Recreation and resources* Blackwell

Patmore J A 1965 The British railway network in the Beeching era. *Economic Geography* **41**:71–81

Patmore J A 1966 The contraction of the network of railway passenger services in England and Wales 1836–1962. *Transactions of the Institute of British Geographers* **38**:105–18

Plane D 1981 Geography of urban commuting fields. *Professional Geographer* **33**, pp 182–8

Quarmby D 1989 Developments in the retail market and their effects on freight distribution. *Journal of Transport Economics and Policy* **23** (1): 75–88

Schaeffer K H, Sclar E 1975 *Access for all: transportation and urban growth* Penguin

Starkie D 1967 *Traffic and industry: a study of traffic generation and spatial interaction* Geographical Paper 3, London School of Economics

TRRL/Gwent C C Joint Working Group 1981 *Accessibility measures in Gwent* Report LR 994, Transport and Road Research Laboratory

Ullman E L 1956 The role of transportation and the basis for interaction. In

Thomas W L (ed) *Man's role in changing the face of the earth* University of Chicago Press, pp 862–88

White H 1963 The reshaping of British Railways: a review. *Geography* **48**:335–7

White H, Senior M 1983 *Transport geography* Longman

Williams A F 1977 Crossroads – the new accessibility of the West Midlands. In *Metropolitan development and change: the West Midlands*. British Association, Saxon House, pp 367–92

Chapter 3

Aloba O 1983 Evolution of rural roads in Nigeria. *Singapore Journal of Tropical Geography* **4**:1–10

Appleton J H 1956 The railway geography of South Yorkshire. *Transactions of the Institute of British Geographers* **22**:159–70

Appleton J H 1965 *A morphological approach to the geography of transport* Occasional Paper 3, University of Hull

Bird J H 1971 *Seaports and seaport terminals* Hutchinson

Black W R 1967 *Growth of the railway network of Maine* University of Iowa, Department of Geography Discussion Paper 5

Black W. R 1993 Transport route location: a theoretical framework. *Journal of Transport Geography* **1**(2):86–94

Ekstrom A, Williamson M 1971 Transportation and urbanisation. In Wilson A G (ed) *Urban and regional planning* Papers in Regional Science 3, Pion, pp 33–46

Fullerton B 1975 *The development of British transport networks* Oxford University Press

Gould P R 1966 *Space searching procedures in geography and the social sciences* University of Hawaii Social Science Research Paper 1

Haggett P, Chorley R 1969 *Network analysis in geography* Edward Arnold

Hoyle B S 1983 *Seaports and development: the experience of Kenya and Tanzania* Gordon and Breach

Hoyle B S (ed) 1990 *Port cities in context* Transport Geography Study Group

Hoyle B S et al (eds) 1988 *Revitalising the waterfront* Belhaven

Kansky K J 1963 *Structure of transport networks* University of Chicago, Department of Geography Research Paper 84

Lachene R 1965 Networks and the location of economic activities. *Papers of the Regional Science Association* **14**:183–96

Taafe E J et al 1963 Transport expansion in underdeveloped countries: a comparative analysis. *Geographical Review* **53**:503–29

Tinkler K J 1979 Graph theory. *Progress in human geography* **3**:85–116

Tolley R S 1973 New technology and transport geography: the case of the hovercraft. *Geography* **58**:227–36

United Nations Statistical Bulletins (1966–92)

Wellington A M 1887 *The economic theory of the location of railways* New York

White H, Senior M 1983 *Transport geography* Longman

Chapter 4

Alonso W 1963 *Location and land-use* Harvard University Press, Cambridge, Mass.

Barwell I, Edmonds G A, Hoare J D, de Veen J (eds) 1985 *Rural transport in developing countries* Intermediate Technology Publications

Burgess E W 1925 The growth of the city. In Park R E et al (eds) *The city* University of Chicago Press, pp 178–84

Christaller W 1933 *Central places in southern Germany* Gustav Fischer, Jena

Daniels P W, Warnes A M 1980 *Movement in cities* Methuen

Dennis R J 1984 *English industrial cities of the nineteenth century* Cambridge University Press

Dicken P 1992 *Global shift* 2nd edn PCP

Greenhut M 1956 *Plant location in theory and practice* University of North Carolina Press

Harris C D, Ullman E L 1945 The nature of cities. *Annals, American Academy of Political and Social Science* **242**:7–17

Herbert D T, Thomas C J 1982 *Urban geography* Wiley

Hoover E M 1948 *The location of economic activity* McGraw-Hill

Howard S 1986 *A role for railways in the inland distribution of maritime containers* Proceedings, Seminar G, PTRC Summer Meeting, pp 21–8

Hoyt H 1939 *The structure and growth of residential neighbourhoods in American cities* Federal Housing Administration, Washington

Isard W 1956 *Location and space-economy* Mass. Institute of Technology Press, Cambridge, Mass.

Lloyd P E, Dicken P 1977 *Location in space* 2nd edn Harper Row

McKinnon A 1991 Transformations of the retail supply chain. In Moyes A (ed) *Companies, regions and transport change* IBG, pp 33–50

Muller P O 1989 Transportation and urban form. In Hanson S (ed) *Geography of urban transportation* Guildford Press, pp 24–48

Mwase N 1987a Railway pricing in developing countries. *Journal of Transport Economics and Policy* **21**(2): 189–214

Mwase N 1987b Zambia, the Tazara and alternative outlets to the sea. *Transport Reviews* **7**(3):191–206

Pearson R, Fossey J 1983 *World deep sea container shipping* Gower

Smith D M 1971 *Industrial location* 1st edn Wiley

von Thunen J H 1875 *The isolated state* (in translation) Hamburg

Weber A 1929 *Alfred Weber's theory of the location of industries* (translated from *Über den standart der industrien* 1909) University of Chicago Press, Chicago

Chapter 5

Beaver S H 1967 Ships and shipping: the geographical consequences of technological progress. *Geography* **52**:133–56

Bird J H 1963 *The major seaports of the United Kingdom* Hutchinson

Bird J H 1967 Seaports and the EEC. *Geographical Journal* **133**:302–27

Bird J H 1968 Traffic flows to and from British seaports. *Geography* **54**(3): 284–302

Bird J H 1971 *Seaports and seaport terminals* Hutchinson

British Airports Authority 1976 *Heathrow Airport – London: master plan report* British Airports Authority

Bruyelle P et al 1994 Channel Tunnel special edition. *Applied Geography* **14**(11): 3–104

Chiu T N, Chu D K-Y 1984 Port development of the Peoples Republic of China. In Hoyle B S, Hilling D (eds) *Seaport systems and spatial change* Wiley, pp 199–216

Chu D K-Y 1989 Planning and development of the container port of Hong Kong. *Third World Planning Review* **11**(3):275–88

Couper A D 1972 *The geography of sea transport* Hutchinson

Cox S 1984 Developments in world sea transport. *Geographical Magazine* **October**:521–7

Dennis N 1993 Paper to IBG Transport Geography Study Group Symposium, London, January

Department of Transport 1987 *Annual digest of port statistics* British Ports Federation

Elliot N R 1969 Hinterland and foreland as illustrated by the port of the Tyne. *Transactions of the Institute of British Geographers* **47**:153–70

Freeman D B 1973 *International trade, migration and capital flows* Chicago Research Paper 146, University of Chicago

Hastings P, Hayles M 1988 Airfreight special report. *Transport* **9**(2):77–81

Hayuth Y 1982 Intermodal transportation and the hinterland concept. *Tijdschrift voor Economische en Sociale Geografie* **73**:13–21

Hilling D 1969 The evolution of the major ports of W. Africa. *Geographical Journal* **135**(3):365–78

Hilling D 1989 Technology and the changing port system of England and Wales. *Geography* **74**:117–27

HMSO 1962 *Report of the Rochdale Committee of enquiry into the major ports of Britain* Cmnd 1824

HMSO 1967 *The third London airport* Cmnd 3259

Hoare A G 1988 Geographical aspects of British overseas trade. *Environment and Planning A* **20**(10):1345–64

Hoyle B S 1968 East African seaports: an application of the concept of Anyport. *Transactions of the Institute of British Geographers* **44**:163–83

Hoyle B S 1988 *Transport and development in Tropical Africa* John Murray

Hoyle B S, Hilling D 1984 (eds) *Seaport systems and spatial change* Wiley

Johnston R J 1976 *The world trade system* Bell

McHale J 1969 *The future of the future* George Braziller, New York

Manchester International Airport 1985 *Development strategy to 1995* Manchester Airport Authority

Matley I M 1976 The geography of international tourism Resource Paper, *Annals American Association of Geographers*

Mayer H M 1973 Geographical aspects of technological change in maritime transportation. *Economic Geography* **49**:145–55

OECD 1983 *Short-sea shipping in the economy of inland Europe* OECD, Paris

Pearce D G 1981 *Tourist development* Longman

Pearce D G 1987 *Tourism today: a geographical analysis* Longman

Pearson R, Fossey J 1983 *World deep sea container shipping* Gower
Robinson D 1989 European ports. *Transport* **10**(3):127–32
Roskill Report 1971 *Report of the Roskill Commission on the third London airport* HMSO
Sealy K R 1966 *The geography of air transport* Hutchinson
Sealy K R 1976 *Airport strategy and planning* Oxford University Press
Sealy K R 1992 International air transport In Hoyle B, Knowles R (eds) *Modern transport geography* Belhaven, pp 233–56
Shneerson D 1981 Investment in port systems. *Journal of Transport Economics and Policy* **15**:201–16
Southern A 1988 Heathrow transport links. *Transport* **9**(2):85–8
Thomas R S, Conkling E 1967 *The geography of international trade* Prentice-Hall, New Jersey
Tolley R S, Turton B J (eds) 1987 *Short-sea crossings and the Channel Tunnel* Transport Geography Study Group
UN 1990a *Annual bulletin of transport statistics for Europe* Paris
UN 1990b *Yearbook of international trade statistics* New York
Whitby R et al 1971 London airport and the Roskill Commission. *Regional Studies* **5**(3):118–83

Chapter 6

(a) Developed countries

Charlesworth G 1984 *A history of British motorways* Telford
Damesick P J et al 1986 M25 – a new geography of development? *Geographical Journal* **152**(2):155–75
Fullerton B 1982 Transport. In House J W (ed) *The UK space* 3rd edn Weidenfeld and Nicolson, pp 356–425
Hall D 1993 Impacts of economic and political transition on the transport geography of central and eastern Europe. *Journal of Transport Geography* **1**:20–35
Hall D (ed) 1993 *Transport and economic development in the new central and Eastern Europe* Belhaven
Hideo Nakamura, Shigeri Morichi (eds) 1989 Japan's transport policy. *Transportation Research A* **23**(1):1–101
Ilbery B W 1986 *Western Europe: a systematic human geography* Oxford University Press, Ch 5, pp 65–83
Lijewski T 1982 Transport in Poland. *Transport Reviews* **2**(1):1–21
Mellor R E 1973 A basic railway passenger network in the US. *Geography* **58**:163–6
Nash C A 1981 Government policy and rail transport in Western Europe. *Transport Reviews* **1**(2):225–50
Scott D 1983 The West German transport system. *Geography* **68**(3):266–70
Smerk G M 1981 A profile on transportation in the US. *Transport Reviews* **1**(2)
Smith J et al 1991 British passenger transport into the 1990s. *Geography* **77**:63–93

Starkie D 1982 *The motorway age: road and traffic policies in post-war Britain* Pergamon

Turton B J 1991a The role of the RMS in the rural road transport sector in Zimbabwe. *Geographical Journal of Zimbabwe* **22**:46–61

Turton B J 1991b The changing transport pattern. In Johnston R J, Gardiner V (eds) *The changing geography of the United Kingdom* Routledge, pp 171–97

Turton B J 1992a British Rail passenger policies. *Geography* **77**(1):64–7

Turton B J 1992b Urban transport problems and solutions. In Hoyle B S, Knowles R D (eds) *Modern transport geography* Belhaven, pp 81–104

White H P et al 1978 Problems of transport in modern Britain. *Geographical Magazine* **50**:386–92; 452–6; 672–8; 763–6; 825–33

(b) Other countries

Cundill M 1985 Road–rail competition for freight traffic in Kenya. *Proceedings, Seminar F, PTRC Summer Meeting*, pp 9–20

Dale E H 1968 Some geographical aspects of African land-locked states. *Annals of the Association of American Geographers* **58**:485–505

Ezeife P C 1984 The development of the Nigerian transport system. *Transport Reviews* **4**(4):305–30

Gibb R A 1987 Effects on the countries of SADCC of economic sanctions against the Republic of South Africa. *Transactions of the Institute of British Geographers* **12**(4):398–412

Gibb R A 1991 Imposing dependence: South Africa's manipulation of regional railways. *Transport reviews* **11**(1):19–39

Griffiths I 1986a The scramble for Africa: Africa's inherited political boundaries. *Geographical Journal* **152**:512–17

Griffiths I 1986b The web of steel. *Geographical Magazine* **58**(10):512–17

Hoyle B S (ed) 1973 *Transport and development* Macmillan

Hoyle B S, Hilling D 1970 *Seaports and development in tropical Africa* Macmillan

Leung C K 1980 *China: railway patterns and national goals* Department of Geography Research Paper 165, University of Glasgow

Lienbach T R 1975 Transportation and the development of Malaya. *Annals of the Association of American Geographers* **65**:270–82

Marsden P 1986 Inland containerisation in developing countries. *Proceedings, Seminars G and F, PTRC Summer Meeting*, pp 11–20

Mwase N 1987 Railway pricing in developing countries. *Journal of Transport Economics and Policy* **21**(2):189–214

Puvanachandram V, Beatty A 1985 Value of freight and other benefits to villages resulting from road improvements in Papua New Guinea. *Proceedings, Seminar F, PTRC Summer Meeting*, pp 31–42

Turton B J 1974 River transport in the less-developed countries. In Hoyle B S (ed) *Spatial aspects of development* Wiley, Ch 17, pp 323–44

Turton B J 1987 Transport in Zimbabwe. *Transport* **8**(4):165–7

Turton B J 1989 Railways and the national economy of Zimbabwe. *Geographical Journal of Zimbabwe* **19**:47–58

Various authors 1989 Transportation in SE Asia. *Transportation Research* **23**(A)

Chapter 7

Adams J 1981 *Transport planning:vision and practice* Routledge

Anon 1989 You'll always be late. *Southern African Economist* **August/September**: 18–19

Buchanan M 1991 *Urban transport trends and possibilities* Rees Jeffreys Discussion Paper 18, Transport Studies Unit, University of Oxford

Cervero R 1989 *America's suburban centres: the land-use transportation link* Unwin-Hyman

Daniels P W, Warnes A M 1980 *Movement in cities* Methuen

Davis R 1992 Congestion: so what's the problem? In Whitelegg J (ed) *Traffic congestion: is there a way out?* Leading Edge, pp 33–50

Dickins I 1988 *An introduction to light rail transit in Europe* Faculty of the Built Environment Working Paper 32, Birmingham Polytechnic

Dimitriou H T 1990 Transport problems of Third World cities. In Dimitriou H T (ed) *Transport planning for Third World cities* Routledge, pp 50–84

Goodwin P B 1969 Car and bus journeys to and from Central London in peak hours. *Traffic Engineering and Control* **December**

Hanson S 1986 Dimensions of the urban transport problem. In Hanson S (ed) *The geography of urban transportation* Guilford Press, New York, pp3–23

Hass-Klau C 1990 *The pedestrian and city traffic* Belhaven

Hillman M, Whalley A 1979 *Walking is transport* Policy Studies Institute

Jones P M 1983 *Understanding travel behaviour* Gower

Kipke B 1991 *Bicycle reference manual for developing countries* GATE: Deutsches Zentrum für Entwicklungstechnologien

Knowles R D, Fairweather L 1991 *The impact of rapid transit* Metrolink Impact Study Working Paper 2, Department of Geography, University of Salford

Mogridge M 1990 *Travel in towns: jam yesterday, jam today and jam tomorrow* Macmillan

Navarro R A et al 1985 *Alternativas de transporte en America Latina: la bicicleta y los triciclos* SKAT, Centro Suizo de Technologia Apropriada, St Gallen, Switzerland

Ogden K W 1992 *Urban goods movement: a guide to policy and planning* Ashgate

Owen W 1972a *Making cities livable* Research Report 126, Brookings Institution, Washington

Owen W 1972b *The accessible city* Brookings Institution, Washington

Plowden S 1980 *Taming traffic* André Deutsch

Preston B 1991 *The impact of the motor car* Brefi

Rao M S V, Sharma A K 1990 The role of non-motorized urban travel. In Dimitriou H T (ed) *Transport planning for Third World cities* Routledge, pp 117–43

Replogle M 1992 Bicycles, rickshaws and carts in Asian cities. *Finance and Development* **September:29**

Richards B 1990 *Transport in cities* Architecture Design and Technology Press

Taylor M A 1968 *Studies of travel in Gloucester, Northampton and Reading* Crowthorne Report LR141, Transport and Road Research Laboratory

Thomson J M 1977 *Great cities and their traffic* Penguin

Turton B J 1992 Urban transport patterns. In Hoyle B S, Knowles R (eds) *Modern transport geography* Belhaven, pp 66–80

Various authors 1991 Transport in world cities. *Built Environment* **17**(2):85–183
White P R 1986 *Public transport: its planning, management and operation* 2nd edn Hutchinson
White P R 1990 Inadequacies of urban public transport systems. In Dimitriou H T (ed) *Transport planning for Third World Cities* Routledge, pp 85–116
Whitelegg J 1985 *Urban transport* Macmillan
Whitelegg J 1992 (ed) *Traffic congestion: is there a way out?* Leading Edge
Williams A F (ed) 1985 *Rapid transit systems in the UK: problems and prospects* Transport Geography Study Group
Zhihao W 1989 Bicycles in large cities in China. *Transport Reviews* **9**(2):171–82

Chapter 8

Atkins S T 1977 Transportation planning – is there a road ahead? *Traffic Engineering and Control* **18**(2):58–62
Bell G et al 1983 *The economics and planning of transport* Heinemann
Bracher T 1988 Radwege – van der Chance zur Illusion. Fahrradplanung aus der Sicht der Radfahrer. In Muller A K *Symposium Fahrradzunkunft* Technische Universität, Berlin
British Medical Association 1992 *Cycling towards health and safety* Oxford University Press
Buchanan C D 1963 *Traffic in towns* HMSO
Cervero R, Hall P 1989 Containing traffic congestion in America. *Built Environment* **15**(3/4):176–84
CROW 1993 *Cycling in the city, pedalling in the polder: recent developmemts in policy and research for bicycle facilities in the Netherlands* Centre for Research and Contract Standardisation in Civil and Traffic Engineering in the Netherlands, Ede
Devon County Council 1991 *Traffic calming guidelines* Devon CC
Dimitriou H T 1990a The urban transport policy process: its evaluation and application to Third World cities. In Dimitriou H T (ed) *Transport planning for Third World cities* Routledge, pp 144–83
Dimitriou H T 1990b Towards a developmental approach to urban transport planning. In Dimitrou H T (ed) *Transport planning for Third World cities* Routledge, pp 379–419
Döldissen A, Draeger W 1990 Environmental traffic management strategies in Buxtehude, West Germany. In Tolley R S (ed) *The greening of urban transport: planning for walking and cycling in western cities* Belhaven, pp 266–86
Earth Resources Research 1993 *Costing the benefits: the value of cycling* ERR
Hägerstrand T 1965 A Monte Carlo approach to diffusion. *European Journal of Sociology* **6**:43–67
Hall P 1988 *Cities of tomorrow: an intellectual history of urban planning and design in the twentieth century* Blackwell
Hall P, Hass-Klau C 1985 *Can rail save the city? Rail rapid transit and pedestrianisation in British and German cities* Gower

Hartman J 1990 The Delft bicycle network. In Tolley R S (ed) *The greening of urban transport: planning for walking and cycling in western cities* Belhaven, pp 193–200

Hass-Klau C 1988 *New life for city centres: planning, transport and conservation in British and German cities* Anglo-German Foundation

Hass-Klau C et al 1992 *Civilised streets: a guide to traffic calming* Environmental and Transport Planning

Headicar P 1991 *Activity, development and 'transport need'* Rees Jeffreys Discussion Paper 22, Transport Studies Unit, University of Oxford

Hillman M 1990 Planning for the green modes: a critique of public policy and practice. In Tolley R S (ed) *The greening of urban transport: planning for walking and cycling in western cities* Belhaven, pp 64–74

Hillman M 1991 *The role of walking and cycling in transport policy* Rees Jeffreys Discussion Paper 8, Transport Studies Unit, University of Oxford

Hudson M 1982 *Bicycle planning: policy and practice* Architectural Press

Hülsmann W 1990 The 'bicycle friendly towns' project in the Federal Republic of Germany. In Tolley R S (ed) *The greening of urban transport: planning for walking and cycling in western cities* Belhaven, pp 218–30

Jacobs J 1962 *The death and life of great American cities* Jonathan Cape

James N, Pharoah T 1992 The traffic generation game. In Roberts J *et al.* (eds) *Travel sickness: the need for a sustainable transport policy for Britain* Lawrence and Wishart, pp 75–87

Jones P 1991 *Restraint of road traffic in urban areas: objectives, options and experiences* Rees Jeffreys Discussion Paper 3, Transport Studies Unit, University of Oxford

Lewis N 1993 *Road pricing: theory and practice* Thomas Telford

McClintock H (ed) 1992 *The bicycle and city traffic* Belhaven

May A 1991 *Integrated transport strategies: a new initiative, or a return to the 1960s?* Rees Jeffreys Discussion Paper 21, Transport Studies Unit, University of Oxford

Ministry of Transport and Public Works 1987 *Evaluation of the Delft bicycle network: final summary report* The Hague

Ministry of Transport, Public Works and Water Management 1991 *Bicycles first: the Bicycle Master Plan* The Hague

Monheim R 1990 The evolution and impact of pedestrian areas in the Federal Republic of West Germany. In Tolley R S *The greening of urban transport: planning for walking and cycling in western cities* Belhaven, pp 244–54

MTRU 1991 *Traffic restraint: five cities, five solutions* Video transcript, MTRU

National Consumer Council 1987 *What's wrong with walking?* HMSO

Newman P W G, Kenworthy J R 1989 *Cities and automobile dependence* Gower

Pas E 1986 The urban transportation planning process. In Hanson S (ed) *The geography of urban transportation* Guilford Press, New York, pp 49–70

Pearman D, Jones M 1985 *Transport interchanges – theory and practice* Summary report, Department of Town Planning, Oxford Polytechnic

Preston B 1991 *The impact of the motor car* Brefi

Proudlove A, Turner A 1990 Street management. In Dimitriou H T (ed) *Transport planning for Third World cities* Routledge, pp 348–78

Ramsay A 1990 A systematic approach to the planning of urban networks for walking. In Tolley R S (ed) *The greening of urban transport: planning for walking and cycling in western cities* Belhaven, pp 159–71

Roberts J 1990 Summary and conclusions for policy. In Tolley R S (ed) *The greening of urban transport: planning for walking and cycling in western cities* Belhaven, pp 287–302

Sherlock H 1991 *Cities are good for us* Paladin

Simpson B 1987 *Planning and public transport in France, Great Britain and West Germany* Longman

Spencer A H, Lin Sien Chia 1985 National policy towards cars: Singapore. *Transport Reviews* **5**(4):301–23

TEST 1987 *Quality streets: how traditional urban centres benefit from traffic calming* TEST

Tolley R S 1990 A hard road: the problems of walking and cycling in British cities. In Tolley R S (ed) *The greening of urban transport: planning for walking and cycling in western cities* Belhaven, pp 13–23

Turton B J 1992 Urban transport problems and solutions. In Hoyle B S, Knowles R D (eds) *Modern transport geography* Belhaven, pp 81–104

Untermann R K 1984 *Accommodating the pedestrian: adapting towns and neighbourhoods to walking and bicycling* Van Nostrand, New York

Untermann R K 1990 Why you can't walk there: strategies for improving the pedestrian environment in the United States. In Tolley R S (ed) *The greening of urban transport: planning for walking and cycling in western cities* Belhaven, pp 172–84

Various authors 1983 Buchanan twenty years after: traffic in towns today. *Built Environment* **9**(2):91–139

Various authors 1992 The compact city. *Built Environment* **18**(4):241–313

Wiedenhoeft R 1981 *Cities for people: practical measures for improving urban environments* Van Nostrand, New York

Young W, Polak J, Axhausen K 1991 *Developments in parking policy and management* Rees Jeffreys Discussion Paper 24, Transport Studies Unit, University of Oxford

Chapter 9

Addus A 1989 Rural transportation in Africa. *Transportation Quarterly* **43**(3):421–33

Airey A 1985a Rural road improvements in Sierra Leone. *Singapore Journal of Tropical Geography* **6**(2):78–90

Airey A 1985b The role of feeder roads in promoting rural change in Sierra Leone. *Tijdschrift voor Economische en Sociale Geografie* **76**:192–201

Airey A 1991 Road construction and health care of rural households in Meru, Kenya. *Transport Reviews* **11**(3):273–90

Banister D 1980 *Transport mobility and deprivation in inter-urban areas* Saxon House, Farnborough

Banister D 1983 Transport and accessibility. In Pacione M (ed) *Progress in rural geography* Croom Helm, pp 136–48

Barwell I, Edmonds G A, Hoare J D, de Veen J (eds) 1985 *Rural transport in developing countries* Intermediate Technology Publications

Brownlea A A, McDonald G T 1981 Health and education services in sparseland Australia. In Lonsdale R E, Holmes J H (eds) *Settlement systems in sparsely populated regions* Pergamon, pp 322–46

Cloke P (ed) 1985 *Rural accessibility and mobility* Institute of British Geographers

Dobbs B 1979 Rural public transport: the economic stranglehold. In Halsall D A, Turton B J (eds) *Rural transport problems in Britain* Transport Geography Study Group, pp 21–33

Halsall D A, Turton B J (eds) 1979 *Rural transport problems in Britain* Transport Geography Study Group

Hine J L et al 1983 *Accessibility, transport costs and food marketing in the Ashanti region of Ghana* Report SR 809, Transport and Road Research Laboratory

HMSO 1965 *Rural bus services: report on local enquiries* HMSO

Hoare A G 1975 Some aspects of rural transport: East Anglia. *Journal of Transport Economics and Policy* **9**:141–53

Howe J D, Tennant B S 1977 *Forecasting rural road travel in developing countries from land use studies* Report LR 754, Transport and Road Research Laboratory

Knowles R 1981 *Island to mainland transport in highland areas* Discussion Paper 12, University of Salford

Millard R S 1967 *Road International* **66**

Moseley M J 1979 *Accessibility: the rural challenge* Methuen

Moseley M J, Packman J 1983 *Mobile services in rural areas* University of East Anglia

Moyes A M 1974 Accessibility to general practitioner services on Anglesey. Paper presented at Transport Geography Study Group conference, Salford University

Moyes A 1988 Travellers' tales in rural Wales. *Geographical Magazine* **June**

Mwase N 1989 Role of transport in rural development in Africa. *Transport Reviews* **9**(3)

Nutley S 1988a Unconventional modes of transport in rural Britain. *Journal of Rural Studies* **4**:73–86

Nutley S 1988b Unconventional modes – appropriate technology. In Tolley R S (ed) *Transport technology and spatial change* Transport Geography Study Group

Nutley S 1990 *Unconventional and community transport in the United Kingdom* Gordon and Breach

Oxley P R 1982 *Effects of withdrawal of rural bus services* Report LR 719 Transport and Road Research Laboratory

Pacione M 1984 Rural geography Ch 16 *Transport and accessibility* Harper and Row, pp 281–97

Patmore J A 1966 The contraction of the network of railway passenger services in England and Wales 1836–1962. *Transactions of the Institute of British Geographers* **38**:105–18

Phillips D, Williams A 1984 Transport and accessibility Ch 6 *Rural Britain: a social geography* Blackwell

Rees G, Wragg R 1975 *A study of the passenger transport needs of rural Wales* Welsh Council

Rutex Working Groups 1979–83 Reports on rural transport experiments *Transport and Road Research Laboratory*

Stanley P A 1975 *Defining minimum socially acceptable levels of mobility in rural areas* Transport Geography Study Group

Strathclyde Regional Council 1978 *Strathclyde regional plan* Strathclyde Regional Council

Thomas St J D 1963 *The rural transport problem* Croom Helm

Watts P F, Stark D C, Hawthorn I H 1978 *British postbuses* Report LR 840 Transport and Road Research Laboratory

Yerrell J S (ed) 1981 *Transport research for social and economic progress.* Vol 1 *Transport planning in developing countries.* Gower, Aldershot

Chapter 10

Adams D 1980 *The restaurant at the end of the universe* Pan

Barrett M 1993 Burning up the troposphere. *Transport Retort* **16**(3):11

Buchan K et al 1993 Taming the truck: freight policy and the environment. *Transport Retort* **16**(1):7–20

Department of Transport 1977 *Report of the Advisory Commission on Trunk Road Assessment* (Leitch Committee) HMSO

Department of Transport 1987 *Transport statistics Great Britain 1976–86* HMSO

Department of Transport 1992 *Assessing the environmental impact of road schemes* Report of the Standing Advisory Committee on Trunk Road Assessment (SACTRA), HMSO

Earth Resources Research (ERR) 1989 *Atmospheric emissions from the use of transport in the UK* Vol. 1 *The estimation of current and future emissions* ERR and World Wide Fund for Nature (WWF)

Earth Resources Research 1991 *Bikes not fumes: the emission and health benefits of a modal shift from motor vehicles to cycling* Cyclists' Touring Club

Earth Resources Research 1993 *Costing the benefits: the value of cycling* Cyclists Touring Club

Farrington J 1992 Transport, environment and energy. In Hoyle B S, Knowles R D (eds) *Modern transport geography* Belhaven, pp 51–66

Heierli U 1993 *Environmental limits to motorisation: non-motorised transport in developed and developing countries.* SKAT, St Gallen, Switzerland

Hodge D 1986 Social impacts of urban transport decisions: equity issues. In Hanson S (ed) *The geography of urban transportation* Guilford Press, New York, pp 301–27

Holman C, Fergusson M, Mitchell C 1991 *Road transport and air pollution* Rees Jeffreys Discussion Paper 25, Transport Studies Unit, University of Oxford

OECD 1988 *Transport and the environment* Paris

OECD 1991 *Freight transport and the environment* Paris

Plowden S 1991 *Present and potential role of appraisal techniques in achieving a balanced transport policy* Rees Jeffreys Discussion Paper 4, Transport Studies Unit, University of Oxford

Seifried D 1990 *Gute Argumente: Verkehr, Geck'sche Reihe* Beck, Munich

Steven H nd *Beurteilung der Auswirkung verschiedener Massnahmen zur Minderung des Verkehrlarms* FIGE GmbH Forschungsinstitut Geräusche und Erschütterungen, Aachen

Transnet 1990 *Energy, transport and the environment* Transnet

Transport and Road Research Laboratory 1992 *Environmental appraisal: a review of monetary evaluation and other techniques* Report CR290, Transport Road and Research Laboratory

Tyme J 1978 *Motorways versus democracy* Macmillan

Various authors 1993 Air transport and the environment. *Transport Retort* **16**(3):6–13

Walsh M P 1990 Trends in vehicle use and emissions control approaches worldwide. In World Wide Fund for Nature (WWF) *UK the route ahead: proceedings of the WWF conference on road transport and the greenhouse effect* WWF, pp 131–46

Whitelegg J 1993 *Transport for a sustainable future: the case for Europe* Belhaven

Chapter 11

Adams J 1981 *Transport planning: vision and practice* Routledge

Adams J 1985 *Risk and freedom* Transport Publishing Projects

Appleyard D 1971 *Social and environmental policies for transportation in the 1970s* University of California, Berkeley

Appleyard D, Lintell M 1969 The environmental quality of city streets: the residents' viewpoint. *Journal of American Planning Association* **35**:84–101. Reprinted in de Boer E 1986 *Transport sociology: social aspects of transport planning* Pergamon, pp 93–120

Davis R 1992 *Death on the streets: cars and the mythology of road safety* Leading Edge

De Jong R 1986 The recapture of the street. In de Boer E *Transport sociology: social aspects of transport planning* Pergamon, pp 77–91

Department of Transport 1991 *Road accidents Great Britain 1990: the casualty report* HMSO

European Conference of Ministers of Transport (ECMT) 1987 *Transport for disabled people: developing accessible transport* HMSO

Gant R 1992 Transport for the disabled. *Geography* **77**(1):88–91

Grieco M, Pickup L, Whipp R 1989 *Gender, transport and employment: the impact of travel constraints* Avebury

Hägerstrand T 1970 What about people in regional science? *Papers, Regional Science Association* **24**:7–21

Hamilton K, Jenkins L 1992 Women and transport. In Roberts J et al *Travel sickness: the need for a sustainable transport policy for Britain* Lawrence and Wishart, pp 57–74

Hillman M (ed) 1993 *Children, transport and the quality of life* Policy Studies Institute

Hillman M, Adams J, Whitelegg J 1990 *One false move . . .: a study of children's independent mobility* Policy Studies Institute

Hillman M, Henderson I, Whalley A 1976 *Transport realities and planning policy: studies of friction and freedom in daily travel* Policy Studies Institute

Jacobs G D, Sayer I A 1983 *Road accidents in developing countries* Report SR 807, Transport and Road Research Laboratory

Knox P 1982 *Urban social geography: an introduction* Longman

Le Corbusier 1929 *The city of tomorrow and its planning* John Rodher, reprinted 1947 by Architectural Press

Levy C 1992 Transport. In Ostergaard L (ed) *Gender and development: a practical guide.* Routledge, pp 94–109

Mackay M 1991 *Effective strategies for accident reductions* Rees Jeffreys Discussion Paper 13, Transport Studies Unit, University of Oxford

Morgan E 1976 *Falling apart: the rise and decline of urban civilization* Souvenir Press

Muller P O 1976 Transportation geography II: social transportation geography. *Progress in Geography* **8**:208–31

Nielsen O H 1990 Safe routes to school in Odense, Denmark. In Tolley R S (ed) *The greening of urban transport: planning for walking and cycling in western cities* Belhaven, pp 255–65

Plowden S, Hillman M 1984 *Danger on the road: the needless scourge* Policy Studies Institute

Pred A, Palm R 1978 The status of American women: a time–geographic view. In Lanegran D A, Palm R (eds) *An invitation to geography*, 2nd edition McGraw-Hill, New York, pp 99–109

Pushkarev B S, Zupan J M 1975 *Urban space for pedestrians* MIT Press, Cambridge, Mass.

Silverleaf A, Turgel J 1991 *Transport safety and security – principles and practices* Rees Jeffreys Discussion Paper 23, Transport Studies Unit, University of Oxford

Skelton N 1982 Transport policy and the elderly. In Warnes A M (ed) *Geographical perspectives on the elderly* Wiley, pp 303–22

Transport and Health Study Group 1991 *Health on the move: policies for health promoting transport* Public Health Alliance, Birmingham

Ullrich O 1990 The pedestrian town as an environmentally tolerable alternative to motorised travel. In Tolley R S (ed) *The greening of urban transport: planning for walking and cycling in western cities* Belhaven, pp 97–112

US Department of Transportation 1973 *The handicapped and elderly market for urban mass transit* Transportation Systems Center, Cambridge, Mass.

Wachs M, Kumagai T 1972 *Physical accessibility as a social indicator* School of Architecture and Urban Planning, University of California. Cited in Button K J, Gillingwater D 1986 *Future transport policy* Croom Helm, Sydney

Whitelegg J 1983 Road safety: defeat, complicity and the bankruptcy of science. *Accident Analysis and Prevention* **15**(2):153–60

Whitelegg J 1987 A geography of road traffic accidents. *Transactions of the Institute of British Geographers* NS **12**:161–76

Whitelegg J 1993 *Transport for a sustainable future: the case for Europe* Belhaven

Chapter 12

Adams J 1992 Towards a sustainable transport policy. In Roberts J et al (eds) *Travel sickness: the need for a sustainable transport policy in Britain* Lawrence and Wishart, pp 320–33

Adams J 1993 No need for discussion: the policy is now in place. In Stonham P (ed) *Local transport today and tomorrow* LTT Ltd, pp 73–7

Banister D, Hall P (eds) 1981 *Transport and public policy planning* Mansell

Bromley R D F, Thomas C J 1993 The retail revolution, the carless shopper and disadvantage. *Transactions of the Institute of British Geographers* NS **18**:222–36

Buchan K 1992 Freight policy. In Roberts J et al (eds) *Travel sickness: the need for a sustainable transport policy for Britain* Lawrence and Wishart, pp 193–203

Buchan K et al 1993 Taming the truck: freight policy and the environment. *Transport Retort* **16**(1):7–20

Button K J, Gillingwater D 1986 *Future transport policy* Croom Helm

Cloke P, Bell P 1990 *Transport deregulation: market forces in the modern world* Wiley

Cooper J 1991 *Freight needs and transport policies* Rees Jeffreys Discussion Paper 15, Transport Studies Unit, University of Oxford

EC Commission 1992 *The future development of the common transport policy: a global approach to the construction of a community framework for sustainable mobility* Com 92 454, Brussels

Ferguson E 1990 Transportation demand management: planning, development and implementation. *Journal of the American Planning Association* **56**:442–56

Fergusson M 1989 *Subsidised pollution: company cars and the greenhouse effect* Greenpeace

Friends of the Earth 1992 *Less traffic, better towns* FoE

Goodwin P et al 1991 *Transport: the new realism* Transport Studies Unit

Hamer M 1987 *Wheels within wheels: a study of the road lobby* Routledge

Hart T 1992 Transport, the urban pattern and regional change, 1960–2010. *Urban studies* **29**(3/4):483–503

Hopkin J 1986 *The transport implications of company-financed motoring* Report 61, Transport and Road Research Laboratory

Janelle D G 1986 Metropolitan expansion and the communication–transport trade-off. In Hanson S (ed) *The geography of urban transportation* Guilford Press, New York, pp 357–85

Jones G 1993 The potential contribution of planning to reducing travel demand. In Stonham P (ed) *Local transport today and tomorrow* LTT Ltd, pp 23–9

Jones P 1993 Changing public attitudes to traffic growth. In Stonham P (ed) *Local transport today and tomorrow* LTT Ltd, pp 39–43

Knowles R (ed) 1989 *Transport policy and urban development* Transport Geography Study Group

Knowles R, Hall D 1992 Transport policy and control. In Hoyle B S, Knowles R (eds) *Modern transport geography* Belhaven, pp 33–50

Kroon M 1990 Traffic and environmental policy in the Netherlands. In Tolley R S (ed) *The greening of urban transport: planning for walking and cycling in western cities* Belhaven, pp 113–33

Ministry of Housing, Physical Planning and Environment 1991 *The right business in the right place: towards a location policy for business and services in the interests of accessibility and the environment* The Hague

Nash C A 1991 *The role of rail in future transport policy* Rees Jeffreys Discussion Paper 12, Transport Studies Unit, University of Oxford

Newman P W G, Kenworthy J R 1989 *Cities and automobile dependence: a sourcebook* Gower

O'Sullivan P 1980 *Transport policy: an interdisciplinary approach* Batsford

Plane D A 1986 Urban transportation: policy alternatives. In Hanson S (ed) *The geography of urban transportation* Guilford Press, New York, pp 386–414

Plowden S 1980 *Taming traffic* André Deutsch

Plowden S 1985 *Transport reform: changing the rules* Policy Studies Institute

Preston B 1991 *The impact of the motor car* Brefi

Roberts J 1989 *User-friendly cities* Rees Jeffreys Discussion Paper 5, Transport Studies Unit, Oxford

Shaw S-L 1993 Hub structures of major US passenger airlines. *Journal of Transport Geography* **1**(1):47–58

Smith J *et al.* 1992 British passenger transport into the 1990s. *Geography* **77**(1):63–93

Tolley R S 1990a *Calming traffic in residential areas* Brefi

Tolley R S 1990b Introduction: trading in the red modes for the green. In Tolley R S (ed) *The greening of urban transport: planning for walking and cycling in western cities* Belhaven, pp 1–9

Transnet 1990 *Energy, transport and the environment* Transnet

Tyson B 1991 *Transport policy: constraints and objectives in metropolitan areas* Rees Jeffreys Discussion Paper 26, Transport Studies Unit, University of Oxford

Various authors 1989 The final gridlock. *Built Environment* **15**(3/4):163–256

Von Weizsäcker E U, Jesinghaus J 1992 *Ecological taxation reform: a policy proposal for sustainable development* Zed

Whitelegg J 1984 The company car factor in the UK as an instrument of transport policy. *Transport Policy and Decision-making* **2**:219–30

Whitelegg J 1988 *Transport policy in the EEC* Routledge

Whitelegg J 1992 Till the pips squeak – ecological taxation reform. In Whitelegg J (ed) *Traffic congestion: is there a way out?* Leading Edge, pp 169–85

Whitelegg J 1993 *Transport for a sustainable future* Belhaven

Chapter 13

Brotchie J et al 1985 *The future of urban form: the impact of new technology* Croom Helm

Filson-Young A B 1904 *The joy of the road* Methuen

Fowles J 1969 *The French lieutenant's woman* Triad

Grahame K 1908 *The wind in the willows* Reprinted by Puffin Books 1983

Hägerstrand T 1967 A Monte Carlo simulation of diffusion. In Garrison W L, Marble D F (eds) *Quantitative geography*: Part 1 *Economic and cultural topics* Northwestern University Studies in Geography 13, Northwestern University Press

Hall P 1991 Moving information: a tale of four technologies. In Brotchie J et al (eds) *Cities of the 21st century* Halsted

Hepworth M, Ducatel K 1991 *Transport in the information age: wheels and wires* Belhaven

Holzapfel H, Sachs W 1987 Speed and prospects for our way of life: how motorisation creates new forms of inequality. *Proceedings*, Seminar A PTRC summer meeting, pp 223–9

Illich I 1974 *Energy and equity* Calder and Boyars

Knowles R 1993 Research agendas in transport geography for the 1990s. *Journal of Transport Geography* **1**(1):3–11

Masefield P G 1976 The challenge of change in transport. *Geography* **61**(4): 206–20

Mozer D 1987 *Transportation, bicycles and development in Africa: progression or regression* International Bicycle Fund, Bellvue, Washington

Muller P O 1986 Transportation and urban form: stages in the evolution of the American metropolis. In Hanson S (ed) *The geography of urban transportation* Guilford Press, New York, pp 24–48

Tolley R S (ed) 1988 *Transport technology and spatial change* Transport Geography Study Group

Transnet 1990 *Energy, transport and the environment* Transnet

Vance J E 1986 *Capturing the horizon: the historical geography of transportation* Harper, New York

Whitelegg J W (ed) 1981 *The spirit and purpose of transport geography* Transport Geography Study Group

Williams A F 1992 Transport and the future. In Hoyle B S, Knowles R (eds) *Modern transport geography* Belhaven

Williams H 1992 *Autogeddon* Jonathan Cape

Index

All references to boxes, figures and tables are in bold type.